Licensing Exams for Refrigeration, Air Conditioning and Heating: 4000 Questions & Answers

James L. Dundas

Business News Publishing Company
Troy, MI

Editor: Joanna Turpin

Production Coordinator: Mark Leibold

Library of Congress Cataloging in Publication Data

Dundas, James L.
Licensing Exams for Refrigeration, Air Conditioning and Heating: 4000 Questions and Answers/James L. Dundas
p. cm.
ISBN 0-912524-74-X
1. Air conditioning-Examinations, questions, etc.
2. Refrigeration and refrigerating machinery-Examinations, questions, etc. 3. Heating-Examinations, questions, etc.
I. Title.

TH7687.5.D86 1992 92-31314
697'.00076-dc20 CIP

Disclaimer

This book is only considered to be a general guide. The author and publisher have neither liability nor can they be responsible to any person or entity for any misunderstanding, misuse or misapplication that would cause loss or damage of any kind, including material or personal injury, or alleged to be caused directly or indirectly by the information contained in this book.

Preface

In the field of heating, refrigeration and air conditioning, competency is a very important subject. Municipalities throughout the country test the skills and knowledge of individuals for the purpose of issuing various types of licenses. This testing includes general knowledge in the subject area as well as local and national codes. The following test questions are very similar in nature to those found on tests throughout the country. However, it is necessary to check with individual municipalities for information concerning local codes and qualifications, which may appear on your particular licensing test. When taking a test, you may not find what you think is the correct answer. In this instance, it is important to select the most correct answer of those provided.

James Lee Dundas
Professor, Macomb Community College
Warren, Michigan

Table of Contents

Refrigeration and Air Conditioning

Level 1

1. An increase in pressure on a liquid will cause the boiling point of the liquid to
   A) decrease.
   B) increase.
   C) remain the same.
2. The temperature above which a vapor cannot be condensed regardless of the pressure placed upon it is
   A) thermodynamic process.
   B) a noncondensable.
   C) critical temperature.
3. A high-side float is almost always found on systems
   A) having one evaporator and one float.
   B) of domestic type only.
   C) having numerous evaporator and liquid controls.
4. If used, a defrost cycle generally occurs when the
   A) system is in normal operation.
   B) system is stopped or on defrost cycle.
   C) demand for refrigeration is at a maximum.
5. The frost line is not affected by the
   A) charge in the system.
   B) superheat setting of the TX valve.
   C) solenoid valve.
6. In a counterflow condenser, the
   A) warmest part of one fluid meets the warmest part of the other fluid.
   B) warmest part of one fluid meets the coldest part of the other fluid.
   C) automatic cooling water valve is installed on the outlet side of the condenser.
7. A substance contains the most heat when it is in the
   A) gaseous state.
   B) liquid state.
   C) solid state.
8. The refrigerant having the highest latent heat is
   A) R-22.
   B) R-12.
   C) ammonia.

9. An agent often used with methyl chloride to call attention to the refrigerant is
   A) SO_2.
   B) acrolein.
   C) methylene chloride.
10. If an accumulator is used, it will be located in the
   A) low side.
   B) high side.
   C) rear bank of tubes.
11. A system using an automatic expansion valve
   A) must be hand operated.
   B) can be thermostatically operated.
   C) can be operated with a hi-low switch.
12. Oil will absorb refrigerant more readily when it is
   A) hot.
   B) warm.
   C) cold.
13. The fusible plug shall be connected directly to
   A) the evaporator.
   B) the pressure vessel.
   C) anything on the system.
14. The use of compressor high-pressure gas in the evaporator for the removal of frost is known as
   A) hot gas defrost.
   B) leak testing.
   C) direct defrost.
15. An instrument that measures specific gravity is a
   A) psychrometer.
   B) gravitational device.
   C) hydrometer.
16. Latent heat is defined as the
   A) heat added by the compressor to the suction gas.
   B) heat associated with a change in temperature that can be measured with a thermometer.
   C) amount of heat required to bring about melting or vaporizing.
17. The purpose of a pressure limiting device is to
   A) maintain a constant suction pressure to the compressor.
   B) stop the compressor in case of overpressure.
   C) maintain constant cooling water pressure to the condenser.
18. Although a cooling tower is a water-conserving device, some make-up water is necessary to compensate for the loss. This percentage of make-up water is

A) 1 to 3%.
B) 6 to 10%.
C) 15 to 20%.

19. An instrument used to measure relative humidity is a
    A) humidifier.
    B) hydrometer.
    C) psychrometer.

20. Dirty condenser tubes could lead to
    A) dirt accumulation on evaporator tubes.
    B) decrease in head pressure.
    C) increase in head pressure.

21. Excessive head pressure in a refrigeration system is usually caused by
    A) condenser water failure.
    B) insufficient refrigerant.
    C) a head gasket that is too thick.

22. How many Btu must be added to change a ton of ice at 32 °F into water at 32 °F?
    A) 500,000
    B) 176,000
    C) 288,000

23. When installing a relief valve on a pressure vessel, it should be placed
    A) above the liquid level.
    B) below the liquid level.
    C) anywhere on the vessel.

24. The pressure-type condenser water valve is actuated by
    A) high-side pressure.
    B) low-side pressure.
    C) intermediate pressure.

25. Refrigerating effect is the
    A) weight of refrigerant circulated per minute.
    B) amount of heat absorbed by the refrigerant.
    C) horsepower per ton of refrigeration.

26. A control device, actuated by changes in evaporator pressure, for regulating the flow of liquid refrigerant into an evaporator is a(n)
    A) thermostatic expansion valve.
    B) automatic expansion valve.
    C) solenoid valve.

27. Heat can be transferred by
    A) radiation, conduction, convection.
    B) radiation, conduction, absorption.
    C) convection, conduction, refraction.

28. A device used to rid a system of noncondensable gases is called a
    A) purger.
    B) classifier.
    C) drier.

29. The reason for a frosted compressor using a TXV could be the result of
    A) the expansion valve passing too much refrigerant.
    B) a restricted liquid line.
    C) a high-pressure control not cutting off.

30. Which of the following involves a change between the liquid and solid states of a substance?
    A) Latent heat of fusion
    B) Latent heat of evaporation
    C) Latent heat of sublimation

31. A heat exchanger in an R-12 system
    A) decreases suction superheat.
    B) increases suction superheat.
    C) reduces flash gas.

32. A controlling device, actuated by changes in superheat of the vapor leaving the unit, used for regulating the flow of refrigerant into a cooling unit is a
    A) thermostatic expansion valve.
    B) hand expansion valve.
    C) superheat control.

33. The transmission of heat through intervening substances without heating those substances is
    A) conduction.
    B) convection.
    C) radiation.

34. A device having a predetermined melting temperature member for pressure relief is a
    A) relief valve.
    B) fusible plug.
    C) fusible link.

35. If a refrigerant comes in contact with an open flame, it breaks down and forms
    A) hydrogen gas.
    B) sulfur dioxide.
    C) phosgene.

36. An atmospheric condenser is similar to a(n)
    A) shell-and-tube condenser.
    B) shell-and-coil condenser.
    C) evaporative condenser without a blower.

37. A thermostatic expansion valve controls evaporator
    A) pressure.
    B) temperature.
    C) refrigerant quantity.

38. The amount of heat necessary to raise the temperature of 1 lb of water 1 °F is known as a
    A) thermometer.
    B) Btu.
    C) thermocouple.

39. The growth of molds or bacteria on foods can be stopped by
    A) lowering the temperature.
    B) lowering the temperature below the freezing point of the product.
    C) storing under 32 °F.

40. The type of condenser that gives the best separation of liquid and vapor is the
    A) double pipe.
    B) evaporative.
    C) shell and tube.

41. A fractional horsepower motor that gives the best starting torque is
    A) induction with one set of windings.
    B) capacitor split-phase motor.
    C) resistance split-phase.

42. The rate of heat interchange (12,000 Btu/hr or 200 Btu/min) is
    A) latent heat.
    B) a ton of refrigeration.
    C) refrigerating effect.

43. A system having the evaporator in contact with the material or space refrigerated is called (an)
    A) flooded system.
    B) indirect system.
    C) direct system.

44. A higher heat transfer rate is usually attained with a
    A) direct system.
    B) dry expansion system.
    C) flooded system.

45. The ripening process of food
    A) releases heat that is known as vital heat.
    B) is caused by bacteria.
    C) is slowed down by warm temperature.

46. A collecting or storage chamber for low-side liquid refrigerant is called a(n)
    A) absorber.
    B) accumulator.
    C) regenerator.

47. The function of the compressor is to
    A) compress the liquid refrigerant so it will vaporize.
    B) circulate the refrigerant.
    C) compress the refrigerant vapor above the condensing point and to help circulate refrigerant.

48. Heat always travels
    A) from a warmer to a colder substance.
    B) from a colder to a warmer substance.
    C) upstream.

49. What can be used for building up a pressure in a refrigeration system for testing purposes?
    A) Air
    B) Oxygen
    C) CO_2

50. What is the purpose of a heat exchanger in a refrigeration system?
    A) Decreasing suction superheat.
    B) Increasing suction superheat.
    C) Reducing flash gas.

51. Methyl chloride is
    A) toxic only.
    B) flammable only.
    C) both toxic and flammable.

52. The rate at which work is done is defined as
    A) energy.
    B) motion.
    C) power.

53. The ability of an oil to block the passage of electrical current is referred to as its
    A) kV rating.
    B) power factor rating.
    C) ampere rating.

54. Silica gel activated alumina are classified as
    A) saturators.
    B) absorbents.
    C) neither A nor B.

55. The expanding fluid most adaptable for use in refrigeration thermometers for temperatures above 38 °F is
    A) alcohol.

B) red oil.
C) mercury.

56. A dehumidifier is a device used to
A) decrease humidity.
B) increase humidity.
C) increase the moisture content of the air.

57. Oil return problems are less common in a system using
A) R-22.
B) R-12.
C) R-114.

58. If heat is added to a gas at constant volume, the pressure of the gas will
A) increase.
B) decrease.
C) double.

59. One mechanical hp represents the completion of 33,000 ft-lb of work in one
A) hour.
B) minute.
C) day.

60. The process by which a substance changes directly from a solid to a gas, without becoming a liquid first is known as
A) condensation.
B) sublimation.
C) gasification.

61. The temperature at which oil no longer flows is called the
A) viscosity.
B) floc point.
C) pour point.

62. Sweat or solder copper alloy fittings are used with
A) soft copper tubing.
B) hard copper tubing.
C) steel tubing.

63. The agent best suited to extinguish a small electrical fire in a machinery room is
A) a foam solution.
B) carbon dioxide.
C) water.

64. A capillary tube best controls the refrigerant flow when used with a
A) back-pressure control switch.
B) temperature control switch.
C) high-pressure cut-out switch.

65. The temperature at which sublimation of solid carbon dioxide occurs at atmospheric pressure is approximately
    A) 0 °F.
    B) -50 °F.
    C) -109 °F.

66. If the pressure in the system is less than atmospheric, it is normally measured in inches of
    A) mercury.
    B) water.
    C) vacuum.

67. What is the recommended size replacement for a capillary tube?
    A) Next largest size
    B) Next smallest size
    C) Same size

68. A substance having a boiling point above 32 °F at atmospheric pressure could
    A) not be used as a refrigerant.
    B) be used as a refrigerant.
    C) be used but wouldn't produce a temperature below 32 °F.

69. The evaporation or condensation temperature of a fluid is the
    A) critical temperature.
    B) saturation temperature.
    C) mean temperature.

70. A heat exchanger connected to the suction and liquid lines decreases the
    A) liquid temperature.
    B) suction temperature.
    C) discharge temperature.

71. A quicker and higher vacuum can be pumped when using a
    A) centrifugal compressor.
    B) reciprocating compressor.
    C) rotary compressor.

72. Of the following substances, which could be considered the most nearly perfect gas?
    A) H_2O
    B) Ammonia
    C) Nitrogen

73. The thickness or thinness of an oil and its resistance to flow at a certain temperature is called
    A) viscosity.
    B) floc point.
    C) pour point.

74. When liquid Carrene 1 is released into the atmosphere, it will ordinarily
    A) change to a solid.
    B) remain a liquid.
    C) change to a gas.

75. What does an anemometer measure?
    A) Temperature
    B) Velocity
    C) Volts
76. Air will absorb more moisture when it is
    A) cold and dry.
    B) hot and saturated.
    C) cold and saturated.
77. The special steel cylinders or drums used for the storage and transfer of refrigerants
    A) should be filled completely.
    B) should not be overfilled.
    C) may be filled completely if provided with a fusible plug.
78. To properly test the accuracy of a refrigeration thermometer, it should be immersed in a
    A) thermometer well.
    B) steam bath.
    C) bath of pure water and ice.
79. To change a refrigerant liquid into a gas at its critical temperature and pressure,
    A) latent heat is not absorbed.
    B) latent heat must be absorbed.
    C) is impossible.
80. One horsepower of mechanical energy is equal to
    A) 746 watts.
    B) 3,413 watts.
    C) 2,513 watts.
81. To convert Fahrenheit degrees into degrees Celsius, use the formula
    A) $°C = 5/9 \times (°F - 32)$.
    B) $°C = 9/5 \times (°F - 32)$.
    C) $°C = 5/9 \times (°F + 32)$.
82. A superheated vapor is a vapor
    A) free of liquid.
    B) saturated with liquid.
    C) at a temperature higher than that corresponding to the pressure.
83. The bulb of a thermostatic expansion valve generally contains
    A) refrigerant.
    B) water.
    C) mercury.

84. As the pressure imposed on a refrigerant increases, the latent heat of evaporation
    A) remains constant.
    B) decreases.
    C) increases.

85. R-11 has the same characteristics as
    A) Carrene 7.
    B) Carrene 1.
    C) Carrene 2.

86. Heat always flows from the substance of a higher temperature to a substance of lower temperature is the definition of the
    A) second law of thermodynamics.
    B) first law of thermodynamics.
    C) third law of thermodynamics.

87. Which of the following refrigerants is the most toxic?
    A) CFC refrigerant
    B) Carbon dioxide
    C) Ammonia

88. Other conditions being the same, ice of the highest quality would be made at a temperature of
    A) 32 °F.
    B) 0 °F.
    C) 16 °F.

89. Other conditions remaining the same, cold air
    A) can absorb less moisture than hot air.
    B) can absorb more moisture than hot air.
    C) cannot contain any moisture.

90. When determining relative humidity with a psychrometer, the humidity can be obtained
    A) directly.
    B) through the use of a psychrometric chart.
    C) in grains per gallon.

91. Evaporative condensers are usually located on building roofs to
    A) maintain a static head on liquid lines.
    B) take advantage of large areas of empty space.
    C) take advantage of the cooling effect obtained by the evaporation of the cooling water.

92. A device used to test the density or strength of brine is called a
    A) calorimeter.
    B) hydrometer.
    C) salinometer.

93. Of the refrigerants listed below, the one that makes the most effective fire extinguisher is
    A) carbon dioxide.

B) sulfur dioxide.
C) R-12.

94. The type of compressor most suitable for handling large quantities of refrigerant vapor is the
    A) reciprocating type.
    B) centrifugal type.
    C) hermetic type.

95. When heat is added to a gas at a constant pressure, the gas will
    A) contract.
    B) expand.
    C) remain constant.

96. The purpose of an oil safety switch used on a refrigeration system is to stop the compressor in case of
    A) excessive head pressure.
    B) insufficient oil pressure.
    C) excessive oil pressure.

97. The standard operating conditions for determining the refrigerating capacity of a machine are
    A) 86 °F condensing and 5 °F evaporating temperatures of the refrigerant.
    B) 212 °F condensing and 32 °F evaporating temperatures of the refrigerant.
    C) 86 °F condensing and 5 °F evaporating temperatures of the brine.

98. 778 ft-lb equals one
    A) horsepower.
    B) Btu.
    C) watt.

99. Other conditions being equal, the moisture-holding capacity of air is greater when the air temperature is
    A) 60 °F.
    B) 32 °F.
    C) 120 °F.

100. Lubricating oil is more miscible in
    A) carbon dioxide.
    B) ammonia.
    C) a CFC refrigerant.

101. The area of a circle equals
    A) d^2 x 3.1416.
    B) d x 0.7854.
    C) d^2 x 0.7854.

102. Upon entering the condenser, the hot gas discharge first gives up its
    A) specific heat.
    B) latent heat.
    C) superheat.

103. When functioning, which of the following pressure relief devices causes the greatest loss of refrigerant?
    A) Rupture disc
    B) Relief valve connected to the receiver
    C) Relief valve connected to the compressor

104. Water vapor or moisture may be removed from air by passing the air
    A) over cooling coils.
    B) over heating coils.
    C) through a steam spray.

105. To convert Celsius degrees into Fahrenheit degrees, use the formula
    A) °F = 5/9 x °C + 32.
    B) °F = 9/5 x °C + 32.
    C) °F = 9/5 x °C – 32.

106. As the temperature of lubricating oil carried over in the evaporator decreases, its viscosity
    A) remains unchanged.
    B) decreases.
    C) increases.

107. The drying agent used in refrigerant driers is most likely
    A) sodium chloride.
    B) calcium chloride.
    C) silica gel.

108. A sensation of dampness in an office or other work area is due to
    A) high relative humidity.
    B) low relative humidity.
    C) high ambient temperature.

109. As the pressure imposed on a refrigerant increases, the latent heat of vaporization
    A) remains constant.
    B) decreases.
    C) increases.

110. As the relative humidity decreases, the difference in the readings of the wet and dry bulb thermometers
    A) increases.
    B) decreases.
    C) remains uniform.

111. Proper maintenance of leather belts requires that they periodically be treated with
    A) shellac.
    B) lubricating oil.
    C) neatsfoot oil.

112. A gauge pressure of 35 psig is equal to an absolute pressure of
    A) 49.7 psia.
    B) 20.3 psia.
    C) 35 psia.

113. Which type of copper is not permitted in the refrigeration system?
    A) Type K
    B) Type L
    C) Type M

114. The ambient temperature of a refrigeration compressor is the
    A) air around the compressor.
    B) temperature of the discharge vapor.
    C) inlet temperature of the compressor cooling water.

115. A refrigerant drier contains more moisture when
    A) cold.
    B) warm.
    C) hot.

116. The fusible plug on a receiver should be located
    A) above the liquid line.
    B) below the liquid line.
    C) either above or below the liquid line.

117. When a refrigerant approaches a perfect gas,
    A) its moisture content increases.
    B) it becomes harder to condense.
    C) it becomes easier to condense.

118. The temperature at which particles of wax in an oil-refrigerant mixture can easily be seen by the naked eye is called the
    A) viscosity.
    B) floc point.
    C) pour point.

119. Sulfur dioxide is
    A) toxic only.
    B) flammable only.
    C) both toxic and flammable.

120. With a CFC refrigerant or methyl chloride, the viscosity of a specified oil should be
    A) lower than that used for other refrigerants.
    B) higher than that used for other refrigerants.
    C) the same as that used for other refrigerants.

121. Ethylene is classed as a
    A) Group I refrigerant.
    B) Group II refrigerant.
    C) Group III refrigerant.

122. If inhaled, excess carbon dioxide causes a(n)
   A) clearness of the head.
   B) decrease in the respiration rate.
   C) increase in the respiration rate.

123. A properly installed heat exchanger will
   A) decrease the flash gas through the compressor valves.
   B) decrease flash gas at the expansion valve.
   C) increase flash gas at the expansion valve.

124. The solubility of water in CFC refrigerants is
   A) not affected by temperature changes.
   B) increased by an increase in temperature.
   C) decreased by an increase in temperature.

125. Acrolein is used as a(n)
   A) aid to better oil distribution.
   B) moisture eliminator.
   C) leak warning agent.

126. A thermodynamic process during which no heat is extracted or added to the system is referred to as
   A) the point of no heat transfer.
   B) an isothermal process.
   C) an adiabatic process.

127. The common unit of heat energy is the
   A) Btu.
   B) Fahrenheit degree.
   C) ft-lb.

128. In 6 hours, a one-ton compressor will handle a load of
   A) 288,000 Btu.
   B) 144,000 Btu.
   C) 72,000 Btu.

129. In a system, the oil separator is located between the
   A) condenser and liquid line.
   B) receiver and liquid line.
   C) compressor and condenser.

130. Of the substances listed below, which is considered a chlorinated refrigerant?
   A) Carbon dioxide
   B) Ammonia
   C) CFC refrigerant

131. Clear water
   A) always makes clear ice.
   B) is considered hard water.
   C) does not necessarily make clear ice.

132. Usually, the dry bulb (db) temperature is
    A) the same as the wet bulb temperature.
    B) lower than the wet bulb temperature.
    C) higher than the wet bulb temperature.

133. The axiom energy can neither be created nor destroyed but can be converted from one form to another pertains to
    A) the second law of thermodynamics.
    B) the first law of thermodynamics.
    C) Boyle's law.

134. The bellows of a high pressure cut-out switch contains
    A) a vacuum.
    B) a refrigerant.
    C) air.

135. The function of a thermometer well is to
    A) store the thermometer expansion fluid.
    B) facilitate thermometer removal without refrigerant loss.
    C) keep the thermometer bulb cold.

136. A sensation of dryness in an office or other work area is due to
    A) high relative humidity.
    B) low relative humidity.
    C) low ambient temperature.

137. The part of the refrigeration system that adds the most heat to the refrigerant is the
    A) compressor.
    B) condenser.
    C) evaporator.

138. In a stem thermometer, the hollow space above the liquid
    A) should be filled with air.
    B) may possibly contain gas vaporized from this liquid.
    C) is under a perfect vacuum.

139. A saturated vapor is a vapor
    A) saturated with liquid droplets.
    B) at a temperature corresponding to the pressure.
    C) at a temperature higher than 5 °F.

140. The K factor of an insulating material defines
    A) Btu loss.
    B) moisture resistance.
    C) combustible qualities.

141. The most perfect insulator would be
    A) a vacuum.
    B) still air.
    C) cork.

142. A thermometer well should be
   A) packed with asbestos.
   B) left dry.
   C) filled with oil.

143. The device that compresses and delivers the refrigerant at a temperature that is adequately above that of the atmosphere to the unit where the heat energy is discarded is called a(n)
   A) heat pump.
   B) compressor.
   C) expansion valve.

144. After leaving the compressor, the high-pressure, high-temperature refrigerant vapor is passed through a(n)
   A) evaporator.
   B) condenser.
   C) receiver.

145. If a thermometer is placed at the outlet of the compressor, it will measure only
   A) latent heat.
   B) superheat.
   C) sensible heat.

146. The high-side float controls the flow of refrigerant
   A) to the evaporator.
   B) from the evaporator.
   C) from the condenser.

147. The refrigerants ammonia and carbon dioxide should be stored
   A) in the direct rays of the sun.
   B) near some source of heat.
   C) away from any source of heat.

148. The high-side of a refrigeration system is
   A) in the highest side of the building.
   B) near some source of heat.
   C) away from any source of heat.

149. When ice melts, its temperature
   A) increases slightly.
   B) increases by degrees.
   C) remains constant.

150. Which refrigerant can be tested for leaks by using a halide leak detector?
   A) R-12
   B) R-717
   C) R-744

151. What is the micron equivalent of 1 inch of vacuum?
    A) 500
    B) 2,400
    C) 25,400

152. Condensed refrigerant travels from the condenser to the expansion valve through the
    A) suction line.
    B) discharge line.
    C) liquid line.

153. A hissing noise at the expansion valve is a good indication that
    A) liquid is passing through.
    B) gas is passing through.
    C) nothing is passing through.

154. CFC refrigerants smell
    A) similar to sulfur dioxide.
    B) very pungent.
    C) slightly sweet.

155. Which of the following refrigerants is lighter than air?
    A) CFC refrigerant
    B) Carbon dioxide
    C) Ammonia

156. The temperature of the liquid line should normally be
    A) the same as the condensing water temperature.
    B) higher than the condensing water temperature.
    C) lower than the condensing water temperature.

157. Of the refrigerants listed below, the one most suitable to air condition a theater is
    A) R-11.
    B) ammonia.
    C) propane.

158. R-12 is composed of
    A) hydrogen and nitrogen.
    B) carbon and oxygen.
    C) fluorine, chlorine and carbon.

159. The unit used for measuring gas pressure is the
    A) pound.
    B) Btu.
    C) psi.

160. Which of the following cannot be used in a mechanical refrigeration system?
    A) Swaged joints on the low side
    B) Flared joints
    C) 50/50 solder joints

161. A vapor that has been heated above its boiling point is referred to as
    A) saturated vapor.
    B) superheated vapor.
    C) sub-heated vapor.
    D) latent heated vapor.

162. The process by which a substance changes from a gaseous to a liquid state is referred to as
    A) evaporation.
    B) absorption.
    C) vaporization.
    D) condensation.

163. The device in which the refrigerant is vaporized is called the
    A) compressor.
    B) vaporizer.
    C) evaporator.
    D) condenser.

164. The device that regulates the flow of liquid refrigerant to the cooling unit so that the evaporator is maintained nearly full of liquid refrigerant is called the
    A) compressor.
    B) receiver.
    C) condenser.
    D) expansion valve.

165. The refrigerant storage tank that holds a surplus of refrigerant and is located below the condenser is called the
    A) receiver.
    B) evaporator.
    C) accumulator.
    D) well.

166. Temperature is measured accurately with a
    A) manometer.
    B) thermometer.
    C) barometer.
    D) thermostat.

167. The odor of ammonia is
    A) pungent.
    B) sweet.
    C) sour.
    D) nothing.

168. The odor of the refrigerant carbon dioxide
    A) is pungent.
    B) is odorless.
    C) smells like phosgene gas.
    D) is sweet.

169. Ammonia in the presence of sulfur vapor produces a
    A) white smoke.
    B) black smoke.
    C) green flame.
    D) blue flame.

170. Carbon dioxide is composed of
    A) hydrogen and nitrogen.
    B) carbon and oxygen.
    C) fluorine, chloride and carbon.
    D) CO_2 and NH_3.

171. The electric motor that drives the compressor in home units is turned on and off by the
    A) receiver.
    B) condenser.
    C) cooling units.
    D) thermostat.

172. After leaving the compressor, the vaporous refrigerant goes to the
    A) receiver.
    B) condenser.
    C) thermostat.
    D) liquid flow control.

173. The device used to activate the compressor in a household refrigerator is a(n)
    A) electric motor.
    B) generator.
    C) heat engine.
    D) air motor.

174. Mercury is used in thermometers primarily because it
    A) expands uniformly.
    B) vaporizes at very high temperatures.
    C) is cheaper than most other suitable liquids.
    D) never freezes.

175. The operation of every type of motor is based on the
    A) principle of induction.
    B) frequency of the power supply.
    C) principle of attraction.
    D) magnetic and electric reactions.

176. The unit used for force is the
    A) pound.
    B) pound of feet.
    C) pounds per square inch.
    D) horsepower.

177. The unit for measuring mechanical power is the
   A) foot-pound.
   B) pound.
   C) Btu.
   D) horsepower.

178. The unit used for measuring electrical power is the
   A) ampere.
   B) horsepower.
   C) watt.
   D) Btu.

179. One mechanical horsepower is defined as the work done at the rate of
   A) 778 Btu per minute.
   B) 550 ft-lb/second.
   C) 3,300 ft-lb/second.
   D) 3,300 ft-lb/minute.

180. The latent heat of vaporization of sulfur dioxide is
   A) 144 Btu.
   B) 168 Btu.
   C) 746 Btu.
   D) 970 Btu.

181. The latent heat of vaporization of water is
   A) 144 Btu.
   B) 168 Btu.
   C) 746 Btu.
   D) 970 Btu.

182. The specific heat of water is
   A) 1.00.
   B) 0.5.
   C) 0.463.
   D) 0.481.

183. One British thermal unit is the amount of heat required to raise the temperature of one
   A) pound of water 1 °F.
   B) gallon of water 1 °F.
   C) cubic foot of water 1 °F.
   D) pound of water 1 °C.

184. The process by which a substance changes from a solid to a liquid state is referred to as
   A) melting.
   B) dissolving.
   C) vaporization.
   D) disintegration of the atom.

185. To melt one ton of ice (2,000 lb) requires
   A) 144 Btu.
   B) 2,000 Btu.
   C) 1,200 Btu.
   D) 288,000 Btu.

186. The heat given up by a substance during the freezing process is called
   A) latent heat of fusion.
   B) latent heat of vaporization.
   C) evolution.
   D) condensation.

187. The refrigerant considered safest for use in domestic refrigeration is
   A) sulfur dioxide.
   B) R-12.
   C) methyl chloride.
   D) ammonia.

188. The refrigerant with which lubricating oil mixes best is
   A) R-12.
   B) SO_2.
   C) ethyl chloride.
   D) methyl chloride.

189. A device that transfers a gaseous substance from a container of low pressure to one of high pressure is called a(n)
   A) evaporator.
   B) receiver.
   C) high-side float.
   D) compressor.

190. The best conduction of heat takes place through a
   A) non-metal object.
   B) ferrous metal.
   C) perfect vacuum.
   D) liquid.

191. R-11 has the greatest corrosive effect on
   A) copper.
   B) steel.
   C) magnesium.
   D) tin.

192. NH_3 has the greatest corrosive effect on
   A) steel.
   B) copper.
   C) aluminum.
   D) lead.

193. The unit of measurement for temperature is
    A) Btu.
    B) ohm.
    C) calorie.
    D) degrees Fahrenheit.

194. What is the moisture condition in air when the dew point is reached?
    A) 40%
    B) 50%
    C) 100%

195. The temperature at which a substance changes from a liquid to a vapor is its
    A) boiling point.
    B) condensing point.
    C) changing point.
    D) latent heat point.

196. Heat that increases temperature of a refrigerant without causing a change in state is called
    A) sensible heat.
    B) latent heat.
    C) specific heat.
    D) vaporizing heat.

197. The quantity of heat in a refrigerator may be measured by a unit called the
    A) degree Fahrenheit.
    B) foot-pound.
    C) calorie/in^2.
    D) Btu.

198. If it were possible to remove all the heat from a refrigerator, its temperature would be
    A) 32 °F.
    B) 0 °F.
    C) -40 °F.
    D) -460 °F.

199. The purpose of the electric motor is to
    A) increase the energy put into it.
    B) generate power.
    C) change mechanical energy into electrical energy.
    D) change electrical energy into mechanical energy.

200. Unless it is of a type approved for more, the maximum number of electrical conductors that may be placed under a screw-type terminal is
    A) one.
    B) two.

C) three.
D) four.

201. In a refrigerator, the food is preserved because the unit
A) puts cold into storage space.
B) removes heat from storage space.
C) allows no heat to get into the storage space.
D) will not permit cold to get out of the storage space.

202. Heat is carried to the refrigeration cooling unit by
A) radiation.
B) convection.
C) conduction.
D) heat coils.

203. The device used to maintain the temperature inside a refrigerator within a certain limit is called a
A) thermometer.
B) pyrometer.
C) governor.
D) thermostat.

204. The amount of heat a substance contains is measured with an instrument called a
A) thermometer.
B) potentiometer.
C) calorimeter.
D) pyrometer.

205. Refrigeration is a process of
A) making ice.
B) preserving food.
C) extracting heat.
D) adding heat.

206. Refrigerant 502 is in Group II.
A) True
B) False

207. Methyl formate is in Group I.
A) True
B) False

208. Refrigerant 30 is in Group I.
A) True
B) False

209. Methyl chloride is in Group II.
A) True
B) False

210. Sulfur dioxide is in Group III.
 A) True
 B) False

211. Refrigerant 500 is in Group I.
 A) True
 B) False

212. Ethyl chloride is in Group II.
 A) True
 B) False

213. Refrigerant 13 is in Group I.
 A) True
 B) False

214. Refrigerant 11 is in Group I.
 A) True
 B) False

215. Carbon dioxide is in Group III.
 A) True
 B) False

216. Butane is in Group II.
 A) True
 B) False

217. Ethane is in Group II.
 A) True
 B) False

218. Isobutane is in Group III.
 A) True
 B) False

219. Propane is in Group II.
 A) True
 B) False

220. Dichlorodifluoromethane is in Group I.
 A) True
 B) False

221. Ammonia is in Group III.
 A) True
 B) False

222. Methylene chloride is in Group I.
 A) True
 B) False

223. Refrigerant 22 is in Group II.
 A) True
 B) False

224. Refrigerant 12 is in Group I.
  A) True
  B) False

225. The chemical name for R-11 is
  A) dichlorodifluoromethane.
  B) chlorodiflouromethane.
  C) trichlorofluoromethane.
  D) methyl chloride.

226. The chemical name for R-12 is
  A) dichlorodifluoromethane.
  B) chlorodiflouromethane.
  C) trichloromonofluoromethane.
  D) methyl chloride.

227. The chemical name for R-22 is
  A) dichlorodifluoromethane.
  B) chlorodiflouromethane.
  C) trichloromonofluoromethane.
  D) methyl chloride.

228. The chemical name for R-30 is
  A) methyl chloride.
  B) methylene chloride.
  C) carbon dioxide.
  D) ammonia.

229. The chemical name for R-40 is
  A) methyl chloride.
  B) sulfur dioxide.
  C) methylene chloride.
  D) ammonia.

230. The chemical name for R-50 is
  A) methyl chloride.
  B) sulfur dioxide.
  C) carbon dioxide.
  D) methane.

231. The chemical name for R-114 is
  A) dichlorodifluoromethane.
  B) dichlorotetrafluoroethane.
  C) trichloromonofluoromethane.
  D) methyl chloride.

232. The chemical name for R-170 is
  A) methyl chloride.
  B) ethane.
  C) carbon dioxide.
  D) methane.

233. The chemical name for R-290 is
   A) methyl chloride.
   B) butane.
   C) propane.
   D) methane.

234. The chemical name for R-600 is
   A) methyl chloride.
   B) butane.
   C) propane.
   D) methane.

235. The chemical name for R-717 is
   A) methyl chloride.
   B) sulfur dioxide.
   C) ammonia.
   D) water.

236. The chemical name for R-718 is
   A) methyl chloride.
   B) sulfur dioxide.
   C) ammonia.
   D) water.

237. The chemical name for R-764 is
   A) methyl chloride.
   B) sulfur dioxide.
   C) carbon dioxide.
   D) ammonia.

238. The chemical name for R-702 is
   A) nitrogen.
   B) neon.
   C) helium.
   D) hydrogen.

239. The chemical name for R-704 is
   A) nitrogen.
   B) neon.
   C) helium.
   D) hydrogen.

240. The chemical name for R-720 is
   A) nitrogen.
   B) neon.
   C) helium.
   D) hydrogen.

241. The chemical name for R-728 is
   A) nitrogen.
   B) neon.

C) helium.
D) hydrogen.

242. The chemical name for R-729 is
A) oxygen.
B) argon.
C) air.
D) hydrogen.

243. The chemical name for R-732 is
A) oxygen.
B) argon.
C) air.
D) hydrogen.

244. The chemical name for R-740 is
A) oxygen.
B) argon.
C) air.
D) hydrogen.

245. The chemical name for R-744 is
A) methyl chloride.
B) sulfur dioxide.
C) carbon dioxide.
D) ammonia.

246. The chemical formula for R-11 is
A) C_3H_3.
B) C_4H_{10}.
C) CCl_3F.
D) CH_2L_2F.

247. The chemical formula for R-12 is
A) $CHClF_2$.
B) CCl_2F_2.
C) CH_3Cl.
D) CH_2L_2F.

248. The chemical formula for R-13 is
A) $CClF_3$.
B) CCl_2F_2.
C) CH_3Cl.
D) CH_2L_2F.

249. The chemical formula for R-14 is
A) $CClF_3$.
B) CF_4.
C) CH_3Cl.
D) CH_2L_2F.

250. The chemical formula for R-22 is
A) $CHClF_2$.
B) CCl_2F_2.
C) CH_3Cl.
D) CH_2L_2F.

251. The chemical formula for R-30 is
A) CH_2Cl_2.
B) C_4HL0.
C) C_3H_3.
D) C_2H6.

252. The chemical formula for R-40 is
A) SO_2.
B) CO_2.
C) CH_3Cl.
D) $CHClF_2$.

253. The chemical formula for R-50 is
A) SO_2.
B) CH_4.
C) CH_3Cl.
D) $CHClF_2$.

254. The chemical formula for R-114 is
A) SO_2.
B) $CClF_2CClF_2$.
C) $CHCl_2CF_3$.
D) $CHClFCF_3$.

255. The chemical formula for R-123 is
A) SO_2.
B) CO_2.
C) $CHCl_2CF_3$.
D) $CHClF_2$.

256. The chemical formula for R-124 is
A) SO_2.
B) CO_2.
C) $CHCl_2CF_3$.
D) $CHClFCF_3$.

257. The chemical formula for R-125 is
A) SO_2.
B) CO_2.
C) $CHCl_2CF_3$.
D) CHF_2CF_3.

258. The chemical formula for R-134a is
A) SO_2.
B) CF_3CH_2F.

C) $CHCl_2CF_3$.
D) CHF_2CF_3.

259. The chemical formula for R-170 is
A) SO_2.
B) C_2H6.
C) $CHCl_2CF_3$.
D) CHF_2CF_3.

260. The chemical formula for R-290 is
A) CO_2.
B) SO_2.
C) C_2H_6.
D) C_3H_8.

261. The chemical formula for R-600 is
A) C_4H_{10}.
B) C_3H_3.
C) C_2H_6.
D) CCl_3F.

262. The chemical formula for R-717 is
A) SO_2.
B) NH_3.
C) CO_2.
D) H_2O.

263. The chemical formula for R-718 is
A) SO_2.
B) NH_3.
C) CO_2.
D) H_2O.

264. The chemical formula for R-744 is
A) SO_2.
B) NH_3.
C) CO_2.
D) H_2O.

265. The chemical formula for R-764 is
A) SO_2.
B) NH_3.
C) CO_2.
D) H_2O.

266. The boiling point for R-11 is
A) 16 °F.
B) 31 °F.
C) 53 °F.
D) 75 °F.

267. The boiling point for R-12 is
   A) 14 °F.
   B) -21.6 °F.
   C) -41.4 °F.
   D) -28 °F.

268. The boiling point for R-13 is
   A) -77 °F.
   B) -93 °F.
   C) -115 °F.
   D) -126 °F.

269. The boiling point for R-14 is
   A) -77 °F.
   B) -105 °F.
   C) -115 °F.
   D) -198 °F.

270. The boiling point for R-22 is
   A) 14 °F.
   B) -21.6 °F.
   C) -41.4 °F.
   D) -28 °F.

271. The boiling point for R-30 is
   A) 95.2 °F.
   B) -105.2 °F.
   C) 104.4 °F.
   D) -28 °F.

272. The boiling point for R-40 is
   A) -11.6 °F.
   B) -28.8 °F.
   C) -5.8 °F.
   D) -48.8 °F.

273. The boiling point for R-50 is
   A) -29 °F.
   B) -59 °F.
   C) -159 °F.
   D) -259 °F.

274. The boiling point for R-113 is
   A) 118 °F.
   B) 46 °F.
   C) -6 °F.
   D) -24 °F.

275. The boiling point for R-13B1 is
   A) -18 °F.
   B) -72 °F.

C) -84 °F.
D) -98 °F.

276. The boiling point for R-114 is
A) -58.4 °F.
B) -21.4 °F.
C) -6.4 °F.
D) 38.8 °F.

277. The boiling point for R-116 is
A) -131 °F.
B) -109 °F.
C) -84 °F.
D) -62 °F.

278. The boiling point for R-123 is
A) 72.2 °F.
B) 82.2 °F.
C) -72.2 °F.
D) -82.2 °F.

279. The boiling point for R-124 is
A) -8.2 °F.
B) -12.2 °F.
C) 8.2 °F.
D) 12.2 °F.

280. The boiling point for R-125 is
A) -25.3 °F.
B) -35.3 °F.
C) -45.3 °F.
D) -55.3 °F.

281. The boiling point for R-134a is
A) -15.08 °F.
B) -25.7 °F.
C) -35.7 °F.
D) -45.7 °F.

282. The boiling point for R-500 is
A) 14 °F.
B) -21.6 °F.
C) -41.4 °F.
D) -28 °F.

283. The boiling point for R-502 is
A) -50 °F.
B) -41.4 °F.
C) -28 °F.
D) -126 °F.

284. The boiling point for R-503 is
A) -50 °F.
B) -41.4 °F.
C) -28 °F.
D) -127.6 °F.

285. The boiling point for R-504 is
A) -71 °F.
B) -41.4 °F.
C) -38 °F.
D) -126 °F.

286. The boiling point for R-704 is
A) -50 °F.
B) -41.4 °F.
C) -452 °F.
D) -320 °F.

287. The boiling point for R-717 is
A) 14 °F.
B) -28 °F.
C) -50 °F.
D) 212 °F.

288. The boiling point for R-718 is
A) -109 °F.
B) 14 °F.
C) -28 °F.
D) 212 °F.

289. The boiling point for R-728 is
A) -50 °F.
B) -41.4 °F.
C) -452 °F.
D) -320 °F.

290. The boiling point for R-744 is
A) 14 °F.
B) 212 °F.
C) -109 °F.
D) -50 °F.

291. The boiling point for R-764 is
A) -109 °F.
B) 14 °F.
C) -28 °F.
D) 212 °F.

292. The color code for R-500 is
A) green.
B) yellow.

C) white.
D) orange.

293. The color code for R-502 is
A) green.
B) yellow.
C) orchid.
D) orange.

294. The color code for R-503 is
A) green.
B) aquamarine.
C) white.
D) orange.

295. The color code for R-504 is
A) green.
B) aquamarine.
C) white.
D) tan.

296. The color code for R-113 is
A) gray.
B) aquamarine.
C) purple.
D) tan.

297. The color code for R-123 is
A) gray.
B) aquamarine.
C) purple.
D) tan.

298. The color code for R-717 is
A) silver.
B) green.
C) white.
D) orange.

299. The color code for R-764 is
A) silver.
B) gray.
C) black.
D) green.

300. The color code for R-744 is
A) silver.
B) gray.
C) black.
D) orange.

301. The color code for R-12 is
    A) green.
    B) yellow.
    C) white.
    D) orange.

302. The color code for R-22 is
    A) green.
    B) yellow.
    C) white.
    D) orange.

303. The color code for R-11 is
    A) green.
    B) yellow.
    C) white.
    D) orange.

304. The liquid refrigerant drops to evaporate temperature
    A) in the liquid line.
    B) in the evaporator.
    C) at the expansion valve.

305. An accumulator
    A) keeps coils flooded and separates liquid from a vapor.
    B) acts as a separator only.
    C) acts as a storage vessel for excess refrigerant.

306. How many dew points can one sample of air have?
    A) One
    B) Two
    C) Three

307. When water freezes, it
    A) expands.
    B) contracts.
    C) does not change in volume.

308. The latent heat of fusion of water is
    A) 32 °F.
    B) 144 Btu.
    C) 0 °F.

309. Good insulation has
    A) a high K factor.
    B) a low K factor.
    C) any K factor.

310. Which has the highest tensile strength?
    A) Copper
    B) Aluminum
    C) Mild steel

311. Melting ice absorbs
    A) sensible heat.
    B) specific heat.
    C) latent heat.

312. Absolute humidity is
    A) actual weight of water in a given amount of air.
    B) dew point.
    C) dry bulb temperature plus the wet bulb temperature.

313. To prevent bacteria and mold in vegetables, the temperature should be
    A) 0 °F.
    B) 45 °F.
    C) -15 °F.

314. Substances exist in how many physical forms (states)?
    A) One
    B) Two
    C) Three
    D) Four

315. Pressures above atmospheric are measured in
    A) degrees Celsius.
    B) degrees Fahrenheit.
    C) psig.
    D) inches of mercury.

316. Atmospheric pressure is
    A) 14.7 feet.
    B) 14.7 psia.
    C) 14.7 psig.
    D) 14.7 inches of Hg.

317. The symbol Hg indicates
    A) halogen inches.
    B) feet of mercury.
    C) hydrogen gallons.
    D) inches of mercury.

318. Compound gauges measure
    A) pressures above and below atmospheric.
    B) pressures above atmospheric only.
    C) pressures below atmospheric only.
    D) density.

319. The lowest temperature believed obtainable is
    A) -400 °F.
    B) -640 °F.
    C) -260 °F.
    D) -460 °F.

320. Entropy is
   A) the heat available measured in Btu per pound per degree change.
   B) something used every day by service technicians.
   C) a condition of critical pressure and temperature.
   D) not used in engineering calculations.

321. Boyle's Law, Charles' Law and the Gas Law all deal with
   A) saturated conditions.
   B) relative humidity.
   C) subcooling.
   D) gases only.

322. Relative humidity is
   A) a point where air will not hold any more moisture than 60%.
   B) not the same for different temperatures.
   C) the percentage of moisture in the air compared to the amount it could hold at the same pressure and temperature.
   D) not affected when air is passed over a cooling coil.

323. The Celsius (C) scale
   A) is presently in common use by service technicians in the U.S.
   B) is a metric system pressure scale.
   C) is a metric system temperature scale.
   D) cannot be converted to Fahrenheit.

324. The gravity pull on an object will give the object a falling acceleration of 32.2 feet/second per
   A) second.
   B) 100-foot fall.
   C) 1-foot fall.
   D) minute.

325. What is mass?
   A) The most common kind of confusion, such as mass confusion
   B) That which is measured in pounds
   C) The property of all matter, for everything has mass
   D) None of the above are correct

326. If food is frozen slowly at or near the freezing temperatures of water, the ice crystals formed will be
   A) small in size.
   B) zero, as there won't be any ice crystals.
   C) large in size.
   D) none of the above

327. Matter can exist as a
   A) solid, vapor or liquid.
   B) solid.
   C) vapor.
   D) liquid.

328. Pressure is
   A) volume per unit area.
   B) space used per unit area.
   C) force per unit area.
   D) shape per unit area.

329. A solid is
   A) any substance that always takes the shape of its container.
   B) any substance that must be contained in a sealed container or it will soon dissipate.
   C) both a and b are correct.
   D) a substance that retains its own shape without support.

330. A perfect vacuum may be expressed as
   A) 0 lb/in^2 absolute.
   B) 0 lb/in^2 gauge.
   C) 25 inches Hg.
   D) 15 inches Hg.

331. One atmosphere equals
   A) 14.7 psig.
   B) 0 psia.
   C) 29.4 psig.
   D) 0 psig.

332. The force supporting a solid is always
   A) downward.
   B) sideways.
   C) upward.
   D) backward.

333. How many square inches are there in one square foot?
   A) 144
   B) 576
   C) 553
   D) 594

334. A mercury barometer is used to measure
   A) pressures in psia below sea level.
   B) pressures in gauge absolute.
   C) atmospheric pressure.
   D) strength of mercury.

335. To convert psig to psia, add
   A) 14.7 to the psia reading.
   B) 14.7 to the psig reading.
   C) 29.92 to the psig reading.
   D) 29.92 to the psia reading.

336. The device used for measuring small pressure differences is the
A) anemometer.
B) pyrometer.
C) psychrometer.
D) manometer.

337. The condenser is usually made of
A) steel and plastic.
B) copper or steel.
C) rubber and copper.
D) a material similar to glass.

338. Condensers can be
A) either air-cooled or water-cooled.
B) air-cooled only.
C) water-cooled only.
D) static only.

339. A sling psychrometer is a
A) wall-mounted recorder that has a built-in sling to catch particles heavier than air (chromets).
B) humidity-measuring device with wet and dry bulb thermometers that is moved rapidly through the air when measuring humidity.
C) strap-type sling for mounting thermometers on the wall.
D) none of the above.

340. Latent heat is the heat
A) required to raise the temperature of a vapor.
B) given off in the compressor as heat of compression.
C) required to change the state of a substance.
D) measured with a thermometer.

341. A suction line on a system is between the
A) compressor and condenser.
B) evaporator and compressor.
C) condenser and metering device.
D) none of the above.

342. What is psia?
A) Gauge pressure minus atmospheric pressure
B) Atmospheric pressure minus gauge pressure
C) Gauge pressure divided by atmospheric pressure
D) Gauge pressure plus atmospheric pressure

343. What is a ton of refrigeration?
   A) A refrigeration system that weighs 2,000 pounds
   B) The refrigeration effect equal to the melting of one ton of ice in twelve hours
   C) The refrigeration effect equal to the removal of 288,000 Btu in twelve hours
   D) The refrigeration effect equal to the melting of one ton of ice in 24 hours

344. What is refrigerant charge?
   A) The electrical charge placed in a refrigerant to increase its heat-absorbing ability
   B) The amount of refrigerant needed in a system
   C) The wholesaler's cost of the refrigerant
   D) How much heat a refrigerant will absorb or give off to change its state

345. A liquid receiver
   A) changes the high-pressure liquid to a low-pressure liquid.
   B) controls the amount of refrigerant flowing into the condenser.
   C) stores excess liquid refrigerant.
   D) increases the volumetric efficiency of the compressor.

346. A high-side float is a
   A) metering device used on a critically charged system that controls the level of the liquid refrigerant in the high-pressure side of the system.
   B) float-type switch that stops the compressor when the pressure on the high side becomes too high.
   C) float-type valve that controls the amount of refrigerant stored in the receiver.
   D) float-type valve that keeps the oil in the crankcase level.

347. What is a foot-pound?
   A) A foot that weighs 16 ounces
   B) A pound that is a foot long
   C) One pound moved a distance of one yard (3 feet)
   D) One pound moved a distance of one foot

348. What is the Rankin scale?
   A) The name given for the absolute centigrade scale
   B) A scale used for measuring the amount of acid in water
   C) A scale used for measuring the amount of acid in refrigerant
   D) The absolute Fahrenheit scale where zero is actually -460 °F

349. What is the Kelvin scale?
   A) The name given to the absolute Fahrenheit scale
   B) A scale for measuring metric weights
   C) The name given to the absolute centigrade scale
   D) None of the above

350. Latent heat of fusion is the heat
   A) released when a substance changes from a liquid to a solid state.
   B) released when a substance changes from a vapor to a liquid state.
   C) absorbed when a substance changes from a liquid to a vapor state.
   D) absorbed in the liquid refrigerant between the receiver and metering device.

351. The compressor's job is to
   A) pump the liquid refrigerant around.
   B) move gas from the low temperature to the high temperature side of the system.
   C) pump heat into the evaporator.
   D) relieve the discharge line of excessive pressure.

352. The condenser
   A) is located between the compressor and the metering device.
   B) is located between the liquid line and the evaporator.
   C) takes heat into the system.
   D) reduces pressure.

353. A receiver is a device
   A) located between the liquid line and the compressor.
   B) not needed on all refrigeration systems.
   C) for getting rid of all heat absorbed into the system.
   D) for storing superheated gas.

354. The evaporator
   A) is located between the compressor and receiver.
   B) is located between the compressor and condenser.
   C) puts the heat into the refrigerant.
   D) gets rid of heat.

355. Refrigerant
   A) is not needed in a mechanical refrigeration system.
   B) is transferred from the inside to the outside of the system.
   C) gets rid of heat by radiation.
   D) transfers heat from one side of the system to the other.

356. The discharge line
   A) connects the metering device to the evaporator.
   B) discharges heat to the metering device.
   C) discharges the gas from the evaporator to the condenser.
   D) is located between the compressor and condenser.

357. Heat enters the evaporator through
   A) radiation only.
   B) radiation, convection and conduction.

C) heat generated by the compressor.
D) the insulation in the outside walls only.

358. The suction line connects
A) the condenser and the receiver.
B) two parts of the system and is the smallest size of line used.
C) the evaporator to the compressor.
D) the metering device to the evaporator.

359. The largest line used on a compression system is the
A) discharge line.
B) suction line.
C) no one line, as the lines are all the same size.
D) liquid line.

360. In a mechanical system, heat flows into the
A) receiver and out of the evaporator.
B) evaporator and out at the condenser.
C) liquid line and out at the receiver.
D) compressor and out at the liquid line.

361. The compressor adds heat to the system by
A) reducing the volume of the gas and raising its pressure.
B) taking the heat of the condenser and compressing it.
C) raising its temperature by specific heat.

362. Most heat flows out of a refrigeration system through the
A) line connecting the condenser to the receiver.
B) suction line.
C) discharge line.
D) liquid line.

363. Fluid leaving the condenser
A) is in a gaseous state.
B) is in a solid state.
C) is slightly superheated.
D) should be slightly subcooled.

364. The condenser and discharge line
A) need help in getting rid of heat.
B) connect the evaporator and suction line.
C) must rid the system of all heat picked up in the cycle.
D) take heat into the system.

365. The line connecting the receiver to the evaporator is the
A) suction line.
B) liquid line.
C) discharge line.

366. A halide torch flame turns what color when it detects a leak?
   A) Red
   B) Blue
   C) Yellow
   D) Green
   E) No color

367. How many basic types of relays are used on domestic units?
   A) One
   B) Two
   C) Three
   D) Four
   E) Five

368. Why is it necessary to be careful when putting refrigerant into a cylinder?
   A) Gas expansion should be considered
   B) Cylinder expansion should be considered
   C) Possible moisture expansion should be considered
   D) The cylinder will burst if full of liquid
   E) Refrigerant is expensive

369. Does the service valve attachment have threads on it?
   A) Yes
   B) No
   C) On some models
   D) Only if it is a high-pressure unit
   E) No valve attachment is used

370. Most service valves are made of
   A) steel.
   B) copper.
   C) brass.
   D) monel.
   E) tool steel.

371. How does an electronic leak detector indicate a refrigerant leak?
   A) Flame color change
   B) Meter reading, light or bell
   C) Color change in the tube
   D) Color trace
   E) Bubbles

372. Liquid refrigerant
   A) burns the eyes and skin.
   B) freezes the eyes and skin.
   C) irritates the eyes and skin.
   D) does nothing to the eyes and skin.
   E) lubricates the eyes.

373. How does the leaking refrigerant reach the halide flame?
   A) It mixes with the fuel
   B) It is drawn through a sniffer tube.
   C) A pump is used
   D) An aspirator is used
   E) The refrigerant is in the air around the flame shield

374. How does one check the unit's electrical power outlet?
   A) Ammeter
   B) Test light
   C) If cabinet light won't work, power is off
   D) If house lights work, the power is all right
   E) Ohmmeter

375. What is the primary purpose of oil in a refrigeration system?
   A) Cooling
   B) Lubrication
   C) Removal of heat
   D) Acting as a catalyst to the refrigerant
   E) Preventing oxidation of parts

376. Considering the areas to be all equal, which of the following would permit the greatest heat gain?
   A) Wood
   B) Glass
   C) Brick

377. An awning over a window is less beneficial on which wall?
   A) South
   B) East
   C) North

378. The lowest temperature at which R-12 will boil is
   A) -21 °F.
   B) below -100 °F.
   C) 0 °F.

379. Dew point is
   A) the point at which air begins to give up its moisture.
   B) 32 °F.
   C) the same as the wet bulb temperature.

380. Relative humidity is
   A) the comfort zone.
   B) grains of moisture in the air.
   C) the amount of moisture in the air compared to 100% saturated air.

381. When the contacts are shorted to ground on a 24-V circuit, which side of the step-down transformer fails?
A) Primary
B) Secondary
C) Neither A nor B

382. Air filters should be cleaned or changed
A) when necessary.
B) every 6 months.
C) every month.
D) annually.

383. The condenser's job is to
A) supply the cycle with heated gas.
B) feed the evaporator gas.
C) condense refrigerant.
D) feed the compressor low-pressure gas.

384. The compressor suction and discharge valves are located
A) in the same place as the service valves.
B) in a safety head.
C) between the king and queen valves.
D) in the valve plate in the compressor.

385. The cooling coil is called a(n)
A) condenser.
B) evaporator.
C) freezer.

386. The purpose of the evaporator is
A) to remove heat.
B) to remove heat of compression.
C) the same as a receiver.

387. A humidistat controls
A) moisture.
B) flash gas.
C) refrigeration.

388. A drier is used for
A) drying oil.
B) holding flash gas.
C) drying refrigerant.

389. Five tons of refrigeration is equal to how many Btu?
A) 120,000
B) 90,000
C) 60,000

390. The two basic types of evaporators are
A) finned and plate.
B) prime surface and finned.
C) direct expansion feed and plate.
D) flooded and direct expansion feed.

391. For comfort cooling, the temperature is usually maintained at what degree below the ambient temperature?
    A) 5 °F
    B) 50 to 60 °F
    C) 10 to 12 °F
    D) 2 to 3 °F

392. The normal cut-out setting of a window unit thermostat is
    A) 40 to 50 °F.
    B) 55 to 60 °F.
    C) 60 to 70 °F.

393. Is there any oil in the condenser?
    A) Yes
    B) Only if the system is water-cooled
    C) Only if there is too much oil in the system
    D) No

394. A strainer screen is supposed to remove
    A) solid impurities.
    B) acid.
    C) oil.
    D) moisture.

395. A sight glass shows refrigerant flow.
    A) True
    B) False

396. Gas vapor temperature is raised by applying pressure.
    A) True
    B) False

397. R-12 comes in a white color-coded cylinder.
    A) True
    B) False

398. A reciprocating compressor can be driven by direct drive.
    A) True
    B) False

399. The two basic types of compressors are open and closed.
    A) True
    B) False

400. An evaporator is that part of the system in which liquid refrigerant is vaporized to produce refrigeration.
    A) True
    B) False

401. An expansion coil is an evaporator constructed of pipe or tubing.
    A) True
    B) False

402. High side indicates the very top of the compressor.
    A) True
    B) False

403. A liquid receiver is for the storage of compressor oil.
    A) True
    B) False

404. Low side indicates the bottom of the compressor.
    A) True
    B) False

405. Pressure is defined as
    A) 14.7 psia.
    B) the weight of air above us.
    C) force per unit area.

406. A low-side float is used on systems having
    A) a critical charge.
    B) one evaporator.
    C) one or more evaporators.

407. The purpose of the heat exchanger is to
    A) prevent sweating of the suction line.
    B) subcool the liquid.
    C) decrease the amount of superheat.

408. Temperature is defined as the
    A) amount of heat measurement.
    B) amount of heat required to cause a change of state.
    C) heat intensity measurement.

409. The most common application of the heat pump is in warmer climates where the coldest temperature reached is
    A) 32 °F or above.
    B) -10 to 0 °F.
    C) 10 to 20 °F.

410. The amount of heat required to change a liquid to a vapor with no change in temperature is called latent heat of
    A) fusion.
    B) vaporization.
    C) sublimation.

411. One must be careful when handling old compressor oil because it
    A) is necessary to keep it clean.
    B) may be hot.
    C) may be acidic.
    D) may spill.

412. How many kinds of capacitors are there?
    A) One
    B) Two

C) Three
D) Four

413. The term DOT stands for
A) Date of Transportation.
B) Department of Transportation.
C) Department of Testing.
D) none of the above.

414. What is another name for a capacitor?
A) Meter
B) Stator
C) Condenser
D) Rotor

415. How are the light switch and light wired?
A) In series with the motor
B) In parallel with the motor
C) Independent of the motor
D) In series with the thermostat

416. The purpose of the freezer door heater is to
A) prevent condensation from forming and freezing.
B) provide enough heat to give an adequate seal to the door.
C) keep the gasket pliable.
D) prevent moisture from dripping on the floor.

417. How many circuits are possible if two wires are fastened to one terminal?
A) One
B) Two
C) Three
D) Four

418. When a dehumidifier is operating in a room, will the temperature change within that room?
A) Yes, it will warm the room a small amount
B) Yes, it will cool the room a small amount
C) No

419. In a heat pump installation, is the direction of refrigerant vapor flow reversed through the compressor when the cycle is reversed from heating to cooling?
A) Yes
B) No
C) Yes, if the unit is a R-717 unit only

420. If provided for a positive displacement compression system, a compound gauge should be located on the
A) high side of the system.
B) low side of the system.
C) receiver.

421. What instrument should be used to check the receptacle outlet?
   A) Ohmmeter
   B) Megohmmeter
   C) Voltmeter
   D) Ammeter

422. How many terminals are used on most hermetic units which operate at 3,400 rpm?
   A) Three
   B) Four
   C) Two
   D) Five

423. What may be housed in the junction box on some units?
   A) Relay
   B) Thermostat
   C) Mullion heater
   D) Defrost control

424. What instrument should be used to check for a short circuit?
   A) Voltmeter
   B) Ohmmeter
   C) Ammeter
   D) Wattmeter

425. How many sets of points does a hot wire relay have?
   A) One
   B) Two
   C) Three
   D) Four

426. A relay is usually located
   A) in the thermostat.
   B) in the motor.
   C) on the condensing unit.
   D) in the capacitor box.

427. How should stranded wire be connected to a screw terminal?
   A) Wrap wire clockwise around screw
   B) Wrap wire counter-clockwise around screw
   C) Solder the wire, then wrap around the screw
   D) Fasten a terminal to wire end

428. What type of refrigerant control is in common use on small domestic freezers?
   A) Capillary tube
   B) Automatic expansion valve
   C) Thermostatic expansion valve
   D) Low-side float

429. Continuity is a(n)
    A) grounded wire.
    B) broken wire.
    C) complete electrical circuit.
    D) open circuit.

430. The resistance across a clean tight terminal should be
    A) 0 ohms.
    B) 50 ohms.
    C) 5,000 ohms.
    D) 50,000 ohms.

431. The common lead is the
    A) left side lead.
    B) right side lead.
    C) middle terminal.
    D) terminal with both a running and starting winding connection.

432. In a complete central air conditioning system, where is the condensing unit usually located?
    A) Inside the building
    B) Outside the building
    C) Inside or outside the building

433. Which of the following is the largest conductor?
    A) No. 8
    B) No. 10
    C) No. 12
    D) No. 14

434. How should a solid wire be wrapped around a terminal screw?
    A) Opposite the direction the screw turns as it is tightened
    B) The same direction the screw is turned as it is tightened
    C) Wrapped twice around the screw
    D) It should be kept straight

435. Temperature expressed in degrees above absolute zero is
    A) absolute temperature.
    B) saturation temperature.
    C) critical temperature.

436. Refrigerant containers must be approved by
    A) API ASME.
    B) ASRE.
    C) ICC.

437. In a mechanical refrigeration system, what is used to circulate the refrigerant?
    A) Condenser
    B) Compressor
    C) Refrigerant
    D) Suction line

438. In a mechanical refrigeration system, what picks up heat?
A) Unloader
B) Crosshead
C) Condenser
D) Evaporator

439. In a mechanical refrigeration system, what carries heat?
A) Refrigerant
B) Bypass line
C) Metering device
D) Motor drive device

440. The refrigerant absorbs heat in a mechanical system when
A) the vapor changes to a gas.
B) the vapor changes to a liquid.
C) it is at its critical points (temperature and pressure).
D) the liquid changes to vapor.

441. The temperature at which a change of state occurs is
A) constant.
B) the critical temperature.
C) dependent on the critical pressure.
D) none of the above

442. Heat flows from
A) a colder body to a warmer one.
B) a warmer body to a colder one.
C) horizontal to vertical.
D) a warm body to a hotter one.

443. When a substance is changing from vapor to liquid, what will happen if the pressure changes?
A) It will be the same as atmospheric pressure
B) The vapor will not vaporize
C) The refrigerant is useless
D) The temperature is changed

444. A refrigeration system must be
A) finger tight.
B) air wrench tight.
C) gas tight.
D) water tight.

445. The two main pressures operating in a system can be termed as
A) outside and inside pressure.
B) high-side and low-side pressure.
C) inside and ambient pressure.
D) crankcase and oil pressure.

446. For heat transfer to occur, there must be
A) a temperature difference between the materials.
B) a similarity in materials.

C) no temperature difference.
D) a difference in atmospheric pressures.

447. Most compressors are built out of
A) copper.
B) brass.
C) cast iron.
D) fiberglass.

448. The most common crankshaft used in compressors is a
A) completely straight one.
B) swash plate.
C) Scotch yoke.
D) crank throw or automotive.

449. Eccentric compressors have
A) extra long pistons.
B) large bearing surfaces on journals.
C) short, fat pistons.
D) small bearing surfaces on journals.

450. Small domestic compressors are usually
A) splash lubricated.
B) pump lubricated.
C) in sizes between 10 and 50 horsepower.
D) supplied with service valves.

451. Compressor valves are usually
A) thin, steel discs.
B) thick, steel discs.
C) a bimetal material.
D) available with handles on them.

452. What refrigerant controls are usually used on commercial air conditioning evaporators?
A) Remote TXV, self-contained capillary tube
B) Remote capillary tube, self-contained TXV
C) Neither A nor B

453. Which of the following is not a form of energy?
A) Chemical action
B) Heat
C) Electricity
D) Condensers

454. Zero gauge pressure is equal to what absolute pressure?
A) 30 inches vacuum
B) 14.7 psia
C) 30 psig
D) 44.7 psia

455. Twenty inches of vacuum is equivalent to what absolute pressure?
A) 30.0 psig
B) 59.7 psia
C) -9.7 psig
D) 5.0 psia

456. Heat is not transmitted by way of
A) insulation.
B) convection.
C) radiation.
D) conduction.

457. Refrigeration
A) transfers heat.
B) produces cold.
C) will stop heat.
D) does none of the above.

458. Heat cannot be converted into
A) light.
B) water.
C) electricity.
D) work.

459. Two atmospheres is equivalent to
A) 29.4 psia.
B) 44.7 psia.
C) 74.7 psia.
D) 14.7 psia.

460. Which part is entirely on the high-side?
A) Compressor
B) Condenser
C) Metering device
D) Evaporator

461. The Celsius temperature equivalent to -40 °F is
A) 0 °C.
B) -20 °C.
C) -40 °C.
D) -60 °C.

462. Temperature is
A) an indication of relative heat or cold.
B) the amount of heat to raise 1 lb of water 1 °F.
C) a quantity that causes an increase in temperature.
D) a feeling of warmth.

463. In a unit comfort cooler, condensate is
A) drained into the condenser.
B) drained into the compressor.
C) neither A nor B.

464. How many degrees Celsius is 75 °F?
   A) 10 °C
   B) 20 °C
   C) 24 °C
   D) 32 °C

465. How should a voltmeter be connected into a circuit?
   A) In series
   B) In parallel
   C) Its prongs are clamped around the insulated wire
   D) It is used only at the power source
   E) It is used with a shunt

466. What size fan motor should be used as a replacement?
   A) 1/10 hp more
   B) 1/10 hp less
   C) One size smaller
   D) The same size
   E) One size larger

467. The definition of specific heat is the amount of heat necessary to raise one pound of
   A) any substance one degree F.
   B) water one degree F.
   C) neither A nor B.

468. How does an increase in evaporator pressure affect the boiling point of the refrigerant?
   A) No change
   B) Raises it
   C) Lowers it

469. In which of the components is the most latent heat dissipated?
   A) Compressor
   B) Condenser
   C) Metering device
   D) Evaporator

470. Is a 3rd class licensed operator permitted to change or remove refrigerant from the system?
   A) Yes
   B) No

471. The thermostatic expansion valve operation is determined by how many pressures?
   A) One
   B) Two
   C) Three
   D) Four

472. On a thermostatic expansion valve, if the inlet pressure is 27 psig, and the spring pressure is 7 psig, what should the bulb pressure be for the valve to be in equilibrium (valve has no equalizer connection)?
    A) 27 psig
    B) 7 psig
    C) 34 psig
    D) 20 psig

473. A capillary tube must be used on a
    A) critically charged system.
    B) flooded system.
    C) receiver system.
    D) multiple evaporator system.

474. Where refrigerant is concerned, name two types of evaporators.
    A) Air-cooled and water-cooled
    B) Direct expansion and flooded
    C) Flooded and water-cooled

475. Pertaining to construction, name three types of evaporators.
    A) Flooded and water-cooled
    B) Air-cooled and water-cooled
    C) Bare tube, plate surface, finned tube

476. The minimum voltage that can be supplied to a 230-V solenoid coil is approximately
    A) 190 V.
    B) 196 V.
    C) 200 V.
    D) 210 V.

477. A window-mounted air conditioner rated at 2-1/3 tons is capable of removing how many Btu per hour?
    A) 27,000
    B) 28,000
    C) 30,000
    D) 32,000

478. What type of air diffuser has a balancing damper for controlling air flow?
    A) Baseboard
    B) High side-wall
    C) Low side-wall
    D) Flush-floor

479. What type of blower is most commonly used in a duct system?
    A) Propeller type
    B) Multiblade, with blades curved forward
    C) Disk type
    D) Multiblade with radical blades

480. The cooling capacities of window-mounted air conditioners range from
   A) 1/3 to 2 tons.
   B) 1/3 to 5 tons.
   C) 2/3 to 5 tons.
   D) 1 to 5 tons.

481. In the refrigeration cycle of an air conditioner,
   A) the fan motor heats the cold air that is picked up by the refrigeration system.
   B) high-pressure liquid refrigerant is increased in pressure by the capillary tube.
   C) the capillary tube acts as a restrictor to keep the high and low sides of the system together.
   D) the fan motor circulates air to remove heat picked up by the refrigerant system.

482. In servicing window-mounted units,
   A) do not use soapy water on aluminum mesh type filters.
   B) clean condenser coils with steel brush.
   C) clean aluminum mesh-type filters with soap and hot water.
   D) clean aluminum mesh-type filters monthly at a maximum.

483. How often should a blower be completely disassembled and inspected for defects?
   A) Once a year
   B) Semiannually
   C) Quarterly
   D) Monthly

484. Which of the following expansion devices does not require a thermostatic motor control for the operation of the condensing unit?
   A) Thermostatic expansion valve
   B) High-side float
   C) Low-side float
   D) Capillary tube

485. In a reach-in refrigerator using an evaporator fan, the thermostat feeler bulb is normally located
   A) in front of the evaporator.
   B) in the refrigerated space.
   C) in evaporator return airstream.
   D) on the evaporator.

486. The most important step in checking continuity with a volt-ohmmeter is to
   A) zero in the motor.
   B) select the proper scale.
   C) disconnect power source.
   D) ensure proper contact between test leads and component being tested.

487. The weight oil used when lubricating a water circulating pump is
   A) SAE 20.
   B) SAE 30.
   C) SAE 40.
   D) SAE 10 W 30.

488. A major cause of poor performance in an air conditioner is
   A) high voltage.
   B) dirty filters.
   C) recoating filters.
   D) using soapy water to clean aluminum mesh-type filters.

489. In testing capacitors, what must be done first?
   A) Disconnect service plug
   B) Check for fully discharged capacitor
   C) Check starting relay
   D) Remove thermostat

490. A fuse that has blown due to a hot conductor touching a neutral conductor is called a(n) ____________ in a defective circuit.
   A) short
   B) cross short
   C) direct short
   D) open

491. In a piston-type compressor, how are the intake valves closed?
   A) With spring tension
   B) On down stroke
   C) By head pressure
   D) By crankcase pressure

492. When flux is applied and spread unevenly on the surfaces being joined, the metal
   A) is not at proper temperature.
   B) is not clean.
   C) has too much copper in the alloy.
   D) has too much iron in the alloy.

493. Always apply refrigerant oil to
   A) fittings and flare before assembly.
   B) tubing after swaging.
   C) tools before swaging.
   D) fittings and flare after joining.

494. When a welding operation is completed, the first and second steps in shutting down the torch are
   A) close oxygen and acetylene cylinder valves.
   B) close acetylene valve, then oxygen valve on torch.
   C) open regulators until the gases cease to flow.
   D) release tension on regulator screws.

495. In an oxygen/acetylene apparatus, the green hose is connected to the
    A) torch needle valve stamped OX.
    B) acetylene regulator.
    C) torch needle valve stamped ac.
    D) acetylene or oxygen regulator.

496. It is necessary to crack the valve on an oxygen cylinder before connecting the regulator to
    A) blow out dirt or other foreign matter.
    B) make certain cylinder is not empty.
    C) blow out moisture.
    D) determine whether valve will operate.

497. In a refrigeration system, the filter-drier should be changed
    A) when suction pressure rises.
    B) as often as necessary.
    C) every 2 months.
    D) every 6 months.

498. A bleeder resistor protects the
    A) motor starting winding.
    B) motor running winding.
    C) capacitor.
    D) relay contacts.

499. Which of the following is true about contacts in a good relay when they are not energized?
    A) Hot wire relay has one set of contacts open and one set closed
    B) Potential relay has one set of contacts open and one set closed
    C) Current relay has one set of contacts open
    D) Current relay has one set of contacts closed

500. Which relay may rattle (as if a part is loose) when the case is tapped?
    A) Potential
    B) Current
    C) Hot wire
    D) Time-delay

501. Sealing the insulation in a refrigerator will keep moisture out of the
    A) box.
    B) shell.
    C) system.
    D) insulation.

502. When brazing copper tubing, how hot should the work be heated before applying the alloy?
    A) To the alloy flow point
    B) To alloy melting point
    C) 1,400 °F
    D) 1,300 °F

503. To make an accurate measurement with an inside micrometer using an extension rod,
   A) use the ratchet stop.
   B) take readings at several points.
   C) hold both feet firmly against the lip.
   D) first check inside micrometer with outside micrometer.

504. Maximum current will flow through the ohmmeter circuit when
   A) there is a minimum amount of resistance between ohmmeter terminals.
   B) the scale indicates 6,000 ohms.
   C) there is maximum resistance to the flow.
   D) the scale indicates INF.

505. Universal motors may be used on dc or
   A) ac.
   B) single-phase ac.
   C) two-phase ac.
   D) three-phase ac.

506. At what temperature will all molecular movement in a substance stop?
   A) 70 °F
   B) 50 °F
   C) 20 °F
   D) Absolute zero

507. In the refrigeration cycle, heat absorbed by the refrigerant takes the form of
   A) absorption heat.
   B) latent heat of evaporation.
   C) latent heat of condensation.
   D) sensible heat that reduces refrigerant temperature.

508. What is the pressure on a person in pounds per square foot at sea level?
   A) 2,117
   B) 1,117
   C) 1,007
   D) 14.7

509. What air is used to cool the condenser of an air-cooled comfort cooler?
   A) Inside air
   B) Outside air
   C) Inside or outside air

510. What is entropy?
   A) Total heat in 10 pounds of a substance
   B) Pressure at which a liquid remains liquid
   C) Critical temperatures
   D) Mathematical constant for calculating energy in a system

511. What heat is added to a refrigerant in the compressor?
   A) Compression
   B) Latent
   C) Specific
   D) Sensible

512. In a refrigeration system, in which of the following components is the vaporized refrigerant liquefied by the removal of heat?
   A) Condenser
   B) Evaporator
   C) Absorber
   D) Compressor

513. Sensible heat is the
   A) hidden heat present in a substance.
   B) movement of molecules within a substance.
   C) heat that changes a substance from liquid to vapor.
   D) heat that can be added or subtracted without a substance changing state.

514. The specific heat of ice is
   A) 1.0.
   B) 0.5.
   C) 1.5.
   D) 144.

515. Most heating, air-conditioning, and ventilating ducts are made from
   A) galvanized sheet iron.
   B) aluminum.
   C) stainless steel.
   D) black iron.

516. Backseating a suction shut-off valve closes the
   A) suction line and compressor port.
   B) compressor and gauge port.
   C) compressor port.
   D) gauge port.

517. How many watts are developed in a circuit having 100 ohms resistance and an amperage draw of 5 amps?
   A) 2,000
   B) 2,500
   C) 2,700
   D) 3,000

518. What color is equipment ground wire insulation?
   A) Yellow
   B) Green
   C) White
   D) Blue

519. While visually checking the wires of a circuit, the electrician notices corrosion at a connection. This is a sign of a
   A) dry condition.
   B) poor connection.
   C) short.
   D) normal condition.

520. What, if anything, occurs in a wiring system when one or more conductors in a circuit is broken or otherwise separated?
   A) A short circuit
   B) An open circuit
   C) A blown fuse
   D) Nothing

521. What results when two wires of opposite polarity touch each other?
   A) Short
   B) Open
   C) Ground
   D) Close

522. Which instrument measures specific gravity of a solution?
   A) Multimeter
   B) Test lamp
   C) Hydrometer
   D) Rectifier

523. To ensure that a deenergized circuit is not shorted and still hot, it is necessary to
   A) remove the fuses.
   B) open the switch.
   C) test it with a voltmeter.
   D) lock the switch to off.

524. The effects of current flow through a conductor are heat,
   A) magnetism, and chemical action.
   B) magnetism, and physical shock.
   C) and magnetism.
   D) and shock.

525. When a conductor is moved across a magnetic field, voltage is induced into the conductor by
   A) relative motion.
   B) mutual induction.
   C) magnetic induction.
   D) electromagnetic induction.

526. In alternating current, the number of cycles per second determines the
   A) voltage.
   B) current.

C) frequency.
D) amperage.

527. The total voltage of four primary cells (1.5 V each) connected in parallel is
A) 0.38 V.
B) 1.5 V.
C) 6 V.
D) 12 V.

528. The formula for computing the current flow through a circuit (Ohm's law) is
A) I= E ÷ R.
B) I= R ÷ E.
C) I= R x E.
D) I= E2 ÷ R2.

529. A 24-V series circuit contains two 6 ohm resistors. What is the current flow through the circuit?
A) 6 ohms
B) 4 ohms
C) 2 ohms
D) 1 ohms

530. Which of the following rules does not apply to a parallel circuit?
A) The applied voltage is the same across each path
B) There is more than one path for circuit flow
C) The total resistance is the sum of the individual resistors
D) Total current flow is the sum of the current in each path

531. What is the total resistance of two 20 ohm resistors connected in parallel?
A) 20 ohms
B) 10 ohms
C) 5 ohms
D) 2 ohms

532. What is the total current flow through a 120-V circuit that contains a 10 ohm resistor in series with two 10 ohm resistors connected in parallel?
A) 20 amperes
B) 15 amperes
C) 10 amperes
D) 8 amperes

533. Polarity must be observed when measurements are made with a(n)
A) dc voltmeter.
B) ac voltmeter.
C) ohmmeter.
D) megohmmeter.

534. When using a voltmeter to troubleshoot an inoperative circuit, first check for voltage at the
   A) first box.
   B) switch.
   C) last outlet.
   D) last transformer.

535. After connecting the test leads of an ohmmeter to a short length of continuous wire, the ohmmeter scale should indicate
   A) 10 ohms.
   B) 100 ohms.
   C) continuity.
   D) infinity.

536. An electrical test instrument that is a combination voltmeter, ohmmeter, and ammeter is called a
   A) megger.
   B) voltmeter.
   C) milliammeter.
   D) multimeter.

537. Fuse pullers are used to
   A) remove fuse clips.
   B) remove large fuses only.
   C) prevent sparkling.
   D) prevent electric shock.

538. A large compressor using R-11 will be
   A) rotary.
   B) reciprocating.
   C) centrifugal.

539. A compressor using R-12 will be
   A) rotary.
   B) reciprocating.
   C) centrifugal.
   D) all of the above.

540. A compressor using R-22 will be
   A) rotary or reciprocating.
   B) reciprocating or centrifugal.
   C) centrifugal.
   D) all of the above.

541. A compressor using R-500 will be
   A) rotary.
   B) reciprocating.
   C) centrifugal.
   D) all of the above.

542. A compressor using R-502 will be
   A) rotary.
   B) reciprocating.

C) centrifugal.
D) all of the above.

543. A compressor using R-503 will be
A) rotary.
B) reciprocating.
C) centrifugal.
D) all of the above.

544. A compressor using R-13 will be
A) rotary.
B) reciprocating.
C) centrifugal.
D) all of the above.

545. A compressor using R-113 will be
A) rotary.
B) reciprocating.
C) centrifugal.
D) all of the above.

546. A domestic refrigerator uses
A) R-11.
B) R-22.
C) R-12.
D) R-502.

547. A domestic food freezer uses
A) R-11.
B) R-22.
C) R-12.
D) R-502.

548. Home air conditioning units use
A) R-11.
B) R-22.
C) R-12.
D) R-502.

549. Cryogenic applications may use
A) R-114.
B) R-22.
C) R-502.
D) R-503.

550. The recommended oil for systems using R-11, R-12 and R-113 with evaporator temperatures above -20 °F is
A) 5 GS.
B) 4 GS.
C) 3 GS.
D) 2 GS.

551. The recommended oil for systems using R-11, R-12 and R-113 with evaporator temperatures below -20 °F is
   A) 5 GS.
   B) 4 GS.
   C) 3 GS.
   D) 2 GS.

552. The recommended oil for systems using R-13, R-22, R-114 and R-502 is
   A) 5 GS.
   B) 4 GS.
   C) 3 GS.
   D) 2 GS.

553. The recommended oil for systems using ethane, propane and isobutane is
   A) 5 GS.
   B) 4 GS.
   C) 3 GS.
   D) 2 GS.

554. The recommended oil for systems using ammonia with evaporator temperatures above -20 °F is
   A) 5 or 3 GS.
   B) 4 or 5 GS.
   C) 3 or 2 GS.
   D) 2 or 4 GS.

555. The recommended oil for systems using ammonia with evaporator temperatures below -20 °F is
   A) 5 GS.
   B) 4 GS.
   C) 3 GS.
   D) 2 GS.

556. What does COP stand for?
   A) Conditioning of performance
   B) Compressor of performance
   C) Component of performance
   D) Coefficient of performance

557. Which refrigerant group requires a gas mask?
   A) Group I
   B) Group II
   C) Group III
   D) All groups

558. What is probably wrong if the cabinet is too cold and the unit runs continuously?
   A) Faulty relay
   B) Out of refrigerant

C) Open circuit
D) Faulty thermostat

559. The condenser surface must be kept clean to
A) prevent overheating the fan motor.
B) keep moisture out of the system.
C) prevent overheating of the heater.
D) keep the head pressure down.

560. What is an indication that there is too much refrigerant in the system?
A) Unit will not start
B) Unit will not freeze
C) Suction line will sweat or frost
D) Suction line will be warm

561. What is the most popular refrigerant control in domestic refrigerators?
A) TXV
B) AEV
C) EPR
D) Capillary tube

562. A filter is needed at the inlet of the capillary tube to remove
A) moisture.
B) oil.
C) air.
D) solid particles.

563. The most common place to install a filter drier is
A) between the compressor and condenser.
B) in the liquid line.
C) between the evaporator and compressor.
D) anywhere.

564. If the unit is running with low-side pressure too high and high-side pressure too low, what is wrong with the unit?
A) Too little refrigerant
B) Bad thermostat
C) Too much refrigerant
D) Bad compressor valves

565. What is the compression ratio if the low-side pressure is 68 lb and the high-side pressure is 243 lb?
A) 3.5
B) 3.1
C) 4.3
D) 4.0

566. What is the compression ratio if the low-side pressure is 6 lb and the high-side pressure is 126 lb?
   A) 4.1
   B) 3.4
   C) 6.7
   D) 21

567. What is the compression ratio if the low-side pressure is 65 lb and the high-side pressure is 260 lb?
   A) 4.0
   B) 3.4
   C) 4.9
   D) 3.5

568. What is the superheat for a unit with a high-side pressure of 260 lb and a low-side pressure of 68 lb; a high-side temperature of 120 °F and a low-side temperature of 40 °F; and an evaporator outlet temperature of 47 °F?
   A) 40 °F
   B) 47 °F
   C) 7 °F
   D) 0 °F

569. What is the subcooling for a unit with a high-side pressure of 260 lb and a low-side pressure of 68 lb; a high-side temperature of 120 °F and a low-side temperature of 40 °F; and a condenser outlet temperature of 90 °F?
   A) 30 °F
   B) 90 °F
   C) 120 °F
   D) 68 °F

570. How is superheat measured?
   A) Low-side pressure divided into high-side pressure
   B) High-side pressure divided into low-side pressure
   C) Low-side pressure converted to temperature, then divided into high-side pressure
   D) Low-side pressure converted to temperature, then divided into high-side temperature
   E) None of the above

571. How is subcooling measured?
   A) High-side pressure divided into low-side pressure
   B) Low-side pressure divided into high-side pressure
   C) High-side pressure converted to temperature then divided into low-side pressure
   D) Low-side pressure converted to temperature then divided into high-side temperature
   E) None of the above

572. 120 V to all legs from neutral is a
    A) star transformer.
    B) delta transformer.
    C) polyphase transformer.
    D) none of the above.

573. A dirty condenser coil will cause
    A) high operating cost.
    B) low operating cost.
    C) oil loss.
    D) none of the above.

574. Which refrigerants are bad for the ozone?
    A) FCF
    B) CHC
    C) CFC
    D) FCC

575. When there is no cooling and the compressor runs continuously, the problem is
    A) air in the system.
    B) no refrigerant.
    C) low oil.
    D) a bad thermostat.

576. Which refrigerant is least dangerous for the ozone?
    A) R-11
    B) R-12
    C) R-22
    D) R-113

577. Some of the sun's ultraviolet radiation is filtered out before it reaches the earth by the
    A) cloud layer.
    B) troposphere.
    C) ionosphere.
    D) ozone layer.

578. The reduction of the ozone in the atmosphere could cause a(n)
    A) increase in plant life growth.
    B) decrease in the Earth's temperature.
    C) increase in marine life growth.
    D) increase in skin cancer and eye cataracts.

579. Recycled refrigerant should only be stored in containers that meet the standards set by
    A) the National Bureau of Standards.
    B) SAE J1141.
    C) DOT CFR Title 49.
    D) Refrigerant Carriers Institute.

580. A container of recycled refrigerant should not be used until it has been checked for
   A) halon.
   B) fluorocarbons.
   C) noncondensable gases (air).
   D) condensed chlorine gas.

581. An Underwriters Laboratories (UL) approval seal on a recovery recycling unit ensures that it is capable of cleaning refrigerant to what SAE standard?
   A) J1988
   B) J1989
   C) J1990
   D) J1991

582. Voltage is
   A) flow.
   B) pressure.
   C) current.
   D) resistance.

583. What causes the needle valve to open in an automatic expansion valve refrigerant control?
   A) Reduction of pressure in the compressor
   B) Reduction of pressure in the high side
   C) Reduction of pressure in the low side

584. The needle valve opens in the thermostatically-controlled expansion valve when the low-pressure side is lower than the pressure in the
   A) condenser.
   B) thermal bulb.
   C) compressor.

585. What substances are most used in absorption-type refrigerators?
   A) R-12 and ammonia
   B) Ammonia and hydrogen
   C) Water, ammonia and hydrogen

586. The main advantage of a cascade system is that it
   A) works with R-11 very well.
   B) has the ability to reach very high temperatures.
   C) has the ability to reach very low temperatures.

587. It is sometimes necessary to use a multiple evaporator system when several cabinets are to be cooled from
   A) two compressors.
   B) a condensing unit.
   C) neither A nor B.

588. What is the basic principle behind the operation of most ice cube makers?
   A) Circulating chilled water over an evaporator surface

B) Circulating R-502 over an evaporator surface
C) Neither A nor B

589. *Expendable refrigerant* cooling is used most often
A) with R-12.
B) with R-504.
C) in transporting frozen foods.

590. What is the refrigerant most commonly used in expendable refrigerant cooling systems?
A) SO_2
B) NH_3
C) Liquid nitrogen

591. In cascade systems, are the refrigerating temperatures usually above or below 0 °F?
A) Above
B) Below
C) Temperature is the same

592. What is another name for an open refrigerating system?
A) Internal drive system
B) External drive system
C) Neither A nor B

593. Hydrogen is used in some absorption systems, because it allows the
A) water to evaporate at a lower pressure.
B) ammonia to evaporate at a higher pressure.
C) ammonia to evaporate at a lower pressure.

594. Name the main parts commonly located in the low-pressure side.
A) Condenser, evaporator and suction line
B) Condenser, compressor and suction line
C) Evaporator, compressor and suction line

595. Name the main parts commonly located in the high-pressure side.
A) Condenser, evaporator and suction line
B) Evaporator, compressor and suction line
C) Condenser, compressor and liquid line

596. Name four types of compressors.
A) Open, closed, gear and centrifugal
B) Open, rotary, gear and centrifugal
C) Reciprocating, rotary, gear and centrifugal

597. Some rotary compressors use check valves in the suction line to keep the high pressure from backing up into the
A) condenser.
B) evaporator.
C) neither A nor B.

598. Condensers should be cleaned to
   A) keep the low side down.
   B) remove heat from the evaporator efficiently.
   C) remove heat from the condenser efficiently.

599. What type of evaporator uses an expansion valve?
   A) Wet evaporator
   B) Dry evaporator
   C) Neither A nor B

600. The control that determines refrigerator cabinet temperature is the
   A) compressor.
   B) evaporator.
   C) motor control.

601. How much clearance is allowed between the piston and the cylinder on small compressors?
   A) 0.01" to 0.05"
   B) 0.001" to 0.005"
   C) 0.0001" to 0.0005"

602. How much lift is allowed in the intake and exhaust valves?
   A) 0.10" to 0.15"
   B) 0.010" to 0.015"
   C) 0.0010" to 0.0015"

603. Why are automatic expansion valves (AEV) usually adjustable?
   A) For different condenser temperatures
   B) For different evaporator temperatures
   C) Neither A nor B

604. A capillary tube reduces pressure by resistance to the flow of
   A) liquid.
   B) gas.
   C) neither A nor B.

605. What basic conditions are necessary to produce refrigeration?
   A) High pressure and liquid refrigerant
   B) Low pressure and liquid refrigerant
   C) Neither A nor B

606. When do the molecules changing from liquid to vapor equal the molecules changing from vapor to liquid?
   A) When the heat input equals the heat output
   B) When the heat output equals the heat output
   C) Neither A nor B

607. The purpose of a compressor is to move the evaporated gas to a
   A) low-temperature, low-pressure level.
   B) high-temperature, high-pressure level.
   C) neither A nor B.

608. Compressor pistons are made of
    A) copper.
    B) brass.
    C) cast iron.

609. Compressor cylinders are cooled by
    A) R-718.
    B) R-12.
    C) air or water.

610. Hermetic motors are usually cooled by
    A) water.
    B) low temperature vapor.
    C) oil.

611. What is sometimes used in a rotary compressor in place of an intake valve?
    A) Stationary blade
    B) Piston
    C) Neither A nor B

612. Does a centrifugal compressor have exhaust valves?
    A) Yes
    B) No
    C) Only in R-502 units

613. On external drive compressors, should the crankshaft seal rubbing surfaces be lubricated?
    A) No
    B) Yes
    C) Only on R-11 units

614. Are some compressors water-cooled?
    A) No
    B) Yes
    C) Only in R-718 units

615. What is the least number of impellers a centrifugal compressor may have?
    A) One
    B) Two
    C) Three

616. How many openings does a compressor service valve have?
    A) One
    B) Two
    C) Three

617. An oil separator removes oil from the
    A) low-side vapor.
    B) discharge vapor.
    C) neither A nor B.

618. A liquid line filter-drier is used to
   A) remove moisture from the low side.
   B) remove dirt and moisture from the liquid refrigerant.
   C) keep the compressor cool.

619. How many pressures or forces influence the needle movement of an automatic expansion valve needle?
   A) Two
   B) Three
   C) Four

620. Why are expansion valves adjustable?
   A) They are not adjustable
   B) So the same valve can be used for different refrigerants
   C) Neither A nor B

621. Thermostatic expansion valve superheat is the temperature difference between the liquid refrigerant in the
   A) condenser and the evaporator temperature.
   B) evaporator and the thermal bulb temperature.
   C) neither A nor B.

622. What mesh screens are used in refrigerant controls?
   A) 20 to 60 mesh
   B) 40 to 80 mesh
   C) 60 to 100 mesh

623. The most common expansion valve body material is
   A) copper.
   B) steel.
   C) brass.

624. The liquid line is usually attached to the expansion valve by a
   A) flare connection.
   B) flare connection or soldered joint.
   C) soldered joint.

625. What is a pilot-operated solenoid valve?
   A) A solenoid that operates a large valve and the pressure from this large valve operates a small valve
   B) A solenoid that operates a small valve and the pressure from this small valve operates a large valve
   C) Neither A nor B

626. How does a capillary tube operate?
   A) High-side pressure
   B) Low-side pressure
   C) Reduces the pressure by friction

627. Does a capillary tube system usually use a liquid receiver?
   A) Yes
   B) No
   C) Only on R-502 units

628. Does a capillary tube need a filter or a screen at its inlet?
   A) Yes
   B) No
   C) Only on R-12 units

629. What type of system needs a check valve?
   A) Reciprocating and rotary
   B) Reciprocating
   C) Rotary

630. A solenoid valve is opened by
   A) low-side pressure.
   B) high-side pressure.
   C) electricity.

631. A thermostatic expansion valve equalizer is used to compensate for the pressure drop through the
   A) condenser.
   B) evaporator.
   C) compressor.

632. What may be the trouble if a capillary tube unit frosts down the suction line?
   A) Too little refrigerant in the system
   B) Too much refrigerant in the system
   C) Air in the unit

633. What is a four-way solenoid valve?
   A) Two solenoid valves
   B) A valve that can reverse the direction of refrigerant flow
   C) Neither A nor B

634. How is a capillary tube usually fastened to 3/8" OD copper tubing?
   A) Tape
   B) Solder
   C) Silver brazing

635. What is the most popular TXV superheat setting?
   A) 1 to 3 °F
   B) 2 to 5 °F
   C) 8 to 10 °F

636. How many field windings are used in a 120 to 240-V repulsion-start induction motor?
   A) One
   B) Two
   C) Three

637. How many terminals does a domestic hermetic compressor usually have?
   A) One
   B) Two
   C) Three

638. How many windings does a single-phase hermetic motor stator have?
   A) One
   B) Two
   C) Three

639. Is the starting capacitor connected in series with the starting winding?
   A) No
   B) Yes
   C) Only on three-phase

640. How is the fan motor connected electrically to the compressor motor?
   A) Series parallel
   B) Series
   C) Parallel

641. How may a shorted capacitor be detected?
   A) Amprobe
   B) Continuity
   C) Neither A nor B

642. How much may the voltage drop at the motor terminals before it causes trouble?
   A) 1 to 3%
   B) 3 to 5%
   C) 5 to 10%

643. What color is the ground wire?
   A) Black
   B) White
   C) Green

644. What is the method used to locate R-502 leaks?
   A) Halide torch
   B) R-718
   C) Oil

645. What is a common head pressure for air-cooled R-12 refrigerating systems?
   A) 70 psig
   B) 127 psig
   C) 250 psig

646. What is the method used to locate R-22 leaks?
   A) Oil
   B) Halide torch
   C) Water

647. What does toxic mean?
   A) A type of oil
   B) Detrimental to the respiratory system
   C) Detrimental to the compressor

648. In an R-12 system, the oil must be free of
   A) wax and moisture.
   B) wax.
   C) moisture.

649. Is it advisable to substitute refrigerants in a system?
   A) Yes
   B) No
   C) Only on R-502 units

650. How does air in the system affect the head pressure?
   A) Air will lower head pressure
   B) Air will increase head pressure
   C) Head pressure remains the same

651. Name two refrigerants that may not be tested with the halide torch.
   A) R-22 and R-12
   B) R-114 and R-502
   C) R-744 and R-717

652. To determine the correct head pressure of a water-cooled condenser, add approximately
   A) 2 to 5 °F.
   B) 10 to 15 °F.
   C) 20 to 30 °F.

653. Group I refrigerants are
   A) toxic.
   B) flammable.
   C) the safest.

654. Group II refrigerants are
   A) toxic.
   B) flammable.
   C) the safest.

655. Group III refrigerants are
   A) toxic.
   B) flammable.
   C) the safest.

656. May carbon dioxide be used as a refrigerant?
   A) Yes
   B) No
   C) Only on small units under 2 tons

657. Can an R-22 system safely have more moisture in it than an R-12 system?
   A) Yes
   B) No
   C) Only on units of 10 tons or more

658. What are some of the safety devices in the electrical circuit of a hermetic domestic refrigerator?
   A) Thermostat
   B) Circuit breaker and overload
   C) Neither A nor B

659. In a butter compartment, the temperature is controlled by
   A) a thermostat, which controls the electrical flow to the heater in the compartment.
   B) hot gas from the compressor.
   C) hot gas from the condenser.

660. What controls the flow of hot gas through the evaporator for defrosting?
   A) Capillary tube
   B) Solenoid valve
   C) Condenser

661. In late model domestic refrigerator, most refrigerant lines are located in the
   A) door jamb.
   B) front.
   C) side.

662. During the defrost cycle, the defrost water is removed
   A) in the evaporator.
   B) to the drain pan near the compressor.
   C) in the condenser.

663. What is the most common refrigerant control system used on domestic frozen food cabinets?
   A) Thermostat
   B) TXV
   C) Capillary tube

664. A freezer should be defrosted once every
   A) 3 months.
   B) 6 months.
   C) year.

665. Why are the suction liquid lines sometimes soldered together?
   A) To heat the liquid line
   B) To cool the liquid line
   C) To cool the condenser

666. What is the best temperature range for a home freezer?
   A) 20 to 30 °F
   B) 10 to 20 °F
   C) 0 to -20 °F

667. What kind of motor control is used on most frozen food cabinets?
   A) Thermostat

B) Capillary tube
C) TXV

668. How does a no-frost refrigerator prevent sweating and frosting?
A) Hot gas from the condenser
B) Heat from the compressor
C) Neither A nor B

669. A drain heater is a heating element at
A) the compressor.
B) the condensate drain.
C) neither A nor B.

670. What are some possible locations of a condensing unit?
A) Base of unit
B) Top of unit
C) Base, top, and remote

671. Why must an evaporator be defrosted?
A) So there will be more room in the evaporator
B) Frost acts as an insulation
C) Neither A nor B

672. Cabinet door joints are made airtight by a
A) gasket.
B) vacuum.
C) clamp.

673. What happens to defrost water in a domestic refrigerator?
A) It is drained to a pan near the compressor
B) It is drained to a pan near the compressor
C) Neither A nor B

674. What type of insulation is used in a domestic freezer?
A) Styrofoam
B) Cork
C) Polyurethane foam

675. Do all hermetic systems have service valves?
A) Yes
B) No
C) Only on 1-hp compressors

676. How many fans does a window-type comfort cooler have?
A) One
B) Two
C) Three

677. Describe a good way to clean the condenser of a domestic refrigerator.
A) Water
B) Vacuum cleaner
C) R-12

678. What is the most popular method of sealing tubing joints when servicing hermetic systems?
A) Silver brazing
B) Solder
C) Neither A nor B

679. How many coils does air pass through in a dehumidifier?
A) One
B) Two
C) Three

680. The compound gauge is usually connected to the
A) high-pressure side.
B) low-pressure side.
C) high- or low-pressure side.

681. The high-pressure gauge is usually connected to the
A) high-pressure side.
B) low-pressure side.
C) high- or low-pressure side.

682. Before a system is opened, the pressures must be balanced to prevent
A) the compressor from overheating.
B) air flow into the system.
C) neither A nor B.

683. What indicates the presence of air in a condenser?
A) A lower than normal head pressure
B) A higher than normal head pressure
C) Neither A nor B

684. Describe one method of adding oil to a system.
A) Pressure forcing into the high side
B) Pressure forcing into the low side
C) Neither A nor B

685. How does a process tube adaptor fasten to a process tube?
A) Silver brazing
B) Solder
C) Clamp on

686. At what temperature does water evaporate at a 29" Hg vacuum?
A) 150 to 212 °F
B) 100 to 150 °F
C) 75 to 80 °F

687. Why is it dangerous to handle oil in a burned out motor compressor?
A) It is very hot
B) It has a high acid content
C) It is very cold

688. What happens to a system if a capillary tube of the same ID, but longer, is installed?
    A) Too much refrigerant in the system
    B) Too little refrigerant in the system
    C) It will be the same

689. What kind of pump is needed to produce a high vacuum?
    A) Two-stage rotary
    B) Reciprocating
    C) Neither A nor B

690. Piercing valves are installed on the
    A) low side only.
    B) high and low side.
    C) high side only.

691. How can one find out if a defrost heating element is burned out?
    A) Unit will be very hot
    B) Test with an ohmmeter
    C) Neither A nor B

692. To clean air and moisture out of the service lines and manifold, purge with
    A) refrigerant.
    B) oil.
    C) R-718.

693. A new liquid line filter-drier should be installed
    A) after a burn out.
    B) if the high-side pressure is too high.
    C) if the low-side pressure is too high.

694. What is the most sensitive leak detector?
    A) Halide torch
    B) Electronic
    C) Soap

695. What is the trouble if the suction line is frosted?
    A) Undercharge
    B) No load on the evaporator
    C) Dirty condenser

696. What happens if the hot gas valve is stuck in the closed position?
    A) System will not defrost
    B) Evaporator will be too hot
    C) Condenser will be too hot

697. Is it necessary to check the service lines and gauge manifold for leaks?
    A) No
    B) Yes
    C) Only on systems using R-717

698. Is the refrigerant charged into the system in liquid or gas form?
   A) Gas in the low side with the unit running
   B) Gas in the high side with the unit running
   C) Liquid in the high side with the unit running

699. Does air contain moisture below 32 °F ?
   A) Yes
   B) No
   C) Only from 0 to 32 °F

700. What three pressures are found when using a pitot tube?
   A) Static pressure, total pressure and velocity
   B) Vapor pressure, dew point pressure and velocity
   C) Vapor pressure, grains of moisture pressure and velocity

Refrigeration and Air Conditioning

Level 2

701. Can some of the return air be used to reduce the humidity of the cooled air?
    A) Yes
    B) No
    C) Only 2%

702. Is it possible to have a hydronic comfort cooling system?
    A) Yes
    B) No
    C) If it's not larger than 100,000 Btu

703. Describe the two ways window air conditioners may be installed.
    A) In a window or through the wall
    B) In a window or in the wall
    C) Neither A nor B

704. If the window comfort cooling unit will not start, first check the
    A) electrical supply.
    B) compressor.
    C) condenser fan.

705. How many capacitors does a window comfort cooler have?
    A) One
    B) Two
    C) Three

706. What is happening if a window unit drips water inside the room?
    A) Floor will get wet
    B) Unit is not level
    C) Both A and B

707. How much fresh air should each person receive?
    A) 2 cfm
    B) 4 cfm
    C) All they can get

708. In an electric filter, what is the voltage of the ionizing wires?
    A) 6,000 V
    B) 12,000 V
    C) 20,000 V

709. How efficient are 1" throwaway filters?
A) 50%
B) 70%
C) 80%

710. Most electronic filters are cleaned by washing them with
A) water.
B) water and any detergent.
C) water and dishwasher detergent.

711. In a heat pump, a capillary tube permits
A) travel of refrigerant one way.
B) the reversal of refrigerant through it.
C) neither A nor B.

712. In a heat pump, how many thermostatic expansion valves are used?
A) One
B) Two
C) Three

713. How many refrigerant connections does a four-way reversing valve have?
A) Two
B) Three
C) Four

714. How many check valves are used in a heat pump system equipped with thermostatic expansion valves?
A) One
B) Two
C) Three

715. In a heat pump system, motor compressor operation is most critical when the outdoor temperature drops to
A) 10 °F.
B) 20 °F.
C) 30 °F.

716. What two electrical devices does a heat pump thermostat control?
A) Compressor and four way valve
B) Compressor and hot gas valve
C) Condenser and hot gas valve

717. What parts of a warm-air furnace may be used with a central comfort cooling system?
A) Gas valve, thermostat and blower
B) Ducts, filters and blower unit
C) Blower, thermostat and ducts

718. What refrigerant controls are generally used on air conditioning evaporators?
A) TXV

B) Capillary tube
C) TXV or capillary tube

719. What type of thermostat must be carefully leveled?
A) Mercury tube
B) All thermostats
C) Neither A nor B

720. What is done to a silver brazed joint before it is heated and pulled apart?
A) Oiled
B) Cleaned and fluxed
C) Cleaned with water

721. What kind of oil is usually found in a motor compressor burnout?
A) Acidic
B) 3 GS
C) 5 GS

722. Why must high-pressure motor cut-outs be used with water-cooled condensing units?
A) Inadequate water supply would give an excessive head pressure
B) To stop the pressure when water flow is too great
C) To stop the pressure when the TXV flow is too great

723. What part of the compressor usually contains the intake valves?
A) Condenser
B) Valve plate
C) TXV

724. According to some refrigeration codes, a self-contained system that has been assembled and tested prior to its installation and which is installed without connecting any refrigerant-containing parts is claimed as a(n)
A) unit system.
B) remote system.
C) indirect system.

725. Solenoid valves are generally installed in the
A) suction line.
B) water line.
C) liquid line before the TXV.

726. The refrigerant methyl chloride
A) can be used for air conditioning.
B) cannot be used for air conditioning.
C) can be used anywhere.

727. TXVs are normally factory set for
A) 5 to 15 degrees of superheat.
B) 22 to 30 degrees of superheat.
C) 0 to 10 degrees of superheat.

728. When run across an open space, refrigeration piping must be installed at a minimum height of
   A) 6 feet.
   B) 7-1/2 feet.
   C) 8 feet.

729. With an automatic expansion valve, the pressure in the evaporator with the machine running
   A) is almost always constant.
   B) varies with the load in the cooling compartment.
   C) varies with temperature leaving the evaporator.

730. Lines to be covered must
   A) be inspected before covering.
   B) be soft copper in conduit.
   C) have all joints brazed with 1,000 °F solder.

731. For refrigerant-containing pressure vessels, all pressure relief devices over 1/2-inch shall be marked with the data required in
   A) Section VIII of the ASME boiler and pressure vessel code.
   B) Underwriters Laboratory.
   C) the safety valve code.

732. When protecting a pressure vessel from overpressure, the relief valve should start to function at
   A) a pressure not exceeding the design working pressure of the vessel.
   B) the refrigerant field leak test pressure of the system.
   C) the pressure corresponding to the cut-off of the high-pressure control.

733. Some refrigeration codes allow the pressure relief discharge on the compressor to be
   A) vented to the low side of the system.
   B) vented to the high side of the system.
   C) dumped into the city sewer.

734. Some refrigeration codes require lines covered with concrete or earth to be
   A) made of soft copper.
   B) encased in rigid conduit.
   C) inspected first.

735. Discharge lines from air conditioners shall
   A) be directly connected to the waste or sewer system.
   B) terminate over and above a trapped and vented plumbing fixture.
   C) be connected to any convenient opening.

736. A pressure vessel shall bear ASME stamping when it exceeds an inside diameter of
    A) 6".
    B) 12".
    C) 18".

737. A gas-tight joint obtained by joining metal parts with metallic mixtures or alloys that melt at temperatures above 1000 °F and less than the melting temperatures of the joined parts is a
    A) leaded joint.
    B) soldered joint.
    C) brazed joint.

738. A fuse plug may be used as the sole means of relieving excessive pressure in a maximum size pressure vessel with a gross volume
    A) of 3 cubic feet or less.
    B) of 3 cubic feet or over, but less than 10 cubic feet.
    C) over 10 cubic feet, but less than 15 cubic feet.

739. When concealing copper tubing, provide removable access plates on
    A) all concealed joints.
    B) only soft-solder joints.
    C) only flared joints.

740. When installing refrigeration equipment on roofs, the responsibility for providing access to the equipment rests with the
    A) contractor.
    B) owner of the equipment.
    C) owner of the building.

741. Pressure limiting devices shall be provided on
    A) water-cooled systems only.
    B) all systems containing more than 20 lb of refrigerant and on all water-cooled systems.
    C) any systems over 2 horsepower.

742. When replacing a vessel on a refrigerating system, it is important that the vessel
    A) be properly identified with necessary approval.
    B) have the same number of openings as the original vessel.
    C) have the gauge metal or thicker.

743. The type of expansion valve that could not be used on a flooded evaporator is a(n)
    A) low-side float.
    B) automatic expansion valve.
    C) thermostatic expansion valve.

744. A liquid seal in a refrigeration system is considered necessary to
    A) stop refrigerant leaks around the shaft.
    B) ensure a flooded evaporator.
    C) prevent hot gas from entering the evaporator.

745. Some refrigeration codes require the discharge of a fusible plug to be on the outside of the building on all systems
   A) used in institutional occupancies.
   B) containing more than 6 lb of Group II or Group III refrigerants.
   C) containing more than 2 lb of Group II or Group III refrigerants.

746. Piping located in an air conditioning system air duct shall have joints constructed to withstand temperatures of
   A) 450 to 1,000 °F.
   B) 1,000 °F or over.
   C) 15,000 °F or over.

747. Soft annealed copper tubing used with Group II refrigerants shall
   A) be inspected after installation.
   B) have rigid or flexible metal enclosures.
   C) be 95/5 solder joints in institutional occupancies.

748. At least 2 gas masks shall be provided when the quantity of refrigerant exceeds
   A) 100 lb of Group II refrigerant.
   B) 100 lb of Group III refrigerant.
   C) 300 lb of refrigerant.

749. Which of the following occupancies has the most stringent code requirements?
   A) Public assembly
   B) Commercial
   C) Institutional

750. Dry bulb temperature is a measure of the
   A) dew point of air.
   B) total heat of air.
   C) sensible heat of air.

751. In the operation of a thermostatic expansion valve, the
   A) pressure of the bulb fluid acts to open the valve.
   B) pressure of the refrigerant in the evaporator acts to open the valve.
   C) spring acts with the evaporator pressure to keep the valve open in relation to the amount of superheat setting.

752. Compression ratio is
   A) initial pressure divided by final pressure.
   B) initial volume divided by final volume.
   C) final volume divided by initial volume.

753. A low-side float valve will maintain a
   A) liquid level in an evaporator, which will correspond to load and pressure.
   B) constant liquid level in the evaporator regardless of the load and pressure.
   C) constant pressure in the evaporator.

754. An external equalizer, if connected to a thermostatic expansion valve,
   A) equalizes the pressure on the inlet and outlet sides of the valve,
   B) compensates for excessive pressure drops in the cooling coils,
   C) equalizes the temperature of the refrigerant with that of the air surrounding the coil,

755. In refrigeration, coefficient of performance concerns the
   A) performance of the system due to expansion.
   B) heat input as compared to the heat transfer.
   C) ratio of refrigeration produced to the work supplied.

756. A high-pressure cut-out shall connect directly to the
   A) condenser.
   B) discharge stop valve.
   C) compressor.

757. In a brewery, carbon dioxide is
   A) used in the refrigeration system.
   B) added to the beer to reduce its temperature.
   C) added to the beer to improve its taste.

758. Ingredients used in the beer brewing process that must be kept in cold storage are
   A) hops and yeast.
   B) barley and yeast.
   C) barley and rice.

759. An excessive charge of refrigerant in a system utilizing a high-side float valve could result in
   A) more efficient cooling.
   B) the stoppage of refrigerant at the expansion valve.
   C) damage to the compressor.

760. Other conditions being equal, more condensing water is required when the source of water supply is
   A) the city water main.
   B) a natural draft cooling tower.
   C) an underground well.

761. A refrigeration pressure relief device may discharge into the low-pressure side providing the low side
   A) capacity is equal to or greater than high side.
   B) is equipped with fusible plugs.
   C) is equipped with a pressure relief device vented to the atmosphere.

762. Proper operating procedure requires a compressor to remove the vapor in the low side of the system
   A) as fast as the evaporator produces it.
   B) faster than the evaporator produces it.
   C) much slower than the evaporator produces it.

763. For a system of given size, a large solenoid valve would be required for the
    A) control line.
    B) liquid line.
    C) suction line.

764. The specific heat of a substance is higher when that substance is
    A) above freezing temperature.
    B) below freezing temperature.
    C) anything but water.

765. The lowest freezing point that can be obtained with a sodium chloride brine is approximately
    A) 32 °F.
    B) 15 °F.
    C) -6 °F.

766. There is less danger of flooding the compressor at the start of the cycle when the solenoid valve is placed in the
    A) liquid line to the evaporator.
    B) suction line from the evaporator.
    C) expansion valve equalizing line.

767. The liquid inlet and outlet lines should be connected to the receiver
    A) with a check valve in each line.
    B) as far apart as possible.
    C) as close together as possible.

768. The speed at which a reciprocating compressor may operate is limited by the
    A) crankshaft rotation speed.
    B) piston speed.
    C) belt speed.

769. The float and mechanism of a high-side float are located in the
    A) surge drum or accumulator.
    B) low-pressure side of the system.
    C) high side of the system.

770. The water enters a vertical shell-and-tube condenser at the
    A) center.
    B) top.
    C) bottom.

771. Recirculation of the brine in an ice-making field
    A) increases the time required for making ice.
    B) decreases the time required for making ice.
    C) considerably decreases the strength of the brine solution.

772. The recommended maximum piston speed of a reciprocating compressor is about
    A) 600 fpm.

B) 750 fpm.
C) 1,000 fpm.

773. Stoppage of flow at the expansion valve caused by moisture in the system is more serious with
A) ammonia.
B) CFC refrigerants.
C) dry ice.

774. A lower condenser pressure requires
A) greater work for the compressor.
B) less work for the compressor.
C) greater power to drive the compressor.

775. The lower the suction pressure on a refrigerant, the
A) less horsepower required to handle a given quantity of gas.
B) less volume of gas required per ton of refrigeration.
C) greater the volume of gas required per ton of refrigeration.

776. A larger clearance on a compressor will
A) decrease volumetric efficiency.
B) increase volumetric efficiency.
C) result in oil slugging.

777. In a refrigeration cycle, the greatest pressure drop occurs as the refrigerant passes through the
A) evaporator.
B) reducing valve.
C) expansion valve.

778. Motors, if used to drive large compressors, should be
A) two phase.
B) single phase.
C) three phase.

779. When at its critical temperature, a refrigerant has
A) a higher liquid than gas density.
B) a higher gas than liquid density.
C) equal densities of gas and liquid.

780. The ratio of the refrigerant gas pumped to the piston displacement of the compressor is called
A) volumetric efficiency.
B) a ton of refrigeration.
C) clearance.

781. Explosions caused by anesthetics in hospital operating rooms can be reduced by
A) high relative humidity.
B) low relative humidity.
C) high static spark condition.

782. High suction pressure and low head pressure indicate the compressor is deficient in
 A) refrigeration gas.
 B) oil charge.
 C) pumping capacity.

783. Leakage at the shaft seal of a centrifugal compressor is generally
 A) of little importance for efficient operation.
 B) inward to the compressor.
 C) outward to the atmosphere.

784. When evaporating coils cool a medium that is circulated in the area to be cooled, the system is called a(n)
 A) indirect system.
 B) flooded system.
 C) direct system.

785. Increasing water flow velocity in a condenser
 A) increases the capacity of the condenser.
 B) decreases the capacity of the condenser.
 C) does not affect the capacity of the condenser.

786. Oil separators on ammonia systems are
 A) always placed as close to the condenser as possible.
 B) next to the compressor because of the higher temperature (in some systems).
 C) never next to the compressor because of the higher temperature.

787. A larger condenser capacity for a given floor area may be obtained with the
 A) vertical shell-and-tube type.
 B) horizontal shell-and-tube type.
 C) vertical shell and coil type.

788. When the load on a residential air conditioner decreases, suction pressure goes
 A) down and discharge pressure goes up.
 B) down and discharge pressure goes down.
 C) up and discharge pressure goes down.

789. Water absorbs more ammonia when the water temperature is at
 A) 50 °F.
 B) 100 °F.
 C) 200 °F.

790. A pump-out compressor
 A) evacuates or transfers refrigerant in one or more parts of a system.
 B) maintains more than one suction pressure and temperature.
 C) separates the refrigerant from the oil.

791. To lower the superheat setting of a TXV, it is necessary to
   A) tighten the spring.
   B) loosen the spring.
   C) add refrigerant to the bulb.

792. What safety device is required on all systems containing more than 20 lb of refrigerant?
   A) Fusible plug
   B) Pressure limiting device
   C) Spring-loaded relief device

793. A pressure relief valve is required on
   A) compressors over 20 hp capacity.
   B) all compressors using Group II or Group III refrigerant.
   C) all positive displacement compressors exceeding 50 cfm of displacement.

794. Increasing the clearance volume of a compressor results in
   A) no change in the volumetric efficiency.
   B) a decrease in the volumetric efficiency.
   C) an increase in the volumetric efficiency.

795. When the temperature of the cooling coil is below the dew point and the temperature of the room is below 32 °F,
   A) moisture will condense and freeze on the coil.
   B) the coils will sweat.
   C) the efficiency of the cooling coil will be increased considerably.

796. The evaporation of the cooling water flowing over an atmospheric condenser
   A) has no effect on the quantity of water handled.
   B) increases the amount of water handled.
   C) decreases the amount of water handled.

797. Gas masks or helmets approved for use shall bear a marking by the
   A) Department of Buildings and Safety Engineering.
   B) Bureau of Mines (U.S. Department of the Interior).
   C) Interstate Commerce Commission.

798. At the existing pressure, vapor at a temperature higher than the saturation temperature is called
   A) superheated vapor.
   B) saturated vapor.
   C) steam vapor.

799. An accumulator
   A) prevents the compressor from becoming vapor-bound.
   B) keeps the expansion coils filled with liquid.
   C) separates the oil from the refrigerant.

800. The main function of a compressor's clearance pocket is to provide
   A) piston lineal clearance.
   B) refrigeration capacity control.
   C) a recess for the piston rod jam nut.

801. An evaporative condenser will operate more efficiently
   A) on a low humidity day.
   B) on a high humidity day.
   C) if the blower is turned off.

802. The refrigeration operator's responsibility to maintain the equipment in a clean and accessible manner
   A) is mandatory in the refrigeration code.
   B) is not mandatory in the refrigeration code.
   C) should be delegated to the janitor.

803. Pumps, when used to agitate brine in an ice field, are the
   A) centrifugal type.
   B) rotary type.
   C) reciprocating type.

804. Subcooling the liquid between the condenser and the evaporator
   A) has no effect on the flash gas at the expansion valve.
   B) increases the flash gas at the expansion valve.
   C) decreases the flash gas at the expansion valve.

805. The condenser that performs with the least amount of water consumption is the
   A) evaporative type.
   B) shell-and-tube type.
   C) double-pipe type.

806. The canisters or cartridges for gas masks that have been used should be renewed
   A) immediately.
   B) within two years.
   C) if the seal cannot be replaced.

807. Piston rings with the inside bore eccentric (making them thicker at one part than another) have the cut for the joint at the
   A) center.
   B) thinnest part.
   C) thickest part.

808. It is considered good practice to connect the automatic water regulating valve to the
   A) city water make-up line.
   B) condenser outlet.
   C) condenser inlet.

809. In case of condenser water failure, which of the following safety devices should operate first?
   A) High-pressure cut-out

B) Low water cut-out switch
C) Pressure relief valve

810. The main advantage of an indirect system over a direct system is
A) increased operating efficiency.
B) overall safety, for the refrigerant can be completely removed from the area being cooled
C) the requirement for fewer parts.

811. The refrigerant vapor in a vertical shell-and-tube condenser
A) flows outside the tubes in the shell.
B) flows inside the tubes.
C) neither A nor B, as this condenser type is no longer used.

812. The high-pressure cut-out shall stop compressor action at a pressure not exceeding
A) 75% of the design working pressure of the system.
B) 90% of the relief device setting.
C) 90% of the pressure relief device setting, 90% of the refrigerant leak test pressure applied, or 90% of the design working pressure of the high side, whichever is smallest.

813. A suitable refrigerant should have a critical temperature and pressure
A) well below the condensing temperature and pressure of the system.
B) well above the condensing temperature and pressure of the system.
C) above pressure and below temperature only.

814. Other conditions being equal, as the latent heat increases, the
A) quantity of refrigerant that must be circulated increases.
B) quantity of refrigerant that must be circulated decreases.
C) sensible heat decreases.

815. Feather valves used in compressors are
A) bevel seated.
B) flat seated.
C) spring seated.

816. An external equalizer, when used in conjunction with a thermostatic expansion valve, is connected to the
A) feeler bulb capillary tube.
B) same side of the power element to which the feeler bulb is attached.
C) side of the power element opposite that having the feeler.

817. The water loss through use of a cooling tower or spray pond runs
A) 0%.
B) between 25 and 50%.
C) between 5 and 10%.

818. Condenser cooling water having a pH value of 7 is considered
   A) alkaline.
   B) acid.
   C) neutral.

819. When sublimated at atmospheric pressure, the latent heat of carbon dioxide is approximately
   A) 246 Btu/lb.
   B) 296 Btu/lb.
   C) 346 Btu/lb.

820. Under certain conditions, it is permissible to charge a unit at the
   A) condenser.
   B) receiver.
   C) liquid line.

821. A compressor having a variable clearance volume would
   A) not be practical.
   B) not affect the capacity of the system.
   C) affect the capacity of the system.

822. The superheat setting is associated with a(n)
   A) automatic expansion valve.
   B) thermostatic expansion valve.
   C) capillary tube.

823. The liquid inlet and outlet of a liquid receiver should be
   A) provided with king valves.
   B) as close together as possible.
   C) as far apart as possible.

824. Of the refrigerants listed below, which would be permissible in a direct system for air conditioning purposes?
   A) Ammonia
   B) CFC refrigerant
   C) Sulfur dioxide

825. A system using a centrifugal compressor would most likely operate at
   A) high pressure.
   B) low pressure.
   C) medium pressure.

826. The maximum allowable working pressure for which a specific part of a system is designed is called the
   A) factor safety.
   B) bursting pressure.
   C) design working pressure.

827. To take advantage of cheaper power rates, a system that operates only 50% of each 24-hour period should operate
   A) condensing.
   B) in the daytime.
   C) at night.

828. The time required for ice-making may be
   A) increased by lowering the suction pressure.
   B) decreased by lowering the suction pressure.
   C) decreased by a rise in suction pressure.

829. To be consistent with safety, a refrigeration operator should operate methyl chloride systems much the same as
   A) R-11.
   B) carbon dioxide.
   C) ammonia.

830. Other conditions being equal, lower refrigeration temperatures require a
   A) smaller compressor capacity.
   B) larger compressor capacity.
   C) higher suction pressure.

831. Compressors equipped with safety heads
   A) do not have discharge valves.
   B) do not have discharge valves, as the safety heads serve this purpose.
   C) have discharge valves in the safety head.

832. There is less cycling of the compressor in a(n)
   A) absorption system.
   B) direct system.
   C) indirect system.

833. Of the metering expansion valves listed below, which is the best for use on multiple systems?
   A) Low-side float
   B) Capillary tube
   C) High-side float

834. Due to solubility, oil of a higher viscosity is required in systems using
   A) CFC refrigerants.
   B) carbon dioxide.
   C) ammonia.

835. The largest size soft annealed copper tubing that may be used for refrigerant piping erected on the premises is
   A) 1" OD.
   B) 1-1/2" OD.
   C) 1-3/8" OD.

836. The freezing point of a refrigerant should be well
   A) below the lowest evaporating temperature of the system.
   B) above the highest evaporating temperature of the system.
   C) above the lowest evaporating temperature of the system.

837. For cleaning and repairing condenser tubes, removable heads are provided for
   A) evaporative condensers.
   B) vertical shell-and-tube condensers.
   C) horizontal shell-and-tube condensers.

838. Increasing the superheat setting of the thermostatic expansion valve will
   A) increase the refrigerating capacity of the unit.
   B) not change the refrigerating capacity of the unit.
   C) decrease the refrigerating capacity of the unit.

839. The discharge piping from a pump-out compressor should be connected to the
   A) atmosphere.
   B) condenser.
   C) condenser as well as the atmosphere.

840. Frost formation on the suction line is most generally caused by
   A) liquid carry over from the evaporator.
   B) a cold, dry atmosphere.
   C) insulation on the suction line.

841. The proper separation of liquid and vapor is best accomplished in a(n)
   A) atmospheric condenser.
   B) shell-and-tube condenser.
   C) double-pipe condenser.

842. Medium and large compressors are rated in
   A) horsepower.
   B) Btu/hour.
   C) operating pressure.

843. Density and latent heat being considered, less condensing is required in systems using
   A) CFC refrigerants.
   B) methyl chloride.
   C) ammonia.

844. When shut down for the winter, the refrigerant in an ice-making plant should be stored in the
   A) cooling coils.
   B) accumulator.
   C) condenser or receiver.

845. Occasionally, when R-12 is added to an R-22 system, the amount is approximately 3% to
   A) make one system more workable by eliminating free oil.
   B) make leaks less expensive.
   C) be able to test for leaks with a halide torch.

846. In a refrigeration system using CFC refrigerant, rusting and corrosion of metal parts is greater with
    A) water dissolved in refrigerant.
    B) water dissolved in oil in the system.
    C) undissolved water.

847. With water-cooled condensers, the requirement per ton of refrigeration is
    A) 1 to 5 gal/min.
    B) 5 to 10 gal/min.
    C) 10 to 15 gal/min.

848. When a vapor is in equilibrium and in contact with the liquid from which it was formed, it is referred to as a
    A) wet vapor.
    B) superheated vapor.
    C) saturated vapor.

849. The main function of the water jacket surrounding the head and cylinder of a compressor is to
    A) pre-cool the refrigerant before it enters the compressor.
    B) keep the cylinder walls cool and facilitate lubrication.
    C) remove superheat.

850. Compressors having no high- to low-side equalizing device for starting purposes require drives with
    A) unloaders.
    B) high starting torques.
    C) low starting torques.

851. The high-side float valve is ordinarily connected to
    A) one evaporator.
    B) more than one evaporator.
    C) the capillary tube.

852. Expansion at the expansion valve actually occurs in the
    A) vapor.
    B) liquid.
    C) metal structure of the valve.

853. Vertical shell-and-tube condensers may be located
    A) only on the inside of buildings.
    B) either inside or outside of buildings.
    C) only on the outside of buildings.

854. All conditions being equal, the pressure differential
    A) must be greater in an indirect system.
    B) will be the same for both direct and indirect systems.
    C) must be greater in a direct system.

855. In a unit, the temperature of the medium being cooled must be
A) above the evaporator temperature.
B) below the evaporator temperature.
C) equal to the evaporator temperature.

856. The larger the compressor size, the
A) slower its speed of rotation.
B) faster its speed of rotation.
C) slower its flywheel rpm speed.

857. In a condenser, the rate of heat transfer from the refrigerant to the water
A) decreases as the velocity of the refrigerant increases.
B) increases as the velocity of the refrigerant increases.
C) means an evaporative condenser is being used.

858. If the evaporator section to which the feeler bulb is attached is completely filled with unevaporated refrigerant, the expansion valve should be
A) open or opening.
B) closed or closing.
C) open fully.

859. A shell-and-tube condenser will operate more efficiently if it is of the
A) counterflow type.
B) parallel flow type.
C) series flow type.

860. If the refrigerant leaving the expansion coils is in a superheated condition, the system is classified as a(n)
A) absorption system.
B) flooded system.
C) dry expansion system.

861. A refrigerant having a high boiling point requires a
A) low pressure or vacuum on the suction side.
B) high pressure on the suction side.
C) reciprocating type of compressor.

862. To recover refrigerant from the lubricating oil in an ammonia system, use an oil
A) chiller.
B) separator.
C) still.

863. The expansion device that is not a valve is the
A) thermostatic type.
B) low-side float.
C) capillary tube.

864. An internal receiver pipe connected to the king valve should
  A) have the open end facing upwards.
  B) have the open end facing downwards.
  C) not be used.

865. When a moisture seal or barrier is used in connection with insulation, it should be placed
  A) adjacent to the cold side.
  B) adjacent to the warm side.
  C) exactly in the center.

866. During operation, the shaft seal on a Carrier centrifugal machine is a(n)
  A) metal-to-metal seal.
  B) metal-to-carbon seal.
  C) oil pressure seal.

867. A field test is performed to prove
  A) system tightness.
  B) that the system is operating properly.
  C) that the system is sized properly.

868. A 10 x 10 compressor refers to one
  A) equaling 100 square inches displacement volume.
  B) having a 10" bore and a 10" stroke.
  C) of 10 horsepower and 10 tons of refrigeration.

869. An air conditioning system serving a hospital should distribute the air so that the
  A) relatively cooler air exhausted from the operating room is recirculated to the convalescence rooms or wards.
  B) air from the operating rooms is not distributed to other rooms.
  C) relative humidity and temperature is as low as possible.

870. When leak-testing a CO_2 compressor system with air pressure, it is considered good practice to
  A) use the CO_2 compressor to build up the necessary air pressure.
  B) use a lubricating oil having a low flash point.
  C) obtain the necessary air pressure with a separate air pump.

871. Saturation temperature corresponding to the critical state of the substances at which the properties of liquid and vapor are identical is known as
  A) absolute temperature.
  B) saturation temperature.
  C) critical temperature.

872. A sizable leak in the float ball of a high-side float valve would cause the valve to
  A) remain closed.
  B) remain open.
  C) open and close intermittently.

873. A higher condensing temperature will require
   A) less water.
   B) a lower condensing pressure.
   C) a higher condensing pressure.

874. A back-pressure valve installed on a system employing a liquid chiller prevents the evaporator from becoming
   A) too cold.
   B) too warm.
   C) saturated with liquid.

875. Which of the following is not classified as a Group II refrigerant?
   A) Ammonia
   B) Sulfur dioxide
   C) Butane

876. What keeps a flat belt from slipping off a pulley?
   A) Pulley is concave
   B) Static electricity developed by the belt
   C) Weight of the belt

877. The proper operation of the high-side float depends chiefly on the
   A) correct amount of refrigerant charge in the system.
   B) size of the condenser.
   C) capacity of the compressor.

878. In a refrigeration system, the expansion valve needs
   A) frequent adjustment.
   B) no adjustment after it is set.
   C) adjusting occasionally as indicated by high pressure.

879. Scale traps on direct expansion systems are located
   A) underneath the scales.
   B) on the suction side of the compressor.
   C) on the discharge side of the compressor.

880. Assuming the compressor is in good condition, what limits the degree of vacuum it will develop?
   A) Speed of compressor
   B) Type of refrigerant used
   C) Clearance volume of the compressor

881. The purpose of a vent or equalizing line between a condenser and receiver is to ensure flow
   A) to the atmosphere in case of an emergency.
   B) from the receiver to the evaporator.
   C) from the condenser to the receiver

882. Heat transfer by conduction is faster in a
   A) superheated vapor.
   B) saturated vapor.
   C) liquid.

883. Of the refrigerants listed below, the one most suitable for ice-making in a hospital is
    A) ammonia.
    B) carbon dioxide.
    C) sulfur dioxide.

884. The brine generally used in ice-making plants is
    A) sodium chloride.
    B) calcium chloride.
    C) barium chloride.

885. Pressure relief valves protecting compressors from overpressure usually vent to
    A) an absorbing solution.
    B) the low side.
    C) the atmosphere.

886. A binary system has two
    A) different refrigerants.
    B) suction pressures.
    C) discharge pressures.

887. An automatic expansion valve
    A) maintains a constant pressure and temperature in the evaporator.
    B) responds well to large changes in heat load.
    C) works well with systems requiring several temperatures.

888. The average weight of an ice cake manufactured by a large plant is
    A) 100 lb.
    B) 300 lb.
    C) 700 lb.
    D) 900 lb.

889. The latent heat of evaporation is highest for
    A) water.
    B) ammonia.
    C) carbon dioxide.
    D) R-500.

890. To protect against overloading single-phase induction motors, it is customary to use
    A) thermal relays.
    B) fuses.
    C) compensators.
    D) fusetrons.

891. Electrical branch circuit conductors supplying an individual motor must have a carrying capacity that exceeds the motor's full load current rating by
 A) 25%.
 B) 50%.
 C) 125%.
 D) 150%.

892. The rating of fuses or overcurrent devices used to protect motors must not exceed the current capacity of the conductors by more than
 A) 5%.
 B) 10%.
 C) 15%.
 D) 25%.

893. The silver-coated contacts in temperature and motor controls should be cleaned with
 A) a crocus cloth.
 B) an emery cloth.
 C) extra fine sandpaper.
 D) an extra fine, smooth-cut file.

894. The differential adjustment in the temperature control unit controls the
 A) defrosting interval.
 B) cycling interval.
 C) amount of refrigerant that can enter the receiver.
 D) cabinet temperature range.

895. The quantity of refrigerant flowing into the high-side float chamber is determined primarily by the
 A) evaporating unit capacity.
 B) condensing unit capacity.
 C) low-side float capacity.
 D) size of the compressor.

896. A pressure-reducing valve placed in the liquid line just before it enters the cooling unit is called a(n)
 A) check valve.
 B) low-pressure valve.
 C) high-side float valve.
 D) expansion valve.

897. The length of the refrigerator's on/off cycles depends upon
 A) the setting of the thermostat.
 B) how much or how little refrigerant is used.
 C) the type of temperature control used.
 D) the type of refrigerant used.

898. The lift of a compressor's suction and discharge valves should be
 A) high.
 B) low.

C) by hand.
D) heat.

899. What type of compressor does an ammonia system generally use?
A) Reciprocating
B) Rotary
C) Centrifugal
D) Gear

900. To detect carbon dioxide leaks, use
A) oil of peppermint.
B) a sulfur stick.
C) an ammonia swab.
D) R-11.

901. Ammonia is composed of
A) hydrogen and nitrogen.
B) carbon and oxygen.
C) fluorine, chloride and carbon.
D) SO_2 and CCl_2F_2.

902. In determining the horsepower of a motor needed to operate a particular refrigerating unit, main consideration should be given to
A) the temperature rise of the motor.
B) operating characteristics of the motor.
C) the slip of the motor.
D) how fast it will stop.

903. The single-phase motor that has the best starting torque is the
A) capacitor type.
B) shaded coil type.
C) shaded pole type.
D) split-phase type.

904. Excessive motor vibration may be caused by
A) overloading the motor.
B) light loading of the motor.
C) the pulley being too tight.
D) worn armature shaft bearings.

905. The latent heat of vaporization of R-12 at 5 °F is
A) 68.2 Btu/lb.
B) 169 Btu/lb.
C) 565 Btu/lb.
D) 970 Btu/lb.

906. White ice is the result of
A) too much ammonia.
B) excessive oxygen.
C) insufficient air agitation.

907. An efficient test for sulfur dioxide leaks is
   A) a sulfur stick.
   B) litmus paper.
   C) an ammonia swab.
   D) R-502.

908. The condenser most likely to be found on the roof is a(n)
   A) shell-and-tube vertical.
   B) air-cooled type.
   C) atmospheric.

909. With a starved evaporator, suction pressure is
   A) high.
   B) low.
   C) inconsequential, as the amount of refrigerant in the evaporator has no bearing on pressure.

910. Leak test an ammonia system
   A) with a sulfur stick.
   B) using a halide torch.
   C) by placing system under vacuum.

911. Leak test a methyl chloride system
   A) with a sulfur stick.
   B) using a halide torch.
   C) by placing the system under vacuum.

912. A dry evaporator
   A) has no refrigerant in it and is fed with a TXV.
   B) is 1/4 to 1/3 full of liquid and is fed with a low-side float.
   C) is about 1/3 full of liquid refrigerant and may be fed with a high-side float.

913. When air is heated, the moisture content of the air
   A) remains the same.
   B) becomes more.
   C) becomes less.

914. Which uses an oil with a high viscosity?
   A) CFC refrigerant
   B) NH_3
   C) CO_2

915. Turning a TXV clockwise
   A) increases the superheat.
   B) decreases the superheat.
   C) decreases the pressure and increases the temperature.

916. The solubility of water in CFC refrigerant is greater at
   A) low temperature.
   B) high temperature.
   C) neither A nor B.

917. If more pressure is applied to a refrigerant, its boiling point
    A) increases.
    B) decreases.
    C) stays the same.

918. Atmospheric condensers use
    A) less water.
    B) more water.
    C) no water.

919. When no heat is added to a process, it is called a(n)
    A) adiabatic process.
    B) isothermal process.
    C) superheat process.

920. At a very low temperature, use
    A) NH_3.
    B) R-12.
    C) R-11.

921. The best refrigerant for -110 °F is
    A) a CFC refrigerant.
    B) NH_3.
    C) CO_2.

922. Which has a lower moisture content?
    A) 55 °F db, 36 °F wb
    B) 35 °F db, 50 °F wb
    C) 50 °F db, 33 °F wb

923. Which heat from a high-pressure refrigerant gas entering the condenser is given up first?
    A) Latent heat
    B) Sensible heat
    C) Superheat

924. The pressure limiting device (safety)
    A) may be field set.
    B) may not be field set.
    C) may be field set by a refrigeration contractor only.

925. An oil separator is located in the
    A) suction line.
    B) liquid line.
    C) discharge lines.

926. A low-pressure oil cut-out stops the compressor on
    A) high oil pressure.
    B) low oil pressure.
    C) low oil level.

927. A drum (cylinder) of refrigerant is
   A) filled to capacity.
   B) partially filled (80%).
   C) filled full because it is practical, it is protected by the fusible plug.

928. An operator must keep the equipment and engine room clean, as required by
   A) state code.
   B) ASME B 9 code.
   C) neither A or B.

929. Oil problems are more common with
   A) R-12.
   B) R-22.
   C) NH_3.

930. The high-pressure cut-out is installed
   A) before the first assembly.
   B) after the first assembly.
   C) without a stop valve.

931. The king valve of a liquid receiver is between the
   A) condenser and the receiver.
   B) receiver and the expansion valve.
   C) accumulator and the compressor.

932. The opening of the king valve (inlet) faces
   A) up from the receiver.
   B) down from the receiver.
   C) across the liquid receiver.

933. Water enters a shell-and-tube condenser at
   A) the top.
   B) the bottom.
   C) neither top nor bottom.

934. What is the maximum (plus or minus) voltage variation allowable at an outlet?
   A) 10%
   B) 15%
   C) 20%
   D) 25%

935. Air passing through a 40 °F evaporator
   A) increases in relative humidity.
   B) decreases in relative humidity.
   C) reaches its dew point.

936. A great deal of humidity in the air creates a(n)
   A) light load on the compressor.
   B) average load on the compressor.
   C) heavy load on the compressor.

937. In regard to its evaporating temperature, the freezing point of a refrigerant must be
   A) higher.
   B) lower.
   C) either higher or lower.

938. Carrene 2 is a
   A) high pressure refrigerant.
   B) low pressure refrigerant.
   C) neither A nor B.

939. Which refrigerant experiences fewer oil problems?
   A) R-12
   B) R-22
   C) R-113

940. As an oil becomes more diluted with refrigerant, its viscosity
   A) increases.
   B) decreases.
   C) remains the same.

941. When an air conditioning system is pumped down and the power shut off, the suction pressure rises very rapidly, indicating the
   A) discharge valves are bad.
   B) suction valves are bad.
   C) system is overcharged.

942. A rise in temperature at the remote bulb of a TXV causes the valve to
   A) open.
   B) close.
   C) hunt.

943. The common name for methylene chloride is
   A) Carrene 7.
   B) Carrene 1.
   C) Carrene 2.

944. Subcooling the liquid before it enters the expansion valve
   A) increases flash gas.
   B) decreases flash gas.
   C) does not affect flash gas.

945. Which refrigerant is flammable?
   A) CH_3Cl
   B) CO_2
   C) SO_2

946. Oil viscosity in a CFC refrigerant system should be
   A) lower than the temperature of refrigerant used.
   B) higher than the temperature of refrigerant used.
   C) the same as that used for the refrigerant.

947. Recovery of refrigerant means
    A) removing refrigerant in any condition from a system and storing it in an external container without necessarily testing or processing it in any way.
    B) cleaning refrigerant for reuse by oil separation and single or multiple passes through devices, such as replaceable core filter-driers, which reduce moisture, acidity and particulate matter.
    C) reprocessing refrigerant to new product specifications by means which may include distillation. Will require chemical analysis of the refrigerant to determine that appropriate product specifications are met.

948. In regard to fire hazards, the safest refrigeration system to use is
    A) direct expansion with propane.
    B) direct expansion with ammonia.
    C) indirect using brine.

949. The expansion valve that responds well to heat loads is the
    A) TXV.
    B) AEV.
    C) capillary tube.

950. The external equalizer line is usually located
    A) at the last coil of the evaporator near the control bulb location.
    B) at the last coil of the evaporator but close to the compressor.
    C) tied into the superheat line of the expansion valve for absolute control.

951. Which expansion valve keeps an evaporator more flooded?
    A) TXV
    B) AEV
    C) Low-side float

952. An accumulator is used to
    A) keep the vapor from going directly to the condenser.
    B) keep the evaporator flooded and separate liquid from vapor.
    C) subcool the liquid.

953. The king valve is located near the receiver to
    A) isolate the condenser from the receiver.
    B) isolate the receiver from the low side of the system.
    C) throttle, so as to control the flash gas in the liquid line.

954. Standard ton conditions mean a
    A) 10 °F evaporator and 86 °F condenser.
    B) 5 °F evaporator and 90 °F condenser.
    C) 86 °F condenser and 5 °F evaporator.

955. Evaporator pipes that are too large
    A) cause oil to flow through the compressor.
    B) help trap oil.
    C) make no difference in oil removal.

956. An evaporator fed with a TXV at the bottom will
   A) operate less efficiently if flooded.
   B) operate more efficiently if flooded.
   C) make no difference.

957. When a TXV thermal bulb loses its charge, the suction and discharge pressures will be
   A) unchanged.
   B) high.
   C) low.

958. A properly licensed refrigeration operator may
   A) charge and evacuate a system.
   B) charge a system.
   C) not charge a system.

959. Finned tubes are likely to be used on
   A) evaporative condensers.
   B) double-pipe condensers.
   C) air-cooled condensers.

960. The purpose of the expansion valve is to
   A) meter the flow of refrigerant according to the temperature.
   B) meter the flow of refrigerant according to the heat load.
   C) bypass the evaporative coil if necessary.

961. A thermostatic process is
   A) when heat is added or extracted.
   B) an adiabatic process.
   C) an isothermal process.

962. A properly installed heat exchanger
   A) increases pressure.
   B) decreases pressure.
   C) does not change pressure.

963. A properly installed heat exchanger
   A) increases flash gas.
   B) decreases flash gas.
   C) does not affect flash gas.

964. Which thermometer is considered the most accurate above -38 °F?
   A) Alcohol
   B) Hg
   C) Spirits

965. Which system is more efficient?
   A) Direct
   B) Indirect
   C) Both are the same

966. When liquid passes through an expansion valve the pressure drops, but there is
   A) little change in total heat.
   B) no change in total heat.
   C) complete change in total heat.

967. What controls the cabinet temperature in most household refrigerators?
   A) Thermostatic expansion valve
   B) Capillary tube
   C) Thermostat in the frozen foods compartment
   D) Thermostat in the perishable foods compartment

968. Drinking water should be at what temperature?
   A) 35 to 45 °F
   B) 45 to 55 °F
   C) 55 to 65 °F

969. The operator shall have the equipment ready for inspection
   A) annually.
   B) biennially.
   C) at all times.

970. Gas masks must be stored
   A) near compressors.
   B) in a cabinet outside machinery room.
   C) near refrigerant storage.
   D) in a safe place.

971. Water will absorb more NH_3 at
   A) 80 °F.
   B) 100 °F.
   C) 200 °F.

972. The water regulating valve is connected to the
   A) low side of the system.
   B) high side of the system.
   C) compressor crankcase.

973. A high-pressure cut-out is connected to the
   A) low side of the system.
   B) high side of the system.
   C) compressor crankcase.

974. Subcooled liquid fed to an expansion valve
   A) decreases the refrigeration effect.
   B) increases the refrigeration effect.
   C) does not change the refrigeration effect.

975. When used to interconnect the high and low sides, a heat exchanger's primary function is to
   A) subcool the suction gas.

B) raise the head pressure.
C) precool the liquid refrigerant.

976. Which of the following is in the main electrical circuit of a domestic unit?
A) Plug-in cord and relay
B) Plug-in cord, relay and motor
C) Plug-in cord, thermostat, relay and motor
D) Plug-in cord, thermostat, relay, motor and accessories

977. What is the adjustable pressure range for R-12 water-regulating valves?
A) 150 to 260 psig
B) 60 to 160 psig
C) 0 to 110 psig
D) 20 to 90 psig

978. To accurately determine the amount of refrigerant required for a system,
A) always use 3 pounds per ton of refrigeration.
B) always use 5 pounds per ton of refrigeration.
C) use the volume of receiver.
D) add the pounds required for each component.

979. The formula for Charles' Law is
A) $P_1 \times T_2 = P_2 \times T_1$.
B) $P_2 \times T_2 - T_1 \times P_1$.
C) $V_1 \times P_2 \div V_2 \times P_1$.
D) both B and C are correct.

980. Water columns are used to measure
A) hose pressure.
B) volume.
C) small pressures, above or below atmospheric.
D) square areas.

981. Density is defined as the
A) pressure exerted on supporting surfaces.
B) length per unit volume.
C) weight per unit volume.
D) same as specific gravity.

982. The Gas Law Formula is
A) $V_1 \times T_2 = V_2 \times T_1$.
B) $P_2 \times T_1 = T_2 \times P_1$.
C) $P_1 \times V_1 = P_2 \times V_2$.
D) $(P_1 \times V_1) \div T_1 = (P_2 \times V_2) \div T_2$.

983. The formula for Boyle's Law is
A) $P_1 \times T_2 = P_2 \times T_1$.
B) $V_1 \times T_2 = V_2 \times T_1$.
C) $P_1 \times V_1 = P_2 \times V_2$.

984. Isothermal expansions and contractions are
   A) coefficients of performance.
   B) actions that take place without a temperature change.
   C) conditions of pipe expansion and contraction.
   D) related to saturated conditions.

985. Adiabatic is
   A) when a liquid expands without any heat loss or gain.
   B) a term used in cryogenics.
   C) when the critical temperature is reached.
   D) when a gas expands or contracts without any heat loss or gain.

986. The valve plate from an open-type or semi-hermetic compressor can be removed without disconnecting the compressor from a charge system by
   A) back-seating both valves.
   B) back-seating the discharge valve and front-seating the suction valve.
   C) front-seating the discharge valve and back-seating the suction valve.
   D) front-seating both valves.

987. If the load on a compressor increases, suction
   A) and discharge pressures will increase.
   B) pressure will decrease, discharge pressure will increase.
   C) pressure will increase, discharge pressure will decrease.
   D) and discharge pressures will decrease.

988. If the outlet valve on a receiver is shut off (assume operating compressor), the
   A) high-pressure control will stop the system.
   B) unit will pump-down.
   C) suction pressure will increase.
   D) valves and/or gaskets will break on the compressor (assume unit has a correctly sized receiver and proper refrigerant charge).

989. When preparing a system to change a cycle component (such as a drier), it is good practice to
   A) maintain normal operating pressure differences between the high and low sides of the system.
   B) pump all the refrigerant into the receiver and shut it off tightly.
   C) bleed refrigerant through the system while the drier is being changed at or about 2 psig pressure.
   D) none of the above.

990. If a suction valve is front-seated on an operating unit,
   A) the compressor is likely to blow up.
   B) it must be done gradually so as not to hurt the compressor.
   C) the discharge pressure will increase to dangerous limits.
   D) it must be done rapidly to prevent the oil from boiling.

991. Which of the following cannot be used in a switching circuit placed in the hot line of a 120-volt, single-phase power supply?
   A) SPST
   B) DPST
   C) Three switches of an electromagnetic relay
   D) Three switches connected in series

992. A control circuit must be low voltage.
   A) True
   B) False

993. What is a pyrometer?
   A) Device used to measure the saturation pressure of refrigerant vapor
   B) Device used to measure specific weights
   C) Unit used for measuring the specific gravity of a liquid
   D) Thermometer used for measuring high temperatures

994. A 34 foot column of water would equal how many inches of mercury?
   A) 50.00
   B) 29.92
   C) 2.992
   D) 299.2

995. On earth, atmospheric pressure on one square inch of surface at sea level is
   A) 14.0 psia.
   B) 0 psia.
   C) 14.7 psia.
   D) 14.7 psig.

996. Temperature work below -250 °F is called
   A) cryogenics.
   B) areas of high boiling point refrigerants.
   C) low temperature work.
   D) hard to come by temperature.

997. Dalton's Law of partial pressure expresses
   A) a condition where two or more gases are present.
   B) heat flow in a refrigeration system.
   C) a condition where new refrigerants are formed.
   D) the law of saturation.

998. Critical pressure of a substance is
   A) the pressure above which all evaporators should operate.
   B) usually an extremely low temperature.
   C) at the same point as the critical temperature.
   D) the point at or above which a liquid cannot be evaporated, regardless of heat applied.

999. Enthalpy is
   A) the means of finding a pressure.
   B) the amount of heat in one pound of a substance.
   C) the same as entropy.
   D) a base temperature.

1000. Pressure drop is the
   A) pressure built up in the cylinder at top dead center.
   B) difference in pressure between the inlet and outlet of an object.
   C) pressure in the receiver after the unit has been shut down for awhile.
   D) pressure in the compressor cylinder at the bottom of the piston stroke.

1001. Mechanical cycle is a
   A) system that operates on the absorption principle.
   B) system that never needs a condensing unit.
   C) refrigeration service technician's mode of transportation.
   D) cycle that is a repetitive series of mechanical events.

1002. A relief valve is
   A) a manually-operated valve that controls the amount of heat load for the compressor.
   B) a safety valve designed to open before a dangerous pressure is reached.
   C) used to release air into the system if additional pressure is needed.
   D) used to reverse the direction of refrigerant flow depending on whether heating or cooling is desired.

1003. Gear and diaphragm compressors
   A) are not commonly used.
   B) are a good way of compressing gas.
   C) use throw eccentric-type crankshafts.
   D) are used on large jobs.

1004. A cooling tower is a
   A) platform the compressor is placed upon to keep it cool while operating.
   B) tower the condenser is placed in to allow better heat transfer for the removal of latent heat of condensation.
   C) device that cools by exposing water to lower temperature air.
   D) device the compressor oil is pumped through to cool it down.

1005. What is lapping?
   A) Smoothing a metal surface to a high degree of refinement or accuracy using a fine abrasive
   B) The proper way of lifting a heavy object off the floor by first placing it in the lap before standing upright
   C) The process of re-packing the service valves
   D) Straightening out the shank on bent reed valves

1006. The valve plate is the part of the compressor
 A) to which the suction valve is connected.
 B) that contains the compressor valves.
 C) to which the discharge service valve is connected.
 D) that contains the valves for the oil pump.

1007. What is a hot gas bypass?
 A) Piping system that moves the refrigerant gas from the condenser into the low-pressure side of the system
 B) Piping system that allows hot gas to flow directly into the conditioned space if it should get too cold
 C) Piping system that directs hot gas directly to the condenser from the compressor
 D) Section of the hermetic-type compressor used to help cool the motor windings

1008. What is used to flame test for leaks?
 A) Halide leak detector
 B) General Electric electronic leak detector
 C) Ammonia swab
 D) Soap and water solution

1009. Where is the liquid line located in the system?
 A) Between the compressor and condenser
 B) Between the evaporator and compressor
 C) None of these answers are correct
 D) Between the condenser and the metering device

1010. Flash gas is a condition of the
 A) metering device.
 B) compressor.
 C) condenser.
 D) receiver.

1011. A gas is superheated as it leaves the
 A) metering device.
 B) discharge line.
 C) condenser.
 D) evaporator.

1012. The pressure in the liquid line is
 A) the same as crankcase pressure.
 B) the same as the suction line.
 C) the same as the evaporator.
 D) usually the same as condenser pressure.

1013. Gas leaving the compressor is
 A) subcooled.
 B) superheated.
 C) saturated.
 D) flash gas.

1014. A saturated condition exists only in the
 A) evaporator and condenser.
 B) suction line and compressor.
 C) receiver and liquid line.
 D) end of the evaporator and the suction line.

1015. As liquid leaves the metering device to the evaporator,
 A) the pressure is reduced and the liquid is superheated.
 B) it is stored in the receiver.
 C) part of the liquid is vaporized, thus cooling the rest of it.
 D) expansion occurs in the condenser.

1016. The discharge service valve is located
 A) usually in a valve plate.
 B) on the inside of the compressor.
 C) on the outside of the compressor.
 D) in the piston.

1017. A receiver is used to
 A) store liquid refrigerant.
 B) supply liquid to the compressor.
 C) connect the line to the evaporator.
 D) measure Btu absorbed in the system.

1018. What type of relay does not need electromagnets?
 A) Magnetic
 B) Voltage
 C) Amperage
 D) Hot wire
 E) Weight

1019. What may be wrong if the relay takes too long before the starting circuit is opened?
 A) Capacitor is shorted
 B) Starting winding is grounded
 C) Voltage is too high
 D) Unit is overloaded
 E) Unit is too cold

1020. What service operations may best be performed with the aid of a high-vacuum pump?
 A) Setting a motor control
 B) Checking the amount of refrigerant in a system
 C) Dehydration of a system
 D) Setting the thermostatic expansion valve

1021. In a hermetic system, what is the most common indication of refrigerant shortage?
 A) Excessive head pressure
 B) Shorter running part of cycle
 C) Lowering of frost line on the evaporator

D) Warmer cabinet
E) Excessive frosting

1022. How may oil be added if the system cannot be made to produce a vacuum?
A) Oil cannot be added
B) Build up pressure on the low side
C) Use a hand pump
D) Remove all refrigerant
E) Correct oil charge is unimportant

1023. What is the most important thing to do if refrigerant shortage is discovered?
A) Charge the unit
B) Stop the unit
C) Overhaul the unit
D) Find the leak
E) Check the quality of oil in the unit

1024. On a system having service valves and on which the vacuum method (using a bottle) is used, what precautions must be taken when adding oil?
A) Always keep the end of the hose submerged in oil
B) Draw a vacuum, open the unit and pour the oil in through a funnel
C) Produce a vacuum and add oil to the high side
D) Remove all refrigerant, then add oil
E) Always purge the high side of air first

1025. What is the purpose of the relay?
A) Permit electricity to flow through the starting winding until the motor reaches 2/3 of its speed and then open the starting winding
B) Permit electricity to flow through the running winding until the motor reaches 2/3 of its speed and then open the running winding
C) Prevent the running winding from overheating
E) Control the box temperature

1026. What permits the service valve attachment to be swiveled to any position?
A) Tapered threads
B) Swivel nut on the valve attachment
C) Attachment fitting has tapered threads
D) It cannot be swiveled
E) Compressible gasket

1027. What kind of threads are used in the gauge opening of a service valve?
    A) Pipe
    B) NF
    C) NC
    D) National extra fine
    E) Any kind of thread

1028. How does the piercing needle provide a leak-proof joint when turned all the way in?
    A) It does not
    B) Synthetic rubber gasket
    C) Needle seats on the walls of the pierced hole
    D) Cap is used
    E) 45° flare

1029. A frozen valve stem can be loosened by
    A) using a larger wrench.
    B) hammering on it.
    C) heating the valve body.
    D) using a pipe wrench.
    E) using oil.

1030. What circuit is energized after the motor reaches operation speed?
    A) Both run and start windings
    B) Only the run winding
    C) Only the start winding
    D) Neither winding
    E) Only the relay circuit

1031. What instrument is often combined with the high vacuum pump as an assembly?
    A) High-vacuum gauge
    B) Drier
    C) Gauge manifold
    D) Filter
    E) Acid indicator

1032. What is considered the best way to heat a service cylinder?
    A) Welding torch
    B) Gas-air torch
    C) Blow torch
    D) Hot water
    E) Hot gas

1033. One of the chief properties of a good refrigerant oil is that it must
    A) separate from the refrigerant at high temperatures.
    B) circulate with the refrigerant.
    C) separate at low temperatures.
    D) have high viscosity.
    E) have high pour point.

1034. Why does heating the cylinder with hot water speed up the charging operation?
   A) Warm gas pumps better
   B) Liquid evaporates faster
   C) Removes the oil from the refrigerant
   D) Keeps the charging line clean
   E) Warms the compressor

1035. For maximum refrigerant flow through a Schrader or dill valve core, the core must be
   A) all the way out.
   B) depressed halfway.
   C) removed.
   D) all the way in.
   E) depressed even with fitting.

1036. The packing nut should be loosened
   A) only if the valve stem turns with difficulty.
   B) whenever the valve is used.
   C) only to replace the packing.
   D) when the packing leaks.
   E) when the valve stem breaks.

1037. A weight-type operated relay must be
   A) carefully leveled.
   B) kept cool.
   C) over capacity.
   D) noise insulated.

1038. When a two-way service valve stem is turned all the way out, it
   A) falls out.
   B) closes the gauge opening.
   C) closes the refrigerant line.
   D) is front-seated.
   E) does nothing.

1039. Condensing mediums must be
   A) in the right temperature range.
   B) a liquid.
   C) a gas.

1040. Some manufacturers use mufflers to
   A) prevent pulsation.
   B) retard noise.
   C) subcool liquid refrigerant.

1041. An oil safety switch operates on
   A) differential between oil pump discharge pressure and compressor suction pressure.
   B) differential between oil pump suction pressure and compressor discharge pressure.
   C) discharge pressure of the oil pump.

1042. A fusible plug will discharge the refrigerant in a system
   A) when the pressure rises more than the pressure stamping on the fusible plug.
   B) when the temperature rises more than the temperature stamping on the fusible plug.
   C) in case of fire.

1043. An evaporator of a given size can extract more heat at
   A) 20 °F evaporator temperature.
   B) 0 °F evaporator temperature.
   C) 40 °F evaporator temperature.

1044. A part of the heat removed by the condenser is called
   A) heat of evaporation.
   B) latent heat.
   C) heat of condensation.

1045. Increasing the superheat setting of a TXV would
   A) increase the frost line.
   B) not affect the frost line.
   C) decrease the frost line.

1046. Crankcase heaters are sometimes used to
   A) prevent the compressor from freezing.
   B) keep the oil viscosity constant.
   C) prevent refrigerant from condensing in the compressor.

1047. Hot gas mufflers are sometimes used on large systems for
   A) removing pulsations.
   B) desuperheating the discharge gas.
   C) preventing oil from leaving the compressor.

1048. When installing a piece of air-cooled refrigeration equipment on the roof, a refrigeration permit
   A) is required.
   B) and plumbing permit are required.
   C) and building permit are required.

1049. The type of valve that has the least resistance to flow is a
   A) gate valve.
   B) globe valve.
   C) shut-off valve.

1050. When an oil separator is used on a system using CFC refrigerant, the oil is
   A) returned to the compressor crankcase.
   B) drained into a receptacle.
   C) returned with the suction vapor.

1051. A water regulating valve shall be installed
   A) upstream from the first component installed.
   B) either at the inlet or the outlet of the condenser.
   C) in the top of the condenser.

1052. Buildings having the greatest cooling requirements would be painted
   A) dark green.
   B) white.
   C) red.

1053. For each person seated at rest, how many Btu per hour are added to the heat load?
   A) 300
   B) 600
   C) 800

1054. Which of the following relative humidity conditions has the greatest heat load?
   A) 10%
   B) 50%
   C) 90%

1055. A pressure relief device is required on the low side of a system
   A) any time the evaporator is located downstream from the heat-producing equipment in an institutional or public assembly occupancy.
   B) any time an accumulator is installed in conjunction with an evaporator.
   C) when the system serves an evaporator directly above the machinery room.

1056. A start kit
   A) equals pressure during off cycle.
   B) increases resistance in the common leg.
   C) increases starting torque.
   D) decreases starting torque.

1057. A start kit for residential equipment contains
   A) start and run capacitors.
   B) a step-up transformer and a run capacitor.
   C) a start capacitor and a potential relay.

1058. When the unit locks out on the high-pressure cut-out, this is a good indication of a restriction in the refrigerant circuit.
   A) True
   B) False

1059. An air conditioning system can accomplish more sensible cooling when the air in the conditioned space has a low moisture content.
   A) True
   B) False

1060. In case of compressor burnout, an oil sample should be taken for acid testing.
   A) True
   B) False

1061. Insulating the suction line prevents
A) excessive superheat.
B) excessive subcooling.
C) condensate damage.
D) damage to the copper line.

1062. The thermostatic expansion valve is designed to maintain constant
A) flow.
B) temperature.
C) pressure.
D) superheat.

1063. The sensing bulb on the TXV is attached to the suction line.
A) True
B) False

1064. Ductwork located in unconditioned areas should be insulated.
A) True
B) Faalse

1065. If a system is properly charged, low outside ambient temperatures do not affect system operation.
A) True
B) False

1066. With a capillary tube system, superheat will vary over a wide range depending on the outside temperature.
A) True
B) False

1067. Where do air and noncondensable gases tend to collect in a system?
A) Condenser
B) Evaporator
C) Compressor
D) Connecting tube

1068. Low suction pressure can be caused by
A) restriction in the system.
B) a faulty thermostatic expansion valve.
C) low air flow through the evaporator.
D) all of the above.

1069. When an internal mechanical failure occurs, the compressor always makes abnormal noises.
A) True
B) False

1070. A tight running compressor will cause the running amps to
A) be higher than normal.
B) be lower than normal.
C) remain the same.
D) change depending upon ambient temperature.

1071. All three-phase equipment must have start capacitors.
 A) True
 B) False

1072. When the contactor buzzes, no current is flowing to the contactor coil.
 A) True
 B) False

1073. Blackened mercury bulbs on a thermostat indicate
 A) thermostat old age.
 B) high amperage draw exists.
 C) line voltage has been applied.
 D) short cycling.

1074. To check out the blower relay at the thermostat, jump the
 A) R & W terminals.
 B) W & G terminals.
 C) Y & R terminals.
 D) R & G terminals.

1075. The cooling anticipator produces false heat when the thermostat calls for cooling.
 A) True
 B) False

1076. When tests indicate questionable starting relay performance, it is good policy to replace both the relay and the starting capacitor.
 A) True
 B) False

1077. Improper capacitor circuit wiring can cause compressor motor burnout.
 A) True
 B) False

1078. An air conditioning unit utilizing a capillary tube would come from the factory with what type of compressor motor?
 A) Shaded pole
 B) Permanent split capacitor
 C) Capacitor start
 D) Capacitor start-capacitor run

1079. In troubleshooting the low voltage control circuit, one can safely assume the transformer is in good condition if the voltage across the secondary winding is 24 volts.
 A) True
 B) False

1080. A standard VOM can be used to determine if a capacitor has shorted internally or shorted to ground.
 A) True
 B) False

1081. The run capacitor will be electrically connected to the compressor at terminals
   A) R & C.
   B) R & S.
   C) S & C.
   D) 2 & 5 on potential relay.

1082. One terminal of a running capacitor is marked to identify it as the
   A) negative terminal.
   B) positive terminal.
   C) terminal that usually shorts to the metal case if the capacitor fails.
   D) load side terminal.

1083. When capacitors are wired in parallel, the equivalent microfarad's capacitance value
   A) increases.
   B) decreases.
   C) remains constant.
   D) differs with start and run capacitors.

1084. When the contactor closes but the motor won't start and hum, the problem is usually in the low voltage circuit.
   A) True
   B) False

1085. Specific gravity is
   A) measured in pounds/in^2.
   B) measured in gallons.
   C) the same as density.
   D) the ratio of weight of an equal volume of water.

1086. The current-type relay has a switch that is normally
   A) open.
   B) closed.
   C) neither A nor B.

1087. If the relay terminals are marked 5, 2, 1, the relay is usually a
   A) current relay.
   B) potential relay.
   C) neither A nor B.

1088. A 2 hp compressor uses what type of relay?
   A) Current
   B) Potential
   C) Neither A nor B

1089. The operating voltage of most commercial defrost timers is
   A) 24 volts.
   B) 230 volts.
   C) 230 volts, three phase.
   D) 115 volts, three phase.

1090. When checking resistance (in ohms) to identify R-S-C, the smallest resistance is from
   A) R to C.
   B) S to C.
   C) R to S

1091. What type of motor uses a potential relay and a start capacitor (one phase)?
   A) PSC
   B) CSIR
   C) CSR
   D) SP

1092. A mechanical shaft seal is necessary in a
   A) semi-hermetic compressor.
   B) hermetic reciprocating compressor.
   C) hermetic rotary compressor.
   D) open-type compressor.

1093. If a drum partially filled with liquid dichlorodifluoromethane is stored in a 32 °F room, the pressure in the drum would be
   A) 10 psig.
   B) 124 psig.
   C) 16 psig.
   D) 30 psig.

1094. A reed valve normally has
   A) 4 springs.
   B) 2 springs.
   C) no springs.
   D) a number of springs determined by the valve size.

1095. The automatic expansion valve is controlled by
   A) evaporator temperature.
   B) evaporator coil superheat.
   C) the rate of refrigerant condensation.
   D) evaporator pressure.

1096. The three operation pressures of a TXV are evaporator pressure,
   A) spring pressure and suction pressure.
   B) bulb pressure and condensing pressure.
   C) spring pressure and political pressure.
   D) spring pressure and bulb pressure.

1097. The thermostatic expansion valve is controlled by
   A) coil temperature.
   B) the difference in gas temperature and the temperature corresponding to the gas pressure.
   C) the difference in gas pressure and the pressure corresponding to the gas temperature.
   D) coil pressure.

1098. When the pressure difference across a capillary tube increases,
A) refrigerant flow increases.
B) refrigerant flow decreases.
C) it gets noisy
D) there is no change in flow rate.

1099. The touching surfaces between a thermostatic expansion valve bulb and the suction line must be
A) clean and bright.
B) clean and tight.
C) downstream of the equalizing line.
D) above the coil.

1100. Liquid slugging is
A) the pounding of liquid refrigerant in the suction line at a point of restriction.
B) a presence of liquid in the condenser causing excessive noise.
C) liquid in compressor clearance space.
D) excessive liquid refrigerant in the receiver.

1101. All refrigeration compressor valves are opened and closed by
A) external springs.
B) inherent spring tension.
C) pressure difference.
D) a camshaft.

1102. Desired fresh food storage temperature range is
A) 0 to 20 °F.
B) 20 to 30 °F.
C) 35 to 45 °F.
D) 50 to 60 °F.

1103. Frozen food storage temperature is usually
A) -40 to -60 °F.
B) 0 to 32 °F.
C) 0 to -10 °F.
D) 35 to 45 °F.

1104. A screen is usually located at the
A) valve outlet.
B) valve seat.
C) bellows section.
D) valve inlet.

1105. If a space measures 2 ft x 4 ft x 8 ft, there is
A) 14 ft^2.
B) 14 ft^3.
C) 64 ft^2.
D) 64 ft^3.

1106. A 100 mesh screen has
A) 10 openings/in^2.
B) 1000 openings/in^2.

C) 100 openings/in^2.
D) 50 openings/in^2.

1107. The pressure of R-12 at 70 °F is approximately
A) 65 psig.
B) 70 psig.
C) 78 psig.

1108. What happens when a capillary tube system is overcharged?
A) Nothing
B) It has a low head pressure
C) It will partially defrost
D) It will sweat and frost back

1109. Where is the liquid receiver usually located in a TXV system?
A) There isn't one
B) In the evaporator
C) At the outlet of the condenser
D) In the suction line

1110. It is possible to tell when a drier without an indicator is absorbing moisture when the
A) drier becomes cold when the unit is operating.
B) drier becomes warm when the unit is operating.
C) head pressure rises.
D) head pressure falls.

1111. If a resistance is 10 ohms and the voltage is 230 volts, what is the current in amperes?
A) 10 A
B) 23 A
C) 100 A
D) 15 A

1112. Air-cooled condensers usually handle how much air as compared to the evaporator coil?
A) More
B) Less
C) Same

1113. On a belt-driven blower, if the size of the motor pulley is decreased, the speed of the blower will
A) increase.
B) decrease.
C) remain the same.

1114. What kind of compressor motor uses one run capacitor and no relay?
A) PSC
B) CS
C) CSR

1115. What kind of compressor motor uses two start capacitors and one run capacitor?
    A) PSC
    B) CS
    C) CSR

1116. Automatic expansion valve systems are usually
    A) dry.
    B) flooded.
    C) very large tonnage.
    D) the same as high-side floats.

1117. A control circuit must be low voltage.
    A) True
    B) False

1118. The reason for using air-cooled, rather than water-cooled, condensing units is their more attractive appearance.
    A) True
    B) False

1119. The components of an air-cooled condensing unit include the TXV or metering device and the evaporator coil.
    A) True
    B) False

1120. A compressor is a specific machine with or without accessories for a given refrigerant vapor.
    A) True
    B) False

1121. A compressor unit is a condensing unit less the condenser and liquid receiver.
    A) True
    B) False

1122. A condenser is a vessel or arrangement of pipe or tubing used for holding gas.
    A) True
    B) False

1123. The pour point of oil is important.
    A) True
    B) False

1124. Crankcase heaters are for heating oil.
    A) True
    B) False

1125. A system would be safe if it were evacuated to a temperature a little lower than evaporator temperature.
    A) True
    B) False

1126. A shaft seal is necessary.
    A) True
    B) False

1127. Solvent is generally used to clean up a system.
    A) True
    B) False

1128. A system may be no more than 8.6% wet.
    A) True
    B) False

1129. Copper tubing is connected to a refrigeration system by using 95/5 solder.
    A) True
    B) False

1130. Moisture has no effect on a system operating above freezing (such as an air conditioner with a 45 °F evaporator).
    A) True
    B) False

1131. Ordinary copper tube differs from refrigeration grade tube.
    A) True
    B) False

1132. A fusible plug is a device for units over 7-1/2 tons.
    A) True
    B) False

1133. A pressure relief device is a pressure-actuated safety valve.
    A) True
    B) False

1134. Chlorodifluoromethane is the word for R-12.
    A) True
    B) False

1135. Dichlorodifluoromethane is the word for R-22.
    A) True
    B) False

1136. Liquid refrigerant returning with the suction gas can prevent proper lubrication.
    A) True
    B) False

1137. Low suction pressure can cause a sealed unit motor to burn out.
    A) True
    B) False

1138. Liquid refrigerant in the crankcase is not harmful in an enclosed crankcase compressor.
    A) True
    B) False

1139. A splash-lubricated compressor can turn in only one direction.
A) True
B) False

1140. The voltage rating of a capacitor indicates the minimum rating at which it should be used.
A) True
B) False

1141. When hot gas enters the evaporator, it
A) becomes warmer.
B) condenses.
C) cools.
D) evaporates.

1142. The actual amount of heat removed by a refrigerating system is known as the
A) net refrigerating effect.
B) volumetric efficiency.
C) compression ratio.

1143. Too much refrigerant will cause the most trouble in a
A) low-side float system.
B) thermostatic expansion valve system.
C) high-side float system.

1144. An increase in the heat load will cause the thermostatic expansion valve to
A) increase refrigerant flow.
B) decrease refrigerant flow.
C) balance.

1145. The compressor compresses the vapor to
A) raise its superheat.
B) raise its pressure and temperature.
C) add the heat of compression.

1146. An evaporative condenser
A) is always used where there is a water shortage.
B) must always be located outdoors.
C) is a combination of an air-cooled condenser, a water-cooled condenser and a cooling tower

1147. The liquid refrigerant that evaporates while the rest of the liquid passing through the metering device is cooled is known as
A) superheat.
B) subcooling.
C) flash gas.

1148. When installing a capillary tube, the only adjustment is its
A) superheat.
B) diameter.
C) length.

1149. What size motor-compressor is used as a replacement?
   A) One size larger
   B) One size smaller
   C) 1/20 hp more
   D) Same size

1150. In condenser operations, the lower the condensing temperature, the
   A) less water will be used.
   B) less efficient the compressor becomes.
   C) lower the condensing pressure.

1151. An undercharged high-side float system will
   A) operate with a starved evaporator.
   B) pass vapor.
   C) remain closed.

1152. The low-side float closes with a
   A) rise in the liquid level.
   B) drop in the liquid level.
   C) rise in the evaporator temperature.

1153. If 970 Btu were added to a pound of water at a temperature of 212 °F at atmospheric pressure, the temperature of the steam would be
   A) 1180 °F.
   B) 212 °F.
   C) 758 °F.

1154. The two basic types of evaporators are
   A) bare pipe and finned.
   B) flooded and dry.
   C) blower coil and gravity.

1155. A capillary tube system has
   A) a condenser receiver.
   B) no receiver.
   C) a receiver.

1156. The automatic expansion valve tries to maintain
   A. the evaporator full of liquid.
   B) a constant pressure.
   C) the flow of refrigerant according to the load.

1157. A shortage of refrigerant will cause a low-side float to
   A) remain closed.
   B) operate at a lower suction pressure.
   C) remain open.

1158. A slightly overcharged low-side float system will
   A) cause flooding.
   B) cause the float to stay open.
   C) have no effect on the evaporator.

1159. Latent heat of fusion is the amount of heat required to
    A) change a liquid to a solid with no change in temperature.
    B) change a solid to a vapor.
    C) raise the temperature of a substance above its boiling point.

1160. A high-side float is used on a system having
    A) any number of evaporators.
    B) one evaporator.
    C) two evaporators of different temperatures.

1161. The thermostatic expansion valve controls the
    A) temperature of the evaporator.
    B) pressure in the evaporator.
    C) flow of refrigerant to the evaporator.

1162. The number of Btu required to raise the temperature of 10 lb of water 300 °F is
    A) 30.
    B) 300.
    C) 3000.

1163. A rise in the liquid level of a high-side float will cause
    A) the float to open.
    B) the float to close.
    C) floodback.

1164. What type of capacitor is usually used for continuous operation?
    A) Run capacitor
    B) Any type
    C) Copper graphite
    D) Oversize capacitor

1165. In a butter conditioner, how is the heating element connected into the system?
    A) Parallel with the other accessories
    B) In series with the other accessories
    C) Independent of the motor
    D) None of the above

1166. How is the starting capacitor connected to the start winding electrically?
    A) In series
    B) In parallel
    C) Either series or parallel
    D) Between the start and run winding

1167. What oil should be used on bronze motor bearings?
    A) SAE 0 to 10
    B) SAE 10 to 30
    C) SAE 30 to 50
    D) SAE 50 to 60

1168. The two thermometers of a sling psychrometer read alike when the relative humidity is
    A) 0%.
    B) 50%.
    C) 100%.

1169. Where would moisture freeze in a window unit using a capillary tube?
    A) On the low-side
    B) On the screen
    C) No place
    D) In the evaporator

1170. Where does the air go after it cools the condenser?
    A) Indoors
    B) Outdoors
    C) Into the evaporator
    D) Into the fresh air intake

1171. Recycling of refrigerant means
    A) removing refrigerant in any condition from a system and storing it in an external container without necessarily testing or processing it in any way.
    B) cleaning refrigerant for reuse by oil separation and single or multiple passes through devices, such as replaceable core filter-driers, which reduce moisture, acidity and particulate matter.
    C) reprocessing refrigerant to new product specifications by means which may include distillation. Will require chemical analysis of the refrigerant to determine that appropriate product specifications are met.
    D) none of the above.

1172. Why is it important to mount the unit in a level position?
    A) For lubrication
    B) To allow good refrigerant flow
    C) To minimize noise
    D) To ensure proper control of condensate

1173. The condenser cooling air comes from
    A) indoors.
    B) outdoors.
    C) the cooling unit.
    D) outdoors only in cool weather.

1174. Which type of evaporating coil is used most often on chest-type freezers?
    A) Shell liner
    B) Plate on shelf
    C) Open coils
    D) Flooded

1175. What is the recommended temperature range for a chest-type food freezer?
- A) 0 to 10 °F
- B) -20 to -10 °F
- C) -30 to -25 °F
- D) -40 to -30 °F

1176. Dirty filters in an air conditioning system cause
- A) high suction pressure.
- B) high discharge pressure.
- C) low suction pressure.
- D) none of the above.

1177. The purpose of a relay is to
- A) protect against an overload.
- B) keep the motor from running too fast.
- C) disconnect the starting winding at the proper moment.
- D) take the place of the thermostat.

1178. What will most thermal relays do when the unit is using too much current?
- A) Nothing
- B) Burn out
- C) Open the circuit
- D) Close the starting circuit

1179. A dehumidifier with a capillary tube must be charged
- A) within plus or minus 1 lb.
- B) with an extra 1/2 lb.
- C) very accurately.
- D) until it frosts back, and then a pound must be purged.

1180. What type of cycling device is often used on a dehumidifier?
- A) Pressure motor control
- B) Humidistat
- C) Rheostat
- D) High-side cut-off

1181. During the off cycle, the pressures in a capillary tube system
- A) decrease a little.
- B) increase a little.
- C) increase on the low-side while the head pressure stays constant.
- D) tend to equalize.

1182. Superheated R-12
- A) must contain liquid globules.
- B) may or may not contain liquid globules.
- C) must be completely free from liquid globules.

1183. A low-side float valve will maintain a
- A) liquid level in the evaporator which will correspond to load and pressure.

B) constant liquid level in evaporator regardless of load and pressure.
C) constant pressure in the evaporator.

1184. What may be wrong if the system frosts back?
A) Undercharge
B) Overcharge
C) Inefficient compressor
D) Moisture in the system

1185. In an upright, frost-free freezer, defrost
A) is disposed of through a normal drain.
B) evaporates in the motor compressor condenser compartment.
C) evaporates by way of a hot gas defrost system.
D) does not accumulate.

1186. Why must relay covers be kept as tight as possible?
A) It is not necessary
B) Dust will cause the points to deteriorate
C) To keep the unit cool
D) To keep the unit warm

1187. When adding a central cooling system to an existing furnace, what controls are needed to correlate heating and cooling functions?
A) Heat-cool thermostat and transformer
B) Heat-cool thermostat and impedance relay
C) Fan relay and pressure switch
D) Fan center and heat-cool thermostat

1188. During the off cycle, the points in a potential relay are
A) closed.
B) midway.
C) open.

1189. What windings do most hermetic motors have?
A) Stator and rotor winding
B) Stator winding only
C) Two stator windings
D) Two rotor windings

1190. If a refrigerator does not operate, what circuit should be checked first?
A) Wall outlet
B) Thermostat motor control
C) Cabinet light circuit
D) Motor relay

1191. When charging an air conditioning system, what are the two most accurate methods to ensure proper refrigerant charge?
A) Weight and temperature-pressure
B) Weight and sight glass
C) Frost line and temperature-pressure
D) Sight glass and frost line

1192. To obtain automatic control of a refrigerator, use a
   A) dualstat.
   B) thermostat.
   C) manual switch.
   D) high-pressure cut-out.

1193. Reciprocating compressors consist mainly of
   A) closely meshed gears.
   B) a closely fitted piston inside a cylinder.
   C) rotating vanes.
   D) anything that turns.

1194. Rotary compressors have
   A) pistons.
   B) impellers.
   C) horizontal pistons only.
   D) one or more blades

1195. Which type of metering device has no moving parts?
   A) Thermostatic expansion valve
   B) Capillary tube
   C) High-side float
   D) Low-side float

1196. What type of relief device conserves refrigerant?
   A) Spring loaded
   B) Rupture disc
   C) Fusible plug
   D) Pressuretrol

1197. A fusible plug should be marked with
   A) its melting temperature.
   B) its pressure setting.
   C) its bursting pressure.
   D) A, S, or E symbol.

1198. In the event of a major fire involving refrigerant equipment, the operator should first
   A) put the fire out.
   B) call building manager.
   C) evacuate the building.
   D) shut down the equipment and call the fire department.

1199. How much heat is required to raise 1 lb of ice from 0 to 32 °F?
   A) 32 Btu
   B) 16 Btu
   C) 144 Btu
   D) 970 Btu

1200. How much heat is required to raise 1 lb of ice at 0 °F to steam at 212 °F?
   A) 16 Btu
   B) 160 Btu

C) 340 Btu
D) 1310 Btu

1201. Additional heat added to steam above 212 °F, at atmospheric pressure, is called
A) superheat.
B) latent heat.
C) heat of fusion.
D) heat of vaporization.

1202. If heat is added to a gas contained in a cylinder, the increase in pressure is proportional to the increase in
A) temperature Fahrenheit.
B) temperature above 100 °F.
C) temperature Celsius.
D) absolute temperature.

1203. In which part is there only liquid?
A) Suction line
B) Liquid line
C) Discharge line
D) Receiver

1204. How does an increase in evaporator pressure affect refrigerant boiling point?
A) No change
B) Raises it
C) Lowers it

1205. How does an increase in liquid line temperature affect refrigerant boiling pressure?
A) No change
B) Raises it
C) Lowers it

1206. In which component is the most latent heat dissipated?
A) Compressor
B) Condenser
C) Metering device
D) Evaporator

1207. What equipment is used to test for a carbon dioxide refrigerant leak?
A) Halide torch
B) Lighted candle
C) Electronic detector
D) R-12 gun

1208. Which type of metering device controls the suction pressure in the evaporator?
A) Low-side float
B) Thermostatic expansion valve
C) Automatic expansion valve
D) Capillary tube

1209. The principal advantage of a hand expansion valve is it
   A) is load oriented.
   B) has simple construction.
   C) controls suction pressure.
   D) uses less refrigerant.

1210. A capillary tube must be used on a
   A) critically charged system.
   B) flooded system.
   C) receiver system.
   D) multiple evaporator system.

1211. The low-side float is used on a
   A) critically charged system.
   B) flooded system.
   C) pump-down system.
   D) direct expansion system.

1212. The low-side float
   A) controls superheat.
   B) controls suction pressure.
   C) minimizes the use of refrigerant.
   D) maintains the liquid level.

1213. Water will boil at 32 °F if the surrounding pressure is
   A) 0.089 psia.
   B) 0.015 psia.
   C) 0.0019 psia.
   D) 0.0001 psia.

1214. Does the heat loss from the condenser equal the heat gain to the evaporator?
   A) It is more
   B) It is less
   C) They are equal
   D) They are equal only after running for a certain time
   E) The heat gain is five times the heat loss

1215. In a TXV system, the liquid receiver is usually located
   A) in the evaporator.
   B) in the suction line.
   C) at the outlet of the condenser.
   D) in the filter.

1216. If a unit runs all the time and the evaporator is warm, the problem is
   A) a bad relay.
   B) a bad thermostat.
   C) a bad connection.
   D) too much refrigerant.
   E) a broken compressor valve.

1217. Which two motor terminals have the largest voltage drop or lowest amperage flow?
   A) Start to common
   B) Start to run
   C) Run to common
   D) They are all the same
   E) Two circuits are equal

1218. How is a motor winding open circuit detected?
   A) Blown fuse
   B) Open circuit breaker
   C) Unit will indicate a short
   D) Neither winding will allow current to pass
   E) One winding will not allow current to pass

1219. The electrical power source must be checked, because the
   A) wrong type wall socket may be used.
   B) outlet may be too far away.
   C) wiring may be overloaded.
   D) black and white wires may be interchanged.

1220. Does a self-contained comfort cooler dehumidify the air?
   A) Only if a humidistat is used
   B) Only if the unit has filters
   C) Only if the air is very warm
   D) Yes
   E) No

1221. The overload motor cut-out protects the
   A) thermostat.
   B) motor.
   C) wiring.
   D) compressor.
   E) filter.

1222. How may one detect if too much refrigerant is in the system?
   A) Take some out
   B) Low-side pressure will be excessive
   C) Head pressure will be above normal
   D) Liquid line will be warm
   E) Suction line will be warm

1223. What is probably wrong if the liquid line frosts?
   A) System is short of refrigerant
   B) Float needle is leaking
   C) System is too cold
   D) Screens in the liquid receiver are partially clogged
   E) Too much refrigerant in the line

1224. As oil becomes colder, it
 A) thickens.
 B) thins.
 C) precipitates.
 D) loses its lubricating properties.

1225. A frozen food cabinet is coldest
 A) in the center of the cabinet.
 B) at the top rear inner wall.
 C) at the top center of the cabinet.
 D) on the coil surface.
 E) all over the cabinet.

1226. A balanced duct system makes sure
 A) all the openings equal in area.
 B) all the ducts equal in length.
 C) all openings discharge the correct amount of air.
 D) the inlet air is equal to the outlet air.
 E) the duct dampers are adjusted until all the air velocities are the same.

1227. What is the best way to clean a condenser in the home?
 A) Carbon tetrachloride
 B) Kerosene
 C) Brush
 D) Vacuum cleaner
 E) Cloth

1228. What is the best indication of a TXV that is stuck closed?
 A) No cooling
 B) Too cold
 C) Partially frosted evaporator coil
 D) Sweating or frosted suction line
 E) Hot liquid line

1229. If a unit does not start, how is it possible to tell if the thermostat is at fault?
 A) Short the thermostat terminals
 B) Connect the power directly to the motor
 C) A faulty thermostat causes continuous running, so it is not the problem
 D) Heat the thermostat bulb
 E) Cool the thermostat bulb

1230. What is wrong if the unit has excessive head pressure with normal low-side pressure?
 A) Too much refrigerant
 B) TXV is stuck closed
 C) Air in the system
 D) Room too cold
 E) Wrong refrigerant

1231. Where is the power bulb located when controlling an air-cooled evaporator?
    A) On the coil
    B) On the liquid line
    C) On the suction line near the coil
    D) On the suction line outside the cabinet
    E) Anywhere on the cabinet

1232. If the power element is located in a warm air stream, the coil will
    A) flood.
    B) starve.
    C) not be affected.
    D) get too cold.
    E) get too warm.

1233. When the liquid line is warmer than the room temperature, there is
    A) an overcharge of refrigerant.
    B) a TXV stuck open.
    C) a lack of refrigerant.
    D) too much oil.
    E) a low-side pressure that is too high.

1234. Does the condenser contain any oil?
    A) Yes
    B) Only if the system is water-cooled
    C) Only if too much oil is in the system
    D) No
    E) Only when the unit is being pumped down

1235. The wiring must be checked to ensure it is
    A) strong enough.
    B) able to carry the voltage imposed.
    C) able to carry the current.
    D) able to carry the running current.
    E) not too large.

1236. Compressor cylinders are usually made out of
    A) aluminum.
    B) copper.
    C) cast iron.
    D) steel.
    E) brass.

1237. How many inches of vacuum should a good compressor be capable of creating?
    A) 36
    B) 28
    C) 24
    D) 15
    E) 10

1238. On open-type compressors, valves are usually located
A) in the cylinder head.
B) in a valve plate.
C) in the piston head.
D) in the tubing.
E) nowhere, as no valves are needed.

1239. The most difficult internal electrical trouble to detect in motors is a
A) short.
B) ground.
C) burned winding.
D) broken line.
E) lack of continuity.

1240. How can air be cleaned directly using electricity?
A) Heat the air
B) Magnetize the air
C) Cool the air
D) Electrical ionization
E) High speed air movement

1241. Do all filters have a pressure drop?
A) Yes
B) No
C) Only if they are dirty
D) Only if high air speeds are used
E) Only the water spray type

1242. Do all air-borne particles travel at the same speed in the duct?
A) No
B) Yes
C) Only if they are the same size
D) Only in round ducts
E) Only if the velocity is low

1243. Are the pressures the same throughout the duct system?
A) No
B) Yes
C) Only if a slow speed fan is used
D) Pressures are higher at the inlet
E) Only if there is no turbulence

1244. A damper
A) improves air flow.
B) retards air flow.
C) streamlines air flow.
D) maintains equal pressures.
E) prevents the fan from overloading.

1245. If all the outlet openings are the same size,
A) each opening will discharge an equal volume of air.
B) each outlet will discharge an equal weight of air.

C) the farthest opening will discharge the least volume of air.
D) the closest opening to the fan will discharge the least volume of air.
E) the closest opening to the fan will discharge all the air.

1246. What is considered a maximum comfortable draft?
A) 5 ft/minute
B) 10 ft/minute
C) 15 ft/minute
D) 20 ft/ minute
E) 25 ft/minute

1247. An instrument used to obtain air velocity values is the
A) hydrometer.
B) speedometer.
C) pedometer.
D) odometer.
E) anemometer.

1248. The velocimeter is popular for determining air velocities, because it
A) does not need the 16 positions.
B) can be mounted in any position.
C) is more accurate than the pitot tube.
D) is excellent for very low velocities.
E) is possible to read the velocities directly.

1249. Can the velocimeter be used to obtain velocities inside a duct?
A) No
B) Yes
C) Yes, if the instrument can be put inside
D) Yes, if the duct is large enough for the person to get inside
E) Yes, but only if rough readings are desired

1250. Does the velocimeter operate on the principle of velocity pressure?
A) Yes
B) No
C) Only at high velocities
D) Only at outlet grilles
E) Only at inlet grilles

1251. What produces the lower reading of the wet bulb?
A) Cooling by evaporation
B) Thermometer is calibrated that way
C) Cloth and water form a cooling solution
D) It doesn't read lower
E) Ice

1252. When obtaining wet and dry bulb temperatures, the thermometers are swirled through the air to
   A) remove excess liquid water.
   B) enable the thermometers to contact more air and then obtain a more average reading.
   C) prevent the vapor saturating the air immediately around the wick.
   D) keep the thermometer mercury from separating.
   E) obtain an average of temperature stratification.

1253. Wet bulb depression is the
   A) cavity in the bulb.
   B) location of the wet bulb below the dry bulb.
   C) difference in readings between the wet and dry bulb thermometers.
   D) inaccuracy of the thermometer.
   E) appearance of mercury in the thermometer.

1254. If the temperature of R-12 is 10 °F, its pressure is
   A) 34.6 psig.
   B) 24.6 psig.
   C) 14.6 psig.
   D) 4.6 psig.

1255. If the temperature of R-12 is 20 °F, its pressure is
   A) 41 psig.
   B) 31 psig.
   C) 21 psig.
   D) 11 psig.

1256. If the temperature of R-12 is 30 °F, its pressure is
   A) 48.4 psig
   B) 38.4 psig.
   C) 28.4 psig.
   D) 18.4 psig.

1257. If the temperature of R-12 is 40 °F, its pressure is
   A) 46 psig.
   B) 40 psig.
   C) 27 psig.
   D) 37 psig.

1258. If the temperature of R-12 is 50 °F its pressure is
   A) 40.7 psig.
   B) 46.7 psig.
   C) 27.7 psig.
   D) 37.7 psig.

1259. If the temperature of R-12 is 60 °F, its pressure is
   A) 40.7 psig.
   B) 66.7 psig.

C) 57.7 psig.
D) 37.7 psig.

1260. If the temperature of R-12 is 70 °F, its pressure is
A) 80.2 psig.
B) 50.2 psig.
C) 60.2 psig.
D) 70.2 psig.

1261. If the temperature of R-12 is 80 °F, its pressure is
A) 64.2 psig.
B) 94.2 psig.
C) 74.2 psig.
D) 84.2 psig.

1262. If the temperature of R-12 is 90 °F, its pressure is
A) 79.8 psig.
B) 89.8 psig.
C) 99.8 psig.
D) 109.8 psig.

1263. If the temperature of R-12 is 100 °F, its pressure is
A) 87.2 psig.
B) 97.2 psig.
C) 107.2 psig.
D) 117.2 psig.

1264. If the temperature of R-22 is 10 °F, its pressure is
A) 52.8 psig.
B) 42.8 psig.
C) 32.8 psig.
D) 22.8 psig.

1265. If the temperature of R-22 is 20 °F, its pressure is
A) 53 psig.
B) 43 psig.
C) 63 psig.
D) 33 psig.

1266. If the temperature of R-22 is 30 °F, its pressure is
A) 54.9 psig.
B) 44.9 psig.
C) 64.9 psig.
D) 34.9 psig.

1267. If the temperature of R-22 is 40 °F, its pressure is
A) 78.5 psig.
B) 68.5 psig.
C) 58.5 psig.
D) 48.5 psig.

1268. If the temperature of R-22 is 50 °F, its pressure is
A) 104 psig.
B) 94 psig.
C) 84 psig.
D) 74 psig.

1269. If the temperature of R-22 is 60 °F, its pressure is
A) 110.6 psig.
B) 101.6 psig.
C) 91.6 psig.
D) 81.6 psig.

1270. If the temperature of R-22 is 70 °F, its pressure is
A) 131.4 psig.
B) 121.4 psig.
C) 110.4 psig.
D) 91.4 psig.

1271. If the temperature of R-22 is 80 °F, its pressure is
A) 153.6 psig.
B) 143.6 psig.
C) 133.6 psig.
D) 123.6 psig.

1272. If the temperature of R-22 is 90 °F, its pressure is
A) 178.4 psig.
B) 168.4 psig.
C) 158.4 psig.
D) 148.4 psig.

1273. If the temperature of R-22 is 100 °F, its pressure is
A) 205.9 psig.
B) 195.9 psig.
C) 185.9 psig.
D) 175.9 psig.

1274. If the temperature of R-500 is 10 °F, its pressure is
A) .7 psig.
B) 9.7 psig.
C) 19.7 psig.
D) 29.7 psig.

1275. If the temperature of R-500 is 20 °F, its pressure is
A) 7.3 psig.
B) 17.3 psig.
C) 27.3 psig.
D) 37.3 psig.

1276. If the temperature of R-500 is 30 °F, its pressure is
A) 16 psig.
B) 26 psig.

C) 36 psig.
D) 46 psig.

1277. If the temperature of R-500 is 40 °F, its pressure is
A) 36.1 psig.
B) 46.1 psig.
C) 56.1 psig.
D) 66.1 psig.

1278. If the temperature of R-500 is 50 °F, its pressure is
A) 67.6 psig.
B) 57.6 psig.
C) 47.6 psig.
D) 37.6 psig.

1279. If the temperature of R-500 is 60 °F, its pressure is
A) 40.6 psig.
B) 50.6 psig.
C) 60.6 psig.
D) 70.6 psig.

1280. If the temperature of R-500 is 70 °F, its pressure is
A) 55.4 psig.
B) 65.4 psig.
C) 75.4 psig.
D) 85.4 psig.

1281. If the temperature of R-500 is 80 °F, its pressure is
A) 72 psig.
B) 82 psig.
C) 92 psig.
D) 102 psig.

1282. If the temperature of R-500 is 90 °F, its pressure is
A) 105.6 psig.
B) 110.6 psig.
C) 120.6 psig.
D) 130.6 psig.

1283. If the temperature of R-500 is 100 °F its pressure is
A) 121.2 psig.
B) 131.2 psig.
C) 141.2 psig.
D) 151.2 psig.

1284. If the temperature of R-502 is 10 °F, its pressure is
A) 31 psig.
B) 41 psig.
C) 51 psig.
D) 61 psig.

1285. If the temperature of R-502 is 20 °F, its pressure is
A) 32.5 psig.
B) 42.5 psig.
C) 52.5 psig.
D) 62.5 psig.

1286. If the temperature of R-502 is 30 °F, its pressure is
A) 45.6 psig.
B) 55.6 psig.
C) 65.6 psig.
D) 75.6 psig.

1287. If the temperature of R-502 is 40 °F, its pressure is
A) 50.5 psig.
B) 60.5 psig.
C) 70.5 psig.
D) 80.5 psig.

1288. If the temperature of R-502 is 50 °F, its pressure is
A) 77.4 psig.
B) 87.4 psig.
C) 97.4 psig.
D) 107.4 psig.

1289. If the temperature of R-502 is 60 °F, its pressure is
A) 96.4 psig.
B) 106.4 psig.
C) 116.4 psig.
D) 126.4 psig.

1290. If the temperature of R-502 is 70 °F, its pressure is
A) 117.6 psig.
B) 127.6 psig.
C) 137.6 psig.
D) 147.6 psig.

1291. If the temperature of R-502 is 80 °F, its pressure is
A) 151.2 psig.
B) 161.2 psig.
C) 171.2 psig.
D) 181.2 psig.

1292. If the temperature of R-502 is 90 °F, its pressure is
A) 177.4 psig.
B) 187.4 psig.
C) 197.4 psig.
D) 104.4 psig.

1293. If the temperature of R-502 is 100 °F, its pressure is
A) 206.2 psig.
B) 216.2 psig.

C) 226.2 psig.
D) 236.2 psig.

1294. If the temperature of R-717 is 10 °F, its pressure is
A) 3.8 psig.
B) 13.8 psig.
C) 23.8 psig.
D) 33.8 psig.

1295. If the temperature of R-717 is 20 °F, its pressure is
A) 13.5 psig.
B) 23.5 psig.
C) 33.5 psig.
D) 43.5 psig.

1296. If the temperature of R-717 is 30 °F, its pressure is
A) 25 psig.
B) 35 psig.
C) 45 psig.
D) 55 psig.

1297. If the temperature of R-717 is 40 °F, its pressure is
A) 48.6 psig.
B) 58.6 psig.
C) 68.6 psig.
D) 78.6 psig.

1298. If the temperature of R-717 is 50 °F, its pressure is
A) 54.5 psig.
B) 64.5 psig.
C) 74.5 psig.
D) 84.5 psig.

1299. If the temperature of R-717 is 60 °F, its pressure is
A) 82.9 psig.
B) 92.9 psig.
C) 102.9 psig.
D) 112.9 psig.

1300. If the temperature of R-717 is 70 °F, its pressure is
A) 104.1 psig.
B) 114.1 psig.
C) 124.1 psig.
D) 134.1 psig.

1301. If the temperature of R-717 is 80 °F, its pressure is
A) 138.3 psig.
B) 148.3 psig.
C) 158.3 psig.
D) 168.3 psig.

1302. If the temperature of R-717 is 90 °F, its pressure is
    A) 165.9 psig.
    B) 175.9 psig.
    C) 188.9 psig.
    D) 205.9 psig.

1303. If the temperature of R-717 is 100 °F, its pressure is
    A) 187.2 psig.
    B) 197.2 psig.
    C) 207.2 psig.
    D) 217.2 psig.

1304. In reclaiming refrigerant, the product specification must meet what standard?
    A) DOT 118.94
    B) DOT 700-88
    C) ARI 700-88
    D) ARI 118.94

1305. The water level is maintained in the sump of a cooling tower by a
    A) bypass valve.
    B) hand valve.
    C) solenoid valve.
    D) float valve.

1306. Which type of metering device controls the suction pressure in the evaporator?
    A) Low-side float
    B) Thermostatic expansion valve
    C) Automatic expansion valve
    D) Capillary tube

1307. Which of the following would most likely cause air-conditioning equipment to operate for a longer period?
    A) Above normal ambient temperature
    B) Low ambient temperature
    C) Slight increase in occupant activity
    D) Decrease in occupant activity

1308. The maximum allowable load on a 15-A circuit is
    A) 16.5 A.
    B) 15 A.
    C) 13.5 A.
    D) 12 A.

1309. The diameter of belt-driven blowers must be
    A) under 20 inches.
    B) under 30 inches.
    C) over 30 inches.
    D) over 40 inches.

1310. Which of these sensing devices can extend when heated?
   A) Helical bimetal element
   B) Bellows
   C) Bimetal strip
   D) Hair

1311. How many Btu are required to raise the temperature of 8 lb of cast iron 4 °F? (The specific heat of iron is 0.119)
   A) 3.8 Btu
   B) 4.8 Btu
   C) 4.9 Btu
   D) 5.0 Btu

1312. The outer dimensions of a building are 10' x 14' and the walls are 8" thick. What is the wall area in square feet that is used in calculating air conditioning?
   A) 148.00
   B) 140.33
   C) 126.67
   D) 120.50

1313. What controls and regulates the air flow from return ducts and intake openings?
   A) Bypass dampers
   B) Branch ducts
   C) Recirculating damper
   D) Multiblade fan

1314. Major repairs and renovations of evaporative air cooling units should be done
   A) every summer.
   B) during winter season.
   C) every six months.
   D) whenever necessary.

1315. The most common type of heat pump in use today is the
   A) air-to-air (refrigerant changeover).
   B) air-to-air (air changeover).
   C) water-to-air (refrigerant changeover).
   D) water-to-water (water changeover).

1316. A refrigerator has an automatic defrost operated by a solenoid and hot gas. What condition indicates the solenoid valve is stuck open?
   A) Compressor running and evaporator cold
   B) Compressor not running and evaporator cold
   C) Compressor running and evaporator warm
   D) Compressor not running and condenser cold

1317. Which one of the following troubles does not indicate low refrigerant?
 A) Noisy compressor
 B) Noisy capillary tube
 C) Oil spot near compressor
 D) Frosted cooling coil

1318. Some compressors are made with an electric heater, which is there to
 A) prevent slugging.
 B) keep moisture from condensing in motor.
 C) keep refrigerant warm.
 D) keep oil warm for lubrication.

1319. The disadvantage of using a line tap on a refrigerator is that
 A) it is only temporary.
 B) it will leak when the gasket dries.
 C) it is hard to remove.
 D) the line must be opened.

1320. In a potential relay, the
 A) coil is operated by heat.
 B) contacts are normally open.
 C) contacts are opened by induced voltage in the motor's start winding.
 D) contacts are opened by induced voltage in the motor's run winding.

1321. To make a valid test of an overload protector for a single-phase motor, which of the following should be observed first?
 A) Relay must be open
 B) Motor must be cool
 C) Protector must be cool
 D) Protector must be disconnected

1322. A king valve is located at the
 A) receiver inlet.
 B) receiver outlet.
 C) condenser outlet.
 D) compressor suction line.

1323. If the thermostat bellows ruptures and loses its charge, it
 A) causes continuous operation of the compressor.
 B) is felt immediately.
 C) causes the compressor to stop quickly.
 D) produces short cycle.

1324. A heat exchanger in the liquid line of a soda fountain
 A) reduces flashing at expansion valve.
 B) reduces capacity of expansion valve.
 C) increases capacity of expansion valve.
 D) ensures refrigerant gas in multiple system having a manifold.

1325. A thermostatic expansion valve charged with R-12 is
    A) yellow.
    B) green.
    C) orange.
    D) white.

1326. A refrigerator with a new capillary shows frost extending too far out on the suction line. To correct this,
    A) bleed off some refrigerant.
    B) lengthen the capillary.
    C) shorten the capillary.
    D) install a larger capillary.

1327. An automatic ice maker in a refrigerator may have two thermostats. What are their functions?
    A) One senses when the ice is made; the other when tray is full
    B) One reduces line pressure; the other prevents ice formation
    C) One checks temperature of the other to keep equal temperatures
    D) One ejects ice cubes when storage tray is full; the other prevents interruption of the circuit in the colder thermostat

1328. If a refrigerator fails to cool or runs continually, the problem is most likely
    A) loose motor mounts.
    B) loose tubing mounts.
    C) broken valves.
    D) a punctured coil.

1329. When flushing a system, use ________________ to get the best results.
    A) refrigerant
    B) carbon dioxide
    C) dry nitrogen gas
    D) dry compressed air

1330. Two different refrigerants may be used in a cascade system
    A) to ensure oil equalizing.
    B) to eliminate heat exchanger between stages.
    C) because there are two separate systems.
    D) because there is no temperature difference needed in heat exchanger.

1331. How can a sight glass be used to observe refrigerant leaving the condenser?
    A) Clear glass indicates partially charged system
    B) Clear glass indicates moisture in system
    C) Bubbles in glass indicate completely discharged system
    D) Bubbles in glass indicate partially charged system

1332. If the voltage of the dc circuit is 12 V and the resistance is a lamp rated at 4 ohms, what is the amperage?
   A) 48 A
   B) 3.3 A
   C) 3 A
   D) 0.33 A

1333. What is the voltage drop across a dc series circuit containing three resistors rated at 5 ohms each, with a circuit flow of 1 ampere?
   A) 0.5 volt
   B) 5 volts
   C) 15 volts
   D) 50 volts

1334. What is the total resistance in ohms in a parallel dc circuit having five resistors rated at 10 ohms each?
   A) 2
   B) 4
   C) 6
   D) 8

1335. An 1,800 rpm motor for an evaporator fan has burned out. The replacement is identical except that the data plate gives the voltage as 120 V, 60 cycles, and the motor has two poles. What is the rpm of the replacement motor?
   A) 1,800 rpm
   B) 3,200 rpm
   C) 3,600 rpm
   D) 3,800 rpm

1336. The capacitor in the run winding of a capacitor-start capacitor-run motor is a(n)
   A) intermittently rated capacitor.
   B) four-pole capacitor.
   C) continuously rated capacitor.
   D) insulated type capacitor.

1337. In an ac circuit, the only time volts times amperage equals true power is when there is a pure
   A) resistive circuit.
   B) inductance circuit.
   C) capacitance circuit.
   D) resistive, inductance, and capacitance circuit.

1338. How can a single-phase wattmeter be used to calculate the total power of a three-phase circuit?
   A) Connect two wattmeters in any two of the three phases and subtract the readings
   B) Connect three wattmeters, one on each of the three phases, and multiply the readings

C) Connect two single-phase wattmeters in any two of the three phases, then add the two readings
D) Connect the wattmeter current coil in one load line and the voltage coil between the line and ground, then multiply the reading by three

1339. When circuit troubleshooting, the first specific check point item is included in the
A) generator.
B) circuit control device.
C) inoperative device.
D) circuit protective device.

1340. Which of the following starts a three-phase induction motor?
A) Squirrel-cage rotor
B) Capacitor in parallel with run winding
C) Speed of rotating magnetic field
D) Torque it exerts when at rest

1341. The capacitor in a permanent split capacitor motor is connected in
A) series with the start winding.
B) series with the run winding.
C) parallel with the start winding.
D) parallel with the start and run windings.

1342. The short-circuited copper band on the field poles in a shaded pole motor is used to
A) create a rotating magnetic field.
B) increase the current in the field coil.
C) delete the rotating magnetic field.
D) decrease the current in the field coil.

1343. The reluctance motor pulls into synchronous speed because of the
A) dc supplied to the rotor.
B) number of poles in the motor.
C) leading power factor.
D) salient poles mounted on the rotor.

1344. Name the three types of water valves.
A) Electric, pressure and thermostat
B) Capillary tube, AEV, and TXV
C) None of the above

1345. Name the various two-temperature valves.
A) Metering, snap action and thermostat
B) Metering, snap action
C) Thermostat, snap action

1346. Water-cooled condensers are advantageous, because they have a
A) higher head pressure than air-cooled units.
B) lower head pressure than air-cooled units.
C) lower initial cost than air-cooled units.

1347. Air-cooled condensers are advantageous, because they have a
A) lower head pressure than water-cooled units.
B) lower initial cost than water-cooled units.
C) higher initial cost than water-cooled units.

1348. What are the advantages of forced circulation evaporators?
A) They need baffles
B) They cool the produce rapidly
C) None of the above

1349. In multiple installations using two-temperature valves, check valves must be placed
A) in the suction lines of the coldest evaporators.
B) in the suction lines of the warmest evaporators.
C) at the compressor.

1350. An evaporator controlled by a two-temperature valve should not have a refrigeration load placed on it of more than
A) 20% of the total refrigeration load.
B) 30% of the total refrigeration load.
C) 40% of the total refrigeration load.

1351. An evaporative condenser saves about 85% water consumption, because it uses
A) specific heat.
B) latent heat.
C) neither A nor B.

1352. Are liquid receivers equipped with safety devices?
A) Yes
B) No
C) Only on water-cooled units

1353. The inner tube of a tube-within-a-tube condenser contains
A) water.
B) refrigerant.
C) oil.

1354. In water-cooled condensers, the counterflow principle means water flow places the
A) cold water near the coolest refrigerant.
B) cold water near the warmest refrigerant.
C) warmest water near the warmest refrigerant.

1355. When defrosting a system, water is sprayed on
A) to melt ice accumulation.
B) the condenser to melt ice accumulation.
C) the entire system.

1356. During defrost, the drain pan and drain pipe must be heated to
A) prevent freezing.
B) keep head pressure up.
C) keep head pressure down.

1357. In a reverse cycle defrost system, the check valve prevents refrigerant from bypassing
    - A) one TXV.
    - B) the condenser.
    - C) neither A nor B.

1358. Why do some hot gas defrost systems reheat the refrigerant before it returns to the compressor?
    - A) To prevent liquid slugging of the compressor
    - B) Hot gas defrost systems are never reheated
    - C) Neither A nor B

1359. Which water valve does not vary the water flow as the refrigeration load changes?
    - A) Pressure
    - B) Electric
    - C) Pressure electric

1360. A float valve is used with a cooling tower to control
    - A) head pressure.
    - B) make-up water.
    - C) neither A nor B.

1361. In a walk-in meat cabinet, does a flash cooler need an automatic defrost system?
    - A) Yes
    - B) No
    - C) Only on water cooled units

1362. A surge tank stores
    - A) heat.
    - B) water.
    - C) gas.

1363. A sweet water bath cools beverages by using
    - A) tap water as a brine.
    - B) CaCl as a brine.
    - C) NaCl as a brine.

1364. Vibration dampers are installed on the
    - A) high side.
    - B) low side.
    - C) high and low side.

1365. To produce flake ice, an auger revolves
    - A) slowly around a cylindrical evaporator.
    - B) at high speed around a cylindrical evaporator.
    - C) neither A nor B.

1366. What shuts off some commercial ice cube makers when the bin is full?
   A) Thermostat
   B) Low-pressure switch
   C) High-pressure switch

1367. During defrost in a reverse cycle commercial ice cube maker the evaporator temporarily becomes the
   A) condenser.
   B) evaporator.
   C) compressor.

1368. An accumulator collects
   A) hot gas.
   B) liquid refrigerant.
   C) oil.

1369. Some systems are pumped down before the electric defrost starts to
   A) reduce the amount of defrost heat needed and prevent liquid refrigerant from entering the compressor.
   B) keep oil out of the evaporator.
   C) both A and B.

1370. A suction line filter-drier should be replaced when the moisture indicator is
   A) yellow.
   B) green.
   C) blue.

1371. A tandem compressor assembly is
   A) two or more compressors connected to the same suction line and discharge line.
   B) the discharge line of the first compressor leading to the suction line of the second compressor .
   C) neither A nor B.

1372. For a commercially-installed, water-cooled unit, the cabinet temperature should be
   A) -10 to 0 °F.
   B) 35 to 40 °F.
   C) 45 to 55 °F.
   D) 55 to 60 °F.

1373. For a commercially-installed, walk-in cooler, the cabinet temperature should be
   A) -10 to 0 °F.
   B) 35 to 40 °F.
   C) 45 to 55 °F.
   D) 55 to 60 °F.

1374. For a commercially-installed florist cabinet, cabinet temperatures should be
   A) -10 to 0 °F.
   B) 35 to 40 °F.
   C) 45 to 55 °F.
   D) 55 to 60 °F.

1375. For a commercially-installed ice cream cabinet, cabinet temperatures should be
   A) -10 to 0 °F.
   B) 35 to 40 °F.
   C) 45 to 55 °F.
   D) 55 to 60 °F.

1376. Why must doors and windows in commercial cabinets be airtight?
   A) To decrease cost of operation
   B) To increase cost of operation
   C) To inconvenience customers

1377. If a thermometer is placed in a walk-in cabinet, where is the average temperature?
   A) At the top
   B) At the bottom
   C) Halfway between the top and bottom

1378. Many commercial refrigeration installations are of the multiple type to permit a
   A) smaller evaporator.
   B) larger evaporator.
   C) neither A nor B.

1379. Open-display, frozen food display case evaporators are defrosted by
   A) hot water.
   B) hot gas or electrical heating coils.
   C) nothing, as there is no need for defrost in this type of case.

1380. What are the two basic types of ice makers?
   A) Cube and flake
   B) Round and flat
   C) Plane and open

1381. In a freezing dispenser, malted milk should be
   A) -10 °F.
   B) 5 °F.
   C) 25 °F.

1382. Are evaporators sometimes mounted outside the refrigerator cabinet?
   A) No
   B) Yes
   C) Only on 2 ton units

1383. What keeps certain parts of a frozen food case from sweating?
A) Electric heating wires
B) Hot gas
C) Condenser heat

1384. Clear ice is manufactured by circulating
A) chilled water over molds.
B) air in the condenser.
C) air in the water.

1385. Why is a high-velocity blower evaporator used when cooling bottled beverages?
A) Cools beverages rapidly
B) Cools beverages slowly
C) Neither A nor B

1386. Electric resistance heat can be used in making ice cubes to keep
A) the evaporator warm.
B) the evaporator cold.
C) neither A nor B.

1387. How many different air streams are common in a commercial frozen food case?
A) One
B) Two
C) Three

1388. An ice maker should be cleaned at least once
A) a month.
B) every 6 months.
C) a year.

1389. In a freezing dispenser for soft ice cream, what two parts are refrigerated?
A) Mixer and drum
B) Mixer and mix storage unit
C) Mix storage unit and drum

1390. Is food heated while being freeze dried?
A) No
B) Yes

1391. It is necessary to level evaporators for the drainage of
A) water.
B) oil.
C) refrigerant.

1392. In a code installation, soft tubing may be used near the
A) condensing unit.
B) evaporators.
C) condensing unit and near the evaporators.

1393. An undersized suction line will
 A) reduce the capacity of the system.
 B) increase the capacity of the system.
 C) have no effect on the capacity of the system.

1394. A suction line should slope down slightly toward the compressor to aid the
 A) refrigerant in returning to the compressor.
 B) oil in returning to the compressor.
 C) refrigerant in returning to the condenser.

1395. If the TXV orifice is too large, the evaporator will be
 A) starved.
 B) flooded.
 C) neither A nor B.

1396. When a hot gas bypass valve is replaced, refrigerant is stored in the
 A) receiver.
 B) condenser.
 C) evaporator.

1397. Can any 1/2-hp hermetic motor compressor replace a worn out 1/2-hp unit?
 A) Yes
 B) No
 C) Only on 120-V systems

1398. When a system is purged, what else leaves the system besides refrigerant?
 A) Air
 B) Air and oil
 C) Air, oil and moisture

1399. When a drier is heated, moisture is
 A) held in the drier.
 B) driven back into the system.
 C) discharged into the system.

1400. Why is nitrogen used when brazing a connection in a refrigerating system?
 A) Nitrogen is not used in brazing
 B) To prevent oxidation
 C) Neither A nor B

Refrigeration and Air Conditioning

Level 3

1401. What is the Btu equivalent of one hp?
   A) 746
   B) 3.41
   C) 2,545.6

1402. A baffle promotes
   A) oil circulation.
   B) air circulation.
   C) refrigerant circulation.

1403. At what temperature should brick ice cream be maintained?
   A) 25 °F
   B) 10 °F
   C) 0 °F

1404. Vapor passing through the compressor is superheated by heat
   A) from the evaporator.
   B) due to compression.
   C) from the condenser.

1405. How many heat leakage surfaces does a cabinet have?
   A) 2
   B) 4
   C) 6

1406. Two basic types of expendable refrigerant systems available today are the
   A) cold plate and spray.
   B) hot plate and cold plate.
   C) hot plate and spray.

1407. One application of a cascade system is to obtain temperatures to
   A) -250 °F.
   B) 70 °F.
   C) 40 °F.

1408. What are the least number of compressors a multistage system will use?
   A) One
   B) Two
   C) Three

1409. The freezing point for R-11 is
   A) 74 °F.
   B) -74 °F.
   C) -168 °F.

1410. The freezing point for R-12 is
   A) -32 °F.
   B) -74 °F.
   C) -252 °F.

4111. The freezing point for R-13 is
   A) -270 °F.
   B) -294 °F.
   C) -200 °F.

1412. The freezing point for R-13B1 is
   A) -270 °F.
   B) -294 °F.
   C) -200 °F.

1413. The freezing point for R-21 is
   A) -137 °F.
   B) -144 °F.
   C) -211 °F.

1414. The freezing point for R-744 at 76.4 psia is
   A) -117 °F.
   B) -109 °F.
   C) -69.9 °F.

1415. The freezing point for R-22 is
   A) -256 °F.
   B) -252 °F.
   C) -109 °F.

1416. The freezing point for R-40 is
   A) -137 °F.
   B) -144 °F.
   C) -211 °F.

1417. The freezing point for R-113 is
   A) -109 °F.
   B) -31 °F.
   C) -252 °F.

1418. The freezing point for R-114 is
   A) -137 °F.
   B) -144 °F.
   C) -211 °F.

1419. The freezing point for R-500 is
   A) -108 °F.
   B) -256 °F.
   C) -254 °F.

1420. The freezing point for R-502 is
    A) -270 °F.
    B) -294 °F.
    C) -200 °F.

1421. The freezing point for R-717 is
    A) -40 °F.
    B) 32 °F.
    C) -108 °F.

1422. The freezing point for R-718 is
    A) -108 °F.
    B) 32 °F.
    C) -40 °F.

1423. The freezing point for R-764 is
    A) -109 °F.
    B) -31 °F.
    C) -103 °F.

1424. Cryogenics ranges in temperature from
    A) -50 to -100 °F.
    B) -150 to -200 °F.
    C) -250 to -459 °F.

1425. The maximum setting for the pressure limiting device shall not exceed
    A) 60% of the setting of the pressure relief valve.
    B) 90% of the setting of the pressure relief valve.
    C) any setting under 300 psig.

1426. The component used solely for lowering the temperature of water utilizing the evaporative principle is a(n)
    A) evaporator.
    B) evaporative condenser.
    C) cooling tower.

1427. A fusible plug is permitted on the
    A) low side of the system only.
    B) high side of the system only.
    C) both high and low side.

1428. Automatic condenser water regulating valves are operated by
    A) hand.
    B) rise in temperature.
    C) rise in head pressure.

1429. A Temprite system using methyl chloride should never use a(n)
    A) oil separator.
    B) back-pressure valve.
    C) aluminum part.

1430. A system containing sulfur dioxide
   A) would never be installed in a hospital.
   B) can be installed in any type of occupancy.
   C) can be installed in some hospitals with certain reservations.

1431. Some refrigeration codes do not require a pressure relief valve or fuse plug on a pressure vessel not exceeding
   A) an internal gross volume of 3 cubic feet.
   B) an internal diameter of 3 inches.
   C) 30 psig working pressure.

1432. A device used to decrease the starting load of the compressor is a(n)
   A) capacitor.
   B) equalizer.
   C) unloader.

1433. An increase in discharge pressure would cause an automatic regulating valve to
   A) open more.
   B) close off.
   C) restrict the flow of water.

1434. Increasing the superheat setting of a TXV will
   A) increase the refrigerating capacity of the unit.
   B) not change the refrigerating capacity of the unit.
   C) decrease the refrigerating capacity of the unit.

1435. The bypass safety valve on a refrigerating system is located between
   A) check valves in the discharge line and the condenser.
   B) suction and discharge valves on the compressor.
   C) the evaporator and compressor.

1436. When required, a check valve is located
   A) in the suction line to the compressor.
   B) between the high and low sides of the compressor.
   C) in the discharge line of the compressor.

1437. A water regulator valve on a refrigerating system shall be located
   A) on the water inlet.
   B) on the water outlet.
   C) anywhere on the system.

1438. A pressure relief device is required on pressure vessels
   A) 3 to 10 cubic feet.
   B) over 6" diameter.
   C) over 50 lb capacity.

1439. It is considered safe practice to install a rupture disc
   A) with no markings on the disc.
   B) on the upstream (inlet) side of the relief valve.
   C) on the downstream (outlet) side of the relief valve.

1440. A device installed in a compressor head and used to protect the compressor from damage due to liquid slugging is a
   A) safety head.
   B) jacketed cylinder.
   C) cylinder head.

1441. When two or more condensers are served in parallel from a common water supply with a shut-off valve on the outlet side of the condenser, a check valve shall be installed
   A) upstream from all supply connections.
   B) downstream from all supply connections.
   C) any place on the water line.

1442. An equalizer connection to a TXV on a large evaporator coil will
   A) tend to starve the coil under a heavy load.
   B) tend to flood the coil under a heavy load.
   C) permit control with 1 or 2 degrees of superheat.

1443. The defrost system that uses an antifreeze solution in the heat exchanges is a
   A) thermobank.
   B) water spray.
   C) hot gas bypass.

1444. A check valve is required in the compressor discharge line
   A) when more than 40 lb of Group II or Group III refrigerant are used.
   B) when more than 100 lb of refrigerant are used.
   C) on all systems.

1445. Cross-heads are used in
   A) vertical, single-acting machines.
   B) both vertical, single-acting and horizontal, double-acting compressors.
   C) horizontal, double-acting machines.

1446. A flooded system containing a flooded evaporator has cooling coils that
   A) are normally full of liquid refrigerant.
   B) are normally immersed in some secondary cooling medium such as brine.
   C) continuously flood back to the compressor, causing liquid slugging.

1447. The item having the greatest effect on evaporative condenser performance is
   A) ambient temperature.
   B) dry bulb (db) temperature.
   C) wet bulb (wb) temperature.

1448. Frost formation on air blast freezers can be prevented by using
    A) hot gas.
    B) electric heaters.
    C) brine spray.

1449. Which of the following compressors does not meet accepted standards?
    A) 6" x 6" at 720 rpm
    B) 3" x 3" at 1,500 rpm
    C) 6" x 6" at 1,500 rpm

1450. Liquid refrigerant that is chilled to evaporator temperature in the expansion coils
    A) expands after its temperature is lowered.
    B) contracts after its temperature is lowered.
    C) doesn't change in volume after its temperature is lowered.

1451. The high-pressure cut-out must be connected to the compressor with
    A) an intervening stop valve.
    B) no intervening stop valve.
    C) an intervening stop valve of the T or lever handle type.

1452. In a centrifugal compressor, the high velocity of the gas leaving the impeller is transformed into static pressure by the
    A) damper.
    B) pressurestat.
    C) stationary diffuser.

1453. The most suitable refrigerant to use in applications of -110 °F is
    A) R-12.
    B) ammonia.
    C) carbon dioxide.

1454. Eliminators, as used on an evaporative condenser, are for the purpose of
    A) directing air flow over tubes.
    B) directing water flow over tubes.
    C) preventing carry over of water with discharge air.

1455. In a vertical shell-and-tube condenser, the water sides of the tubes can be cleaned mechanically
    A) only when the unit is out of service.
    B) when the unit is in service.
    C) on weekends only.

1456. Lower head pressures and discharge temperatures can be obtained by use of
    A) compound compression.
    B) single-stage compression.
    C) evaporative condensers.

1457. Vertical shell-and-tube condensers are generally used to condense
   A) ammonia.
   B) carbon dioxide.
   C) CFC refrigerants.

1458. Wet compression on a booster compressor is indicated by
   A) frost on the suction line.
   B) a cold discharge line.
   C) a warm discharge line.

1459. When condensing water is hard and has scale-forming tendencies, the most practical type of condenser is a
   A) double-pipe type.
   B) vertical shell-and-tube type.
   C) horizontal shell-and-tube type.

1460. The lineal clearance of a compressor piston should be
   A) greater at the head end of the piston.
   B) greater at the crank end of the piston.
   C) equal for both sides of the piston.

1461. Steam engines are primarily designed to drive a
   A) horizontal double-acting compressor.
   B) vertical single-acting compressor.
   C) centrifugal compressor.

1462. The oil separator for an ammonia system should be placed near the
   A) outlet from the compressor.
   B) inlet to the condenser.
   C) outlet from the condenser.

1463. In an ammonia system, gaskets on flanged joints on the high-pressure side, should be
   A) rubber.
   B) lead.
   C) asbestos composition.

1464. Within the compressor cylinder, leaks past valves and piston rings can be pinpointed by
   A) using a sounding tube.
   B) taking an indicator diagram.
   C) overhauling the compressor.

1465. Compressor capacity control is best achieved by using
   A) a compound compressor.
   B) a booster compressor.
   C) valve unloaders.

1466. To maintain a proper heat balance in a brewery refrigeration plant,
   A) only electric prime movers should be considered.
   B) both steam and electric prime movers should be considered.
   C) only steam prime movers should be considered.

1467. The eliminator used in conjunction with systems having centrifugal compressors is located in the
A) condenser.
B) evaporator (cooler).
C) compressor.

1468. The formation of ice on the brine side of the cooling coils in an ice-making plant
A) decreases the time required for ice manufacturing.
B) increases the time required for ice manufacturing.
C) has no effect on the time required for ice manufacturing.

1469. The quality of ice in an ice-making plant is
A) not affected by the rate of cooling.
B) best maintained at or near a specified brine temperature.
C) increased for every Fahrenheit degree of decrease in brine temperature.

1470. An economizer, as used in a refrigeration system, is identified with
A) centrifugal compressors.
B) booster systems.
C) flash intercoolers.

1471. Spherical heads on compressors
A) increase volumetric efficiency.
B) permit greater valve area.
C) increase pressure developed.

1472. Gas binding difficulties in condenser operation are experienced to a greater degree with the
A) evaporative type.
B) vertical shell-and-tube type.
C) atmospheric type.

1473. The suction valves in a vertical single-acting ammonia compressor are usually located in the
A) piston.
B) safety head.
C) cylinder wall.

1474. The striking clearance in a vertical, single-acting ammonia compressor is approximately
A) 1/64 inch.
B) 1/4 inch.
C) 1/2 inch.

1475. An intercooler used between the cylinders of compound compressors is similar in appearance to a
A) shell-and-tube or surface condenser.
B) double-pipe condenser.
C) evaporative condenser.

1476. When making ice, the primary purpose for blowing air into the cans is to
    A) shorten the time required for freezing.
    B) aid in the removal of impurities.
    C) make the ice much lighter.

1477. In ammonia plants, the condenser tubes should be composed of
    A) steel.
    B) copper.
    C) brass.

1478. An intercooler is used in conjunction with
    A) two coolers.
    B) two-stage or compound compressors.
    C) single-stage compressors.

1479. Moisture eliminators are generally found on
    A) evaporative condensers.
    B) shell-and-tube condensers.
    C) double-tube or double-pipe condensers.

1480. Assuming that both types of compressors are serving the same system, the stuffing box that has to withstand the greater pressure is on the
    A) vertical single-acting machine.
    B) horizontal double-acting machine.
    C) centrifugal machine.

1481. The heat transfer rate per square foot of evaporator surface is usually greater with
    A) a dry expansion system.
    B) a flooded system.
    C) both saturated and superheated vapor in the system.

1482. A stop valve, if located between a pressure relief device and a vessel, is
    A) not permissible under any circumstances.
    B) permissible.
    C) permissible, providing two or more valves are connected to the stop valve in parallel and only one can be rendered inoperative for testing and repair.

1483. To obtain a desired temperature with the greatest efficiency, the back pressure of the system should be maintained
    A) as high as possible.
    B) as low as possible.
    C) below atmospheric.

1484. Indicator diagrams can be taken to check the performance of the
    A) compressor.
    B) expansion valves.
    C) automatic purge device.

1485. A thin film of calcium carbonate, if deposited uniformly over the water side surfaces of a condenser, will
   A) cause an acid condition in the water.
   B) retard corrosion.
   C) accelerate corrosion.

1486. The stuffing box for the enclosed vertical compressor is subject to
   A) suction pressures.
   B) discharge pressures.
   C) either suction or discharge pressures.

1487. When using the same cooling water for the condenser and compressor, the automatic cooling water valve should be located at the
   A) inlet to the compressor.
   B) inlet to the condenser.
   C) outlet from the condenser.

1488. Corrosion of the water side of a condenser is due to the presence of
   A) scale.
   B) dissolved oxygen and carbon dioxide.
   C) soda ash.

1489. Surging in a centrifugal compressor is more apt to occur at
   A) heavy load periods.
   B) light load periods.
   C) periods where noncondensable gases are present.

1490. When starting a motor-driven centrifugal compressor, the suction damper (when provided) should be
   A) closed.
   B) opened completely.
   C) locked in the open position.

1491. The water flowing in a vertical shell-and-tube condenser makes a
   A) single pass.
   B) double pass.
   C) triple pass.

1492. A system using CFC refrigerants must be vented to the outside of the building when the system
   A) contains over 100 lb of refrigerant.
   B) contains over 50 lb of refrigerant.
   C) is over 15 tons.

1493. Where a compressor operates with two suction pressures, the
   A) lowest suction pressure enters the compressor first.
   B) highest suction pressure enters the compressor first.
   C) compressor must be the centrifugal type.

1494. When a pressure relief valve and rupture disc are used in series,
   A) the rupture disc is placed closest to the vessel protected.
   B) the relief valve is placed closest to the vessel protected.
   C) either may be placed closest to the vessel protected.

1495. Liquid refrigerant injected into the suction line ahead of the compressor will
   A) increase the operating temperature of the compressor.
   B) not affect the operating temperature of the compressor.
   C) reduce the operating temperature of the compressor.

1496. With a two-temperature system, the solenoid valve is generally located at the
   A) liquid line to lower temperature box.
   B) liquid line to higher temperature box.
   C) suction line of high temperature box.

1497. Dual pressure relief devices are required when the pressure vessel is
   A) 10 ft^3 or more.
   B) 3 ft^3 or more.
   C) over 10 feet long.

1498. Which of the following compressors is best suited for intermittent (cycling) operation?
   A) Vertical double acting
   B) Centrifugal
   C) Vertical single acting

1499. For use in ultra-low temperature systems, the most practical type of thermostatic expansion valve is one employing a
   A) gas charge.
   B) liquid charge.
   C) cross charge.

1500. An indicator card having a suction line starting below the normal suction pressure line and meeting the normal suction line before completion of the stroke indicates the
   A) suction valve spring may be too stiff.
   B) suction valve spring may be too weak or loose.
   C) compressor discharge is not high enough.

1501. The booster compressor discharges into the
   A) suction side of the main compressor.
   B) discharge side of the main compressor.
   C) inlet side of the condenser.

1502. Centrifugal compressor capacity is often governed by
   A) a low-side float control.
   B) changing the condenser pressure.
   C) the use of clearance pockets.

1503. An automatic expansion valve functions best when used in conjunction with a
   A) temperature control switch (thermostat).
   B) back-pressure control switch.
   C) high-pressure cut-out switch.

1504. Indicator diagrams are taken while the compressor is
A) down for overhaul.
B) stopped.
C) running.

1505. When a calcium chloride brine begins foaming, it might indicate
A) lack of aeration.
B) a brine that is too strong.
C) corrosion.

1506. On an ammonia system, the scale trap is located preferably on the
A) suction line close to the evaporator.
B) discharge line close to the compressor.
C) suction line close to the compressor.

1507. In an ammonia system, liquid line piping is
A) Schedule 20.
B) Schedule 40.
C) Schedule 80.

1508. A safety collar is used in conjunction with a
A) discharge valve.
B) built-up piston.
C) suction valve.

1509. Leaking piston rings and valves are indicated by
A) abnormally low discharge temperatures.
B) abnormally high discharge temperatures.
C) increased frost on the suction lines.

1510. High discharge temperatures can be caused by
A) liquid ammonia in the suction side.
B) the expansion valve being open too wide.
C) accumulation of scale or dirt in the compressor cooling water jackets.

1511. To recharge weak brine solution,
A) add concentrated brine solution at inlet of system.
B) dump calcium chloride into the tank.
C) suspend a sack containing calcium chloride at the outlet of the system.

1512. In an ice field, brine temperature variation should not exceed
A) 2 to 5 °F.
B) 0 to 2 °F.
C) 5 to 10 °F.
D) 10 to 15 °F.

1513. The high side of a system usually has a purge valve to
A) remove oil from the system.
B) add oil to the system.
C) add refrigerant to the system.
D) remove air from the system.

1514. When cleaning the tubes of a vertical shell-and-tube condenser by mechanical means, it is considered good practice to
   A) keep the water flowing through the tubes.
   B) stop the flow of water through the tubes.
   C) use ammonia as the cleaning agent.
   D) use SO_2 as the cleaning agent.

1515. If a small part of the system requires a much lower temperature than the remainder of the system, it is best to utilize a
   A) centrifugal compressor.
   B) booster compressor.
   C) pump-out compressor.
   D) pump-in compressor.

1516. A trunk-type piston, compared to the ordinary type, is
   A) the same length.
   B) much shorter.
   C) much longer.
   D) made of lead.

1517. Vertical, single-acting compressors normally discharge gas
   A) at the top of the cylinder.
   B) at the bottom of the cylinder.
   C) through the safety head.
   D) at the side of the unit.

1518. Temperature is a term used to designate the
   A) amount of energy in the substance.
   B) intensity of energy in the substance.
   C) amount of heat in the substance.
   D) intensity of heat in the substance.

1519. A CO_2 system in which the water enters the condenser at 93 °F will
   A) not operate on the liquid cycle.
   B) not operate as a refrigerant system.
   C) operate but inefficiently.

1520. A spherical piston head is used to
   A) increase volumetric efficiency and strengthen the piston.
   B) ease the installation of piston rings.
   C) keep down the turbulence of the compressed gas.

1521. What causes brine to foam?
   A) Weak brine and air
   B) An NH_3 leak
   C) Too much agitation

1522. Saturated NH_3 has
   A) few liquid droplets.
   B) no liquid particles.
   C) some liquid and some gas.

1523. A vertical two-piston compressor is
- A) single stage.
- B) two stage.
- C) three stage.

1524. On a centrifugal compressor, the pressure on the seal is
- A) outside the seal.
- B) inside the seal.
- C) both inside and outside the seal.

1525. A thin line of bromide on the pipes
- A) leaves scale deposits.
- B) corrodes the metal.
- C) leaves lime deposits.

1526. An indicator chart is hooked to a
- A) compressor.
- B) suction line.
- C) TXV.

1527. What temperature is best for making ice?
- A) 0 °F
- B) 16 °F
- C) 28 °F

1528. The bearing surface of a bellows type of shaft seal
- A) rubs the shaft.
- B) is separated by a thin film of oil.
- C) rotates with the shaft.

1529. The compressor more likely to use a bellows-type shaft is
- A) horizontal.
- B) vertical single acting.
- C) rotary.

1530. With a back-pressure control, the refrigeration unit
- A) starts on pressure drop, stops on pressure rise.
- B) stops on pressure rise, starts on pressure drop.
- C) stops on pressure drop, starts on pressure rise.

1531. Which relieves more pressure?
- A) Relief valve
- B) Rupture disc
- C) Fusible plug

1532. The shaft seal that is subject to the highest pressure on the compressor is
- A) horizontal double acting.
- B) vertical single acting.
- C) a rotary compressor.

1533. In starting a CO_2 system, the
- A) suction and discharge valves should be open.

B) discharge valves should be open.
C) H_2O should be shut off to the condenser.

1534. The vapor passing through the discharge valve of the compressor is
A) slightly above the condensing pressure.
B) slightly below the condensing pressure.
C) at the condensing pressure.

1535. Leaking discharge valves on a compressor will cause
A) high head pressure.
B) the unit to short cycle.
C) low suction pressure.

1536. A bellows-type shaft seal
A) moves diagonally only.
B) rotates with the shaft.
C) is fixed and stationary.

1537. Piston rings are sometimes bored on an eccentric
A) for longer life.
B) to make the rings heavier.
C) to make the rings expand against the cylinder wall.

1538. The type of pipe used on an ammonia system for the low side is
A) Schedule 20.
B) Schedule 40.
C) Schedule 60.

1539. In an institutional occupancy, refrigerant piping
A) may pass between the floors.
B) may not pass between the floors.
C) either A or B.

1540. On a centrifugal compressor, the shaft seal is
A) metal-to-metal.
B) metal-to-carbon.
C) an oil seal.

1541. Evaporator condensers are placed on the roof because
A) there is more space.
B) the static pressure is greater.
C) the evaporation is better.

1542. In a cabinet, the vapor barrier is
A) next to the inner wall.
B) next to the outer walls.
C) midway between the two walls.

1543. High pressure in the evaporator is such that when the refrigerant boils at 15 °F,
A) all the evaporator will be at this temperature.
B) none of the evaporator will be at this temperature.
C) some of the evaporator will be above and some below this temperature.

1544. A low boiling temperature of the refrigerant in the evaporator may be obtained by
A) a vacuum.
B) raising the pressure.
C) slowing down the compressor.

1545. NH_3 vapor at saturation is a
A) wet vapor with liquid droplets in it.
B) dry vapor.
C) dry vapor that is exactly at evaporating temperature.

1546. The true compression curve is likely to be closer to the
A) power curve.
B) isothermal curve.
C) adiabatic curve.

1547. Brine with the highest specific heat is
A) NaCl.
B) CaCl.
C) H_2O.

1548. CH_3Cl and CFC refrigerants require oil of
A) high viscosity.
B) low viscosity.
C) any viscosity.

1549. In ice making, the flow of brine is
A) through the shell.
B) through the tubes.
C) around the shell.

1550. Which is most corrosive?
A) NaCl
B) Sodium bromide
C) CaCl

1551. What type of pump is used to circulate brine to a system?
A) Centrifugal
B) Rotary
C) Reciprocating

1552. A salimeter reads
A) relative humidity.
B) dew point temperature.
C) density of the brine.

1553. C_2Cl and C_2Br are
A) extractors.
B) adsorbers.
C) absorbers.

1554. The clearance between the piston and the head of an NH_3 compressor is
    A) 1/64".
    B) 1/4".
    C) 1/2".

1555. At saturating evaporating pressure, refrigerants for comparison purposes are taken at
    A) 5 °F.
    B) 0 °F.
    C) 36 °F.

1556. Which would not normally be used on a multiple evaporator system?
    A) Low-side float
    B) TXV
    C) Capillary tube

1557. In appearance, an accumulator looks like a(n)
    A) shell-and-tube condenser.
    B) vertical shell-and-tube condenser.
    C) evaporative condenser.

1558. Liquid refrigerant expanded into the suction side would cause
    A) high head pressure.
    B) low head pressure.
    C) a decrease in superheat of the suction gas.

1559. Which of the following devices will pass leaking vapors?
    A) Low-side float
    B) High-side float
    C) Thermostatic expansion valves

1560. With a high-side float, a rising float ball
    A) closes the valve.
    B) opens the valve.
    C) has no effect on the valve.

1561. In a system with a low-side float, too much refrigerant
    A) has no effect on the evaporator.
    B) causes flooding.
    C) starves the evaporator.

1562. With a TXV, a rise in superheat
    A) has no effect on refrigerant flow.
    B) increases the refrigerant flow.
    C) decreases the refrigerant flow.

1563. With an automatic expansion valve, an increase in pressure
    A) opens the valve wider.
    B) decreases evaporator capacity.
    C) closes the valve.

1564. At saturated condensing temperature, refrigerants for comparison purposes are figured at
A) 5 °F.
B) 0 °F.
C) 86 °F.

1565. In ice-making, the brine level should be
A) below the water level in the can.
B) above the water level in the can.
C) 18" below the water level in the can.

1566. Finned tubes are not likely to be used on an evaporative condenser if
A) they may retard heat transfer.
B) cleaning will be more difficult.
C) they can cause pressure drop.

1567. To control capacity on a VSA compressor, use
A) synchronous motors.
B) clearance pockets.
C) an inlet vane in the suction line.

1568. Smaller condenser size is needed for
A) NH_3.
B) CFC refrigerants.
C) R-40.

1569. Eliminators are used on a condenser to
A) prevent H_2O carry over.
B) diffuse the air.
C) prevent a drain draft.

1570. The refrigerant that mixes best with oil is
A) NH_3.
B) CFC refrigerant.
C) CO_2.

1571. A cross head will be found on a(n)
A) VSA compressor.
B) HDA compressor.
C) horizontal compressor.

1572. On an NH_3 compressor, the device used to separate the oil is a(n)
A) oil still.
B) separator.
C) oil chiller.

1573. Between the compressor and condenser, the
A) superheat is given up first.
B) latent heat is given up first.
C) sensible heat is given up first.

1574. If an NH_3 compressor at 4 psi is changed to operate at 24 psi, the amount of refrigeration would be
A) the same.

B) twice as much.
C) six times as much.

1575. Frost accumulating on the evaporator coil of an air conditioning unit is caused by
A) high superheat across the coil.
B) excess cfm across the coil.
C) low cfm across the coil.

1576. A stop valve may be installed in the liquid line if
A) a receiver is used.
B) a relief valve is used.
C) the installer wants to.

1577. Stop valves may be used ahead of the relief valve
A) if a dual relief valve is used.
B) if a receiver is ASME.
C) never.

1578. In a safety head, the machine suction valves are in the
A) safety head.
B) valve plate.
C) neither A nor B.

1579. In a direct system,
A) more refrigerant is in the shell for air conditioning purposes.
B) more refrigerant is in the tubes for industrial purposes.
C) either A or B can be correct.

1580. A thin layer of carbonate on the inside of the pipe would
A) be corrosive.
B) restrict heat transfer.
C) make the pipe last longer.

1581. Which type of condenser gives the higher efficiency for less floor space?
A) Atmospheric
B) Shell and tube
C) Vertical shell and tube

1582. Which is the most common compressor on the market?
A) 6 x 6 at 750 rpm
B) 3 x 3 at 1,500 rpm
C) 6 x 6 at 1,500 rpm

1583. According to code, all pressure relief devices over 1/2 inch in size must be
A) approved by any nationally recognized testing laboratory.
B) constructed to meet ASME code requirements.
C) constructed of the same material as the vessel.

1584. What are the two basic types of commercial ice makers?
   A) Cube and flake
   B) Dispenser and activated
   C) Cold evaporator and warm evaporator

1585. The heavier the brine,
   A) the higher the specific heat.
   B) the less the specific heat.
   C) neither A nor B.

1586. With a compound compressor, the larger cylinder discharges to the
   A) high side of the system.
   B) low side of the system.
   C) small cylinder.

1587. The tubes on a horizontal shell-and-tube condenser are held rigid and made gas tight by
   A) welding.
   B) rolling and flaring.
   C) threading.

1588. A single cylinder acting compressor (horizontal) needs at least
   A) 2 valves.
   B) 4 valves.
   C) 8 valves.

1589. The head pressure in a CO_2 system would probably be
   A) 72 atmospheres.
   B) 35 atmospheres.
   C) 1020 atmospheres.

1590. The correct brine temperature for ice-making is
   A) 14 to 15 °F.
   B) 10 to 15 °F.
   C) 0 to 10 °F.

1591. In an R-12 system, what is the suction line superheat of an evaporator if the expansion valve bulb reads 47 °F and the evaporator pressure is 26.1 psig?
   A) 50 °F
   B) 10 °F
   C) 12 °F
   D) 20 °F

1592. In an R-22 system, if the condenser water temperature is 75 °F at the inlet, the compressor discharge pressure should be
   A) 75 to 100 psig.
   B) 125 to 150 psig.
   C) 160 to 190 psig.

1593. Overload heaters should be selected at what percent of full load amperage?
   A) 100%

B) 115%
C) 125%
D) 150%

1594. Determine the re-heat coil capacity in heating air from 50 °F db, 49 °F wb to 62 °F db, 54 °F wb using 1000 cfm.
A) 12,960 Btu
B) 14,521 Btu
C) 16,819 Btu
D) 18,314 Btu

1595. What is the smallest amount of condenser water at 85 °F entering temperature that could be used on a Trane 10-ton unit?
A) 14.0 gpm
B) 18.5 gpm
C) 24.0 gpm
D) 36.0 gpm

1596. If 1/3 of the air supplied to an air conditioning unit is outside air at 95 °F db, 75 °F wb, and 2/3 is return air at 80 °F db, 67 °F wb, what is the dry bulb temperature of the mixture?
A) 83 °F
B) 85 °F
C) 87 °F
D) 90 °F

1597. If a compressor draws 4 kW using 220 volt, 3 phase current, what is the draw per phase?
A) 10 amps
B) 10.5 amps
C) 15 amps
D) 17.5 amps

1598. Shortage of refrigerant is not indicated by
A) high superheat temperature.
B) high liquid line temperature.
C) low suction temperature.
D) solid flow of refrigerant in the sight glass.

1599. Make-up water is commonly available from the city water supply at
A) 30 to 75 ft head.
B) 70 to 125 ft head.
C) 175 to 300 ft head.

1600. In an average installation, the water tower lowers the condenser water temperature within how many degrees of local wet bulb temperature?
A) 5 to 15 °F
B) 15 to 25 °F
C) 25 to 35 °F

1601. In the average tower, increasing the water flow from 3 to 4 gpm increases the lower capacity
   A) 5%.
   B) 10%.
   C) 15%.
   D) 20%.

1602. The conversion factor for changing pressure in psi to feet of water is
   A) 3.21.
   B) 2.13.
   C) 3.12.
   D) 2.31.

1603. A stationary blade compressor is a
   A) compressor that has a stationary fan blade.
   B) compressor that has the blades inside the rotor and rotates with it.
   C) reciprocating compressor mounted securely to its base.
   D) rotary pump that uses a stationary blade and rides against the rotor.

1604. The capacity of an ammonia compressor operating at 24 psi suction, as compared to the same compressor operating at 4 psi, is
   A) about the same.
   B) twice as much.
   C) six times as much.

1605. What is a good vacuum, measured in millimeters of mercury column?
   A) 5 to 10
   B) 25 to 35
   C) 40 to 50
   D) 55 to 65
   E) 70 to 80

1606. What is the basis of operation of the potential relay?
   A) Increase in voltage as the unit approaches and reaches its rated speed
   B) Decrease in amperage as the unit approaches and reaches its rated speed
   C) Difference in magnetic impulses
   D) Potential difference between the two magnetic forces
   E) Operation of a magnetic coil connected to the solenoid

1607. The low-side pressure should be kept within reasonable limits when charging refrigerant into the low side to
   A) keep from overworking the compressor.
   B) prevent liquid pumping.
   C) speed up the operation.
   D) prevent freezing the cylinder.
   E) prevent rupturing the cylinder.

1608. Alcohol fuel must be kept very clean to
    A) obtain a good flame color.
    B) enable the fuel to go through the small passage.
    C) prevent corrosion.
    D) build up enough pressure to keep the flame burning.
    E) keep the alcohol from exploding.

1609. Under most conditions, the preferable method of capacity control for a centrifugal compressor is
    A) suction line throttling dampers.
    B) condenser temperature control.
    C) variable speed drive.

1610. In order to place a system on a pump-down cycle, which of the following devices is necessary?
    A) Solenoid valve
    B) Thermostat
    C) Pressure control

1611. All oil safety switches should be the type that are
    A) automatically reset.
    B) manually reset.
    C) internally reset.

1612. In a CFC refrigerant system, what size system must be vented to the outside?
    A) Any system over 15 tons
    B) A system containing over 100 lb of refrigerant
    C) Any system containing over 31 lb of refrigerant

1613. An increase in compression ratio would
    A) decrease the efficiency of the system.
    B) have no effect on the efficiency of the system.
    C) increase the efficiency of the system.

1614. In which part of the system is oil more soluble?
    A) Liquid line
    B) Suction line
    C) Discharge

1615. A compressor would be required to pump a larger volume of vapor with an evaporator temperature of
    A) 0 °F.
    B) 40 °F.
    C) 20 °F.

1616. System A has a 5 °F evaporator temperature and an 86 °F condenser temperature. System B has a 40 °F evaporator temperature and an 86 °F condenser temperature. To produce one ton of refrigeration, the compressor would have to handle
A) the same weight of refrigerant in both systems.
B) more refrigerant in system A.
C) more refrigerant in system B.

1617. A water-cooled condensing unit having a rise of 15 °F in the condensing water requires how many gallons of water per minute for each ton of refrigeration produced?
A) 1 gallon
B) 2 gallons
C) 3 gallons

1618. System A has a 5 °F evaporator temperature and an 86 °F condenser temperature. System B has a 40 °F evaporator temperature and an 86 °F condenser temperature. The compression ratio would be
A) the same in both systems.
B) greater in system A.
C) greater in system B.

1619. Dual pressure relief devices are required when
A) two compressors discharge into one condenser.
B) Group II or Group III refrigerants are used.
C) the condenser exceeds 10 cubic feet.

1620. In a system with an evaporator temperature below zero, it is good practice to install a(n)
A) oil separator.
B) oil trap.
C) crankcase heater.

1621. Compressor heaters are installed by the manufacturer to prevent
A) frosting of the compressor.
B) the oil from congealing.
C) refrigerant from condensing in the compressor.

1622. The principal time when a compressor heater should be in operation is
A) during the off cycle.
B) during the running cycle.
C) when adding refrigerant.

1623. A condenser that uses both water and air is called a(n)
A) dual purpose condenser.
B) evaporative condenser.
C) cooling tower.

1624. Oil traps are installed in the suction line when the condensing unit is
A) air-cooled.

B) installed below the evaporator.
C) installed above the evaporator.

1625. Assuming the bodies of a 10-ton and a 5-ton expansion valve are of the same size, the power element would be
A) interchangeable.
B) larger on the 5-ton expansion valve.
C) larger on the 10-ton expansion valve.

1626. Under what condition may bushings be used in fittings?
A) When the reduction is two or more pipe sizes
B) When the reduction is one or more pipe sizes
C) On any fitting

1627. ASME stamping is mandatory when a pressure vessel exceeds
A) 6" ID.
B) 12" ID.
C) 10 ft^3.

1628. When condenser water temperature increases, the compression ratio
A) decreases.
B) increases.
C) remains the same

1629. What type of compressor uses a damper as a means of capacity control?
A) Rotary
B) Centrifugal
C) Reciprocating

1630. Under which of the following temperatures would one pound of saturated air contain the greatest amount of moisture?
A) 25 °F
B) 60 °F
C) 80 °F

1631. Under which of the following conditions would a compressor be most efficient?
A) 40 lb suction psig
B) 25 lb suction psig
C) 150 lb discharge psig
D) 125 lb discharge psig

1632. All conditions being equal, a given compressor removes more heat when the refrigerant used is
A) R-12.
B) R-22.
C) ammonia.

1633. Liquid injection into the suction line is sometimes necessary to cool the compressor
  A) under full load.
  B) under unloaded conditions.
  C) upon condenser water failure.

1634. A relief valve on an ammonia system must be discharged
  A) into the room.
  B) to the atmosphere.
  C) above the roof level.

1635. A properly installed refrigeration line should be
  A) level.
  B) pitched toward the compressor.
  C) pitched in the direction of flow.

1636. When installing refrigerant lines, it is recommended that oil traps be used
  A) on all vertical lines.
  B) when the suction lines rise 8 feet or more.
  C) when the condenser is on the roof.

1637. The greatest permissible pressure drop allowed is in the
  A) discharge line.
  B) suction line.
  C) liquid line.

1638. The approximate pressure drop in the vertical rise of a liquid line in a CFC refrigerant system is
  A) 0.50 lb/ft.
  B) 0.75 lb/ft.
  C) 1.50 lb/ft.

1639. What type of control is usually used for loading or unloading a compressor automatically?
  A) Pressure switch
  B) Thermostat
  C) Manual valve

1640. The minimum velocity of refrigerant in a hot gas line should be
  A) 1,500 ft/min.
  B) 500 ft/min.
  C) 1,200 ft/min.

1641. When a refrigerant line rises vertically, the minimum velocity, compared to a horizontal line, should be
  A) increased.
  B) the same.
  C) decreased.

1642. The heat gain due to solar heat is greatest on the south wall of a building
  A) at the equator.

B) at 30 °F latitude.
C) at 40 °F latitude.

1643. In a commercial occupancy, an installation permit is not required on
    A) a new self-contained system of 1 hp or less, 6 lb or less of refrigeration.
    B) any system 1 hp or less.
    C) any self-contained system.

1644. When a warm system starts after a long idle period, it operates under high load conditions until the system pulls down to normal operating temperatures and pressures.
    A) True
    B) False

1645. With an increased heat load on the evaporator coil, the TXV will
    A) completely close.
    B) blow a fuse.
    C) allow more refrigerant to flow.
    D) de-energize the compressor.

1646. With a TXV, the superheat should be
    A) 0 to 5 °F.
    B) 5 to 15 °F.
    C) 15 to 25 °F.
    D) 25 to 35 °F.

1647. A stuck compressor can be caused by
    A) a loss of oil.
    B) oil trap in low-side system.
    C) internal overheating.
    D) all of the above.

1648. The three terminals on a single-phase compressor are
    A) main, auxiliary and ground.
    B) primary, secondary and intermediate.
    C) start, run and common.
    D) normally open, normally closed and common.

1649. The internal overload of the compressor senses
    A) low voltage at the compressor.
    B) amperage draw of the control circuit.
    C) excessive fan motor amp draw.
    D) excessive amperage in the compressor motor circuit.

1650. The compressor does not run, but the condenser fan is operating. Reading out the compressor the results are: R to S, 5 ohms; R to C , open; and S to C, open. The problem is a(n)
    A) faulty compressor contactor.
    B) open run winding.
    C) open internal overload.
    D) open start winding

1651. High-pressure safety switches generally close their contacts at
   A) 300 to 400 psi.
   B) 400 to 500 psi.
   C) 500 to 600 psi.
   D) none of the above.

1652. Compressor motor winding resistance is generally in the range of
   A) 0 to 10 ohms.
   B) 100 to 200 ohms.
   C) 1,000 to 2,000 ohms.
   D) all of the above.

1653. The purpose of a time delay fuse is to avoid nuisance blowing during starting, when the in-rush is much higher than the normal running current.
   A) True
   B) False

1654. A low voltage reading across an energized contactor from L to T side indicates
   A) an open contactor.
   B) excessive resistance.
   C) everything is fine.
   D) none of the above.

1655. A pressure relief device is required on a shell-and-tube evaporator when the refrigerant is in
   A) the shell.
   B) the tubes.
   C) either the shell or the tubes.

1656. The coil on a potential relay is placed between what terminals on the compressor?
   A) R and C
   B) C and S
   C) S and R

1657. A relay can be used with one or more capacitors.
   A) True
   B) False

1658. The water valve on a packaged air conditioner controls
   A) head pressure.
   B) suction pressure.
   C) discharge temperature.

1659. Comparing tower cfm with evaporative condenser cfm, the tower cfm is
   A) less than an evaporative condenser.
   B) equal to an evaporative condenser.
   C) greater than an evaporative condenser.

1660. The recommended pressure drop for tower piping is _____ psi per 100 feet of pipe length.
 A) 1 to 2
 B) 2 to 3
 C) 3 to 4
 D) 3 to 8

1661. The pH value of water in a tower should be kept between
 A) 3.5 to 4.5.
 B) 4.5 to 5.5.
 C) 5.5 to 6.5.
 D) 6.5 to 7.5.

1662. Nominal tonnage ratings for evaporative condensers on air conditioning units are based on what condensing temperature?
 A) 95 °F
 B) 100 °F
 C) 105 °F
 D) 110 °F

1663. Nominal tonnage ratings for evaporative condensers on air conditioning units are based on a suction temperature of
 A) 35 °F.
 B) 40 °F.
 C) 45 °F.
 D) 50 °F.

1664. Nominal tonnage ratings for evaporative condensers used for air conditioning units are based on an entering wet bulb temperature of
 A) 75 °F.
 B) 76 °F.
 C) 77 °F.
 D) 78 °F.

1665. The heat of compression is
 A) carried away in the evaporator.
 B) sometimes wasted.
 C) less on water-cooled compressors.
 D) carried away in the cooling medium leaving the condenser.

1666. Natural draft condensers are most frequently found in
 A) residential cooling.
 B) large industrial plants.
 C) household refrigerators.
 D) room air conditioners.

1667. The horizontal shell-and-tube condenser is more efficient than the vertical shell-and-tube condenser because
 A) it is mounted in a horizontal position.
 B) it is easier to clean.
 C) the water passes through it more than once.
 D) it takes advantage of more air cooling.

1668. The capacity of an evaporative condenser depends on the
   A) fan horsepower.
   B) amount of heat that exiting air is capable of absorbing.
   C) entering air wet bulb temperature.
   D) temperature of the entering air.

1669. Prime surface evaporators operating below 32 °F must frequently be
   A) cleaned.
   B) purged.
   C) painted.
   D) defrosted.

1670. Secondary refrigerants are usually
   A) Prestone.
   B) alcohol.
   C) water or brine.
   D) expensive.

1671. An under-capacity valve
   A) tends to flood the evaporator.
   B) tends to starve the evaporator.
   C) will work well at heavy loads.
   D) none of the above.

1672. Valve seats are usually made of
   A) the same materials as the needles.
   B) a material that can be adjusted.
   C) softer material than the needles.
   D) harder material than the needles.

1673. Automatic expansion valves used to replace capillary tubes
   A) are the simplest and safest way to replace a capillary tube.
   B) are not a good idea.
   C) require no provision for pressure difference.
   D) must have a high starting torque motor compressor.

1674. A leaking needle and seat on an automatic expansion valve will
   A) flood over the evaporator if a receiver is used.
   B) possibly not admit liquid to compressor.
   C) not react the same as a bleeder or bypass type.
   D) have no effect on the system at all.

1675. Moisture in an automatic expansion valve system
   A) does not affect the system at all.
   B) acts the same whether the compressor is running or not.
   C) does not cause a restriction in the system.
   D) acts the same as a plugged screen.

1676. To raise the temperature on an automatic expansion valve system,
   A) lessen spring tension.
   B) increase spring tension.

C) increase running time on compressor.
D) do not manually adjust.

1677. In adjusting an automatic expansion valve,
A) less tension on the spring admits more refrigerant to the evaporator.
B) less tension on the spring admits less refrigerant to the evaporator.
C) evaporator temperature will not change.
D) the compressor will cycle more often.

1678. A system using an AEV is cycled by the
A) float level.
B) low-side pressure.
C) thermostat.
D) pressure control.

1679. An automatic expansion valve is operated by
A) high-side pressure.
B) low-side pressure.
C) liquid-live pressure.
D) atmospheric pressure.

1680. If the voltage is 120 V and the current is 13 amps, the resistance (in ohms) is
A) 12.2.
B) 10.04.
C) 9.23.
D) 7.11.

1681. The watt input per ton of cooling is __________ for air-cooled units as compared to water-cooled units.
A) higher
B) lower
C) the same

1682. The cfm per ton on a packaged air conditioner is normally
A) 300.
B) 400.
C) 500.
D) 600.

1683. The sensible heat ratio on most packaged air conditioners is approximately
A) 50.
B) 60.
C) 75.
D) 95.

1684. The capacitance formula is $2{,}650 \times \frac{\text{amps}}{\text{volts}} = \text{mfd.}$
 A) True
 B) False

1685. The power input Kwh and horsepower per ton rating for an air-cooled condensing unit is higher than for a water-cooled condensing unit.
 A) True
 B) False

1686. The operating cost of an air-cooled condensing unit is higher than the operating cost of a water-cooled condensing unit using city water under comparable conditions.
 A) True
 B) False

1687. In figuring operating costs, the running time of the unit is used in the calculations, not the full load hours.
 A) True
 B) False

1688. The water cost of a cooling tower is due to bleed off and the evaporation of water.
 A) True
 B) False

1689. Air-cooled condensing units located outside the building usually use a propeller type fan.
 A) True
 B) False

1690. The maximum permissible quantities of Group II refrigerants for indirect systems are 600 lb for commercial and 300 lb for residential.
 A) True
 B) False

1691. In refrigeration systems containing Group II or III refrigerant, sweat joints on copper tubing shall be soldered.
 A) True
 B) False

1692. Refrigerant pipe joints erected on the premises shall be exposed for visual inspection prior to being covered or enclosed.
 A) True
 B) False

1693. Refrigerant piping crossing an open space that affords passageway in any building shall be not less than 7-1/2 feet above the floor unless against the ceiling of such space.
 A) True
 B) False

1694. Aluminum, zinc or magnesium shall not be used in contact with methyl chloride in a refrigerating system.
    A) True
    B) False
1695. Magnesium alloys shall not come into contact with any halogenated refrigerant.
    A) True
    B) False
1696. Discharge of pressure-relief devices and fusible plugs on all systems containing more than 6 lb of Group II or III refrigerant shall be to the outside of the unit.
    A) True
    B) False
1697. A brazed joint is a gas-tight joint obtained by the joining of metal parts at temperatures higher than 1,000 °F but less than the melting temperatures of the joined parts.
    A) True
    B) False
1698. Brine is any liquid used for the exchange of heat without a change in its state.
    A) True
    B) False
1699. Companion (or block) valves are pairs of mating stop valves.
    A) True
    B) False
1700. A compressor cannot be unloaded by throttling the flow of suction gas.
    A) True
    B) False
1701. A broken discharge reed can be caused by high head pressure.
    A) True
    B) False
1702. A loose rod bearing will always result in a compressor motor burnout.
    A) True
    B) False
1703. Suction line length can be the cause of compressor motor burnout.
    A) True
    B) False
1704. A worn wrist pin causes the suction pressure to run higher than it normally would.
    A) True
    B) False

1705. A leaking discharge reed will prevent a compressor from pulling down as low a vacuum as it normally would.
   A) True
   B) False

1706. A leaking suction reed will prevent a compressor from pulling as deep a vacuum as it normally would.
   A) True
   B) False

1707. The higher the suction pressure, the higher the compressor capacity.
   A) True
   B) False

1708. Copper plating is more likely to be found in air conditioning compressors than in low-temperature compressors.
   A) True
   B) False

1709. A suction line filter-drier is less likely to cause trouble in a low temperature compressor than in an air conditioning compressor.
   A) True
   B) False

1710. Changing an R-12 sealed unit compressor to an R-22 unit would increase the current draw by approximately 40%.
   A) True
   B) False

1711. A centrifugal compressor equalizes as soon as it stops turning.
   A) True
   B) False

1712. A ruptured disc, which will blow with high head pressure to bypass from the high-side back into the low side, is an excellent protective device to guard the compressor against abnormal head pressure.
   A) True
   B) False

1713. A direct drive compressor must have not more than .010" misalignment between compressor shaft and motor shaft to prevent damage to the main bearings, seal and couplings.
   A) True
   B) False

1714. The lower the suction gas superheat, the less efficient the compressor.
   A) True
   B) False

1715. The dome of a Scotch yoke compressor is on the high-pressure side of the system.
   A) True
   B) False

1716. Acid or moisture can be detected in a sealed unit compressor by using a megohmmeter.
    A) True
    B) False
1717. Ammonia compressors cannot be used with other refrigerants.
    A) True
    B) False
1718. If a compressor uses individual cylinder unloading, the unloading is always accomplished by holding the suction valve open by the force of a spring.
    A) True
    B) False
1719. Moisture indicators check the amount of frost.
    A) True
    B) False
1720. Ammonia discharge (where ammonia is used) may be into a tank of water that shall be used for no purpose except ammonia absorption.
    A) True
    B) False
1721. Sulfur dioxide discharge (where sulfur dioxide is used) may be into a tank of absorptive brine.
    A) True
    B) False
1722. Pressure vessels over 3 ft^3 but less than 10 ft^3 have no relief device.
    A) True
    B) False
1723. Moisture cannot affect metering device operation as long as there is no more than 8.6% water in the system.
    A) True
    B) False
1724. A hot gas line is for holding the compressor's refrigeration.
    A) True
    B) False
1725. Hot gas from a condensing unit can be used to defrost an evaporator.
    A) True
    B) False
1726. The position of the muffler is important.
    A) True
    B) False
1727. Design working pressure is the maximum allowable working pressure for which a specific part of a system is designed.
    A) True
    B) False

1728. A non-positive displacement compressor is a compressor in which increase in vapor pressure is attained without changing the internal volume of the compression chamber.
A) True
B) False

1729. A positive displacement compressor is a compressor in which increase in vapor pressure is attained by changing the internal volume of the compression.
A) True
B) False

1730. A pressure-imposing element is any device or portion of the equipment used for the purpose of increasing the oil pressure.
A) True
B) False

1731. A pressure limiting device is a pressure-responsive mechanism designed to automatically increase the refrigerant vapor pressure.
A) True
B) False

1732. Unloading a cylinder in a compressor increases the head pressure.
A) True
B) False

1733. Bypassing some of the discharge gas back into the suction line decreases the discharge temperature.
A) True
B) False

1734. The greatest problem in two-stage systems using two compressors with R-12 is overheating.
A) True
B) False

1735. Dropping the suction pressure on a straight air-cooled system, a standard system with a cooling tower, or on an evaporative condenser system will drop the head pressure.
A) True
B) False

1736. Dropping the suction pressure lowers the current draw on any refrigeration system.
A) True
B) False

1737. Dropping the suction pressure lowers the compressor discharge temperature on any type of system.
A) True
B) False

1738. A muffler will often be more effective in preventing vibration of the refrigerant lines than will a bellows type vibration eliminator.
    A) True
    B) False

1739. Vibration eliminator fittings should always be installed parallel to the compressor crankshaft.
    A) True
    B) False

1740. A compressor with a leaky discharge valve may have a backlash when it stops.
    A) True
    B) False

1741. A compressor that must have its belts pulled up tightly to prevent slippage so that no bow is seen in the off side of the belts when running does not have enough grooves in the pulleys.
    A) True
    B) False

1742. A system does not need to follow normal rules of piping for proper oil return if a high-side oil separator is installed.
    A) True
    B) False

1743. Pulling head bolts down tighter on one head of a V-type compressor than on the other head can damage the compressor.
    A) True
    B) False

1744. Bolting a compressor firmly to a heavy concrete block will prevent vibration and result in longer compressor life.
    A) True
    B) False

1745. The advantages of using a centrifugal fan on an air-cooled condensing unit are quieter operation and the allowance of air duct installation.
    A) True
    B) False

1746. The preferable location of the receiver on an air-cooled condensing unit is below the level of the condenser-coil outlet.
    A) True
    B) False

1747. On most remote air-cooled condensing units, the condenser blower runs all the time and the compressor cycles on head pressure.
    A) True
    B) False

1748. A good pressure setting for the relief valve on an air-cooled condensing unit using R-22 is 300 psi where the normal operating condensing temperature does not exceed 125 °F.
 A) True
 B) False

1749. It is reasonable to expect that the running power input in kilowatts for an air-cooled condensing unit will not exceed 1/2 kW per ton.
 A) True
 B) False

1750. On an air-cooled, self-contained, air conditioning unit, the cfm handled by the condenser fan is often double or higher than the cfm handled by the evaporator.
 A) True
 B) False

1751. According to ASHRAE, the pressure drop for the refrigerant in suction lines should not exceed 2 psi.
 A) True
 B) False

1752. According to ASHRAE, the pressure drop in refrigerant liquid lines should not exceed 4 psi.
 A) True
 B) False

1753. An evaporative condenser can handle the gas discharge from
 A) one compressor only.
 B) centrifugal compressors only.
 C) several compressors, providing it has the capacity.

1754. In a low-side float valve, the float ball and mechanism are
 A) in the high-pressure side of the system.
 B) in the low-pressure side of the system.
 C) completely outside the system.

1755. The function of an oil regenerator is to
 A) return oil carry over to the regenerator.
 B) purify the oil.
 C) separate the refrigerant from the oil.

1756. A snap-action, two-temperature valve is usually used
 A) where a small temperature difference is desired.
 B) where a large temperature difference is desired.
 C) in small and large cabinet combination installations.
 D) anywhere in a multiple installation.

1757. Where a cooling pond is used, the water is cooled primarily by
 A) evaporation.
 B) convection.
 C) radiation.

1758. The expansion valve most adaptable for varying load conditions is the
   A) thermostatic expansion valve.
   B) automatic expansion valve.
   C) high-side float.

1759. In all condenser operations, the heat gained by the condensing medium must equal the
   A) amount of refrigerant condensed.
   B) heat given up by the refrigerant.
   C) amount of water used.

1760. Superheat vapor is
   A) wet vapor at its condensing temperature.
   B) dry vapor at its boiling point.
   C) vapor that has been heated above its boiling point.

1761. R-12 vapor enters the condenser at 117.4 psig and 100 °F. If it leaves the condenser at 85 °F, its pressure is
   A) 91.9 psig.
   B) 117.4 psig.
   C) 154 psig.

1762. An excessive pressure drop in an evaporator will cause a thermostatic expansion valve to
   A) operate at a higher temperature.
   B) operate at a lower temperature.
   C) equalize.

1763. Vapor that is in contact with the liquid from which it is evaporated, or is at its evaporating temperature, is
   A) superheated vapor.
   B) saturated vapor.
   C) dry vapor.

1764. When using a recorder, how long should it be connected?
   A) 6 hours
   B) 8 hours
   C) 24 hours

1765. If the specific heat of beef is 0.77, the amount of heat to be removed from 100 lb of beef in cooling it from 75 to 35 °F is
   A) 7,700 Btu.
   B) 400 Btu.
   C) 3,080 Btu.

1766. The purpose of the equalizing line between the condenser and the receiver is to
   A) compensate for pressure drop.
   B) prevent flash gas.
   C) balance pressure.

1767. When the compression ratio of the compressor increases, its capacity
    A) increases.
    B) decreases.
    C) remains the same.

1768. In a counterflow condenser, the coldest water
    A) meets the coldest refrigerant.
    B) meets the warmest refrigerant.
    C) enters at the top.

1769. A higher heat transfer per ft^2 of evaporator is obtained with a
    A) dry expansion system.
    B) flooded evaporator.
    C) finned evaporator.

1770. Subcooling the liquid refrigerant is desirable because
    A) it prevents liquid from returning to the compressor.
    B) it reduces the amount of flash gas.
    C) the compressor operates at a lower head pressure.

1771. 746 watts of electrical energy, or 2,545 Btu/hour of heat energy, is equal to one
    A) ton of refrigeration.
    B) horsepower.
    C) kilowatt.

1772. With a high suction pressure, the capacity of the compressor is
    A) not changed.
    B) decreased.
    C) increased.

1773. An externally-equalized thermostatic expansion valve
    A) maintains constant evaporator pressure.
    B) keeps the evaporator filled with liquid.
    C) overcomes pressure drop in an evaporator.

1774. The normal cfm/ton of cooling for residential units is
    A) 200.
    B) 400.
    C) 650.

1775. In case of condensing water failure, which of the following safety devices should operate first?
    A) High-pressure cut-out switch
    B) Low water cut-out switch
    C) Pressure relief valve

1776. If the condensing water flow stops, which of the following safety devices should be the second to go into operation if the first safety device fails?
    A) High-pressure cut-out switch
    B) Low water cut-out switch
    C) Pressure relief valve

1777. What should be done if a fan blade is bent out of line?
    A) Straighten it
    B) Replace it
    C) Operate it "as is"
    D) If possible, bend the other blades to match

1778. The water in an evaporative condenser is cooled chiefly by
    A) conduction.
    B) evaporation.
    C) convection.

1779. If a gas mask is discovered with a broken seal on the canister, the
    A) seal should be renewed.
    B) canister should be renewed after a two-year period.
    C) canister should be renewed at once.

1780. In the refrigeration cycle, the motor current is highest at the time the
    A) system starts.
    B) condenser pressure is the highest.
    C) low-side pressure is the highest.
    D) low-side pressure is the lowest.

1781. What is one advantage of an oil separator?
    A) Keeps the oil in the compressor
    B) Low cost
    C) Easy to service
    D) Traps the moisture

1782. If the oil separator float collapses,
    A) nothing will happen.
    B) the valve will stay open.
    C) the valve will stay closed.
    D) the liquid refrigerant will short circuit into the crankcase.

1783. An oil separator must be mounted
    A) in any position.
    B) level.
    C) suspended from the condenser line.
    D) below the compressor.

1784. What could be wrong if a voltage exists between the motor-compressor housing and a water pipe when the power is turned on?
    A) Open circuit
    B) Closed circuit
    C) Short
    D) Ground

1785. In a large ice-making plant, the effect of any re-adjustment of the expansion valve would
    A) not be noticeable.
    B) not be immediately noticed.
    C) be immediately noticed.

1786. Where should the check valve be located when used in two-temperature installations?
   A) In the warmest coil suction line
   B) In the coldest coil suction line
   C) In the liquid line
   D) At the compressor

1787. If not used, the canisters (or cartridges) for gas masks should be renewed every
   A) two years.
   B) three years.
   C) five years.

1788. A lower condenser pressure will require
   A) greater work for the compressor.
   B) less work for the compressor.
   C) greater power to drive the compressor.

1789. Lower head pressures and temperatures can be obtained by use of
   A) compound compression.
   B) single-stage compression.
   C) evaporative condensers.

1790. Leakage at the shaft seal of an R-11 centrifugal compressor is usually
   A) of little importance to efficient operation.
   B) inward to the compressor.
   C) outward to the compressor.

1791. In laying-up a large unit for the winter months, it is considered good practice to store the refrigerant in the
   A) accumulator.
   B) evaporator.
   C) condenser and/or the receiver.

1792. Adding more calcium chloride to the brine will
   A) lower its freezing point under certain conditions.
   B) always lower its freezing point.
   C) have no effect on its freezing point.

1793. If the temperature of a cooling coil is below the dew point of the air passing through it, the
   A) air will be free of any moisture.
   B) air will evaporate any moisture on the cooling coil.
   C) moisture will condense out of the air.

1794. If the unit rumbles as it stops, what may be wrong?
   A) Nothing
   B) Wall is not strong enough
   C) Filters are clogged
   D) Unit may not be mounted level

1795. Of the conditions listed below, the one having the most adverse effect on the proper operation of a low-side float would be
   A) too much refrigerant in the system.
   B) the correct amount of refrigerant in the system.
   C) not enough refrigerant in the system.

1796. A discharge line oil trap or separator is utilized in a refrigeration system having a
   A) high volumetric efficiency.
   B) centrifugal compressor.
   C) reciprocating compressor.

1797. How is the defrost water kept from freezing in the drain pan and drain tubes?
   A) There is no danger of freezing
   B) It drains away too fast to freeze
   C) By the heat in the liquid line
   D) By electrical resistance heaters

1798. In the evaporative condenser, the tubes are usually
   A) finned.
   B) bare.
   C) studded.

1799. A refrigerant that is flammable is
   A) methyl chloride.
   B) sulfur dioxide.
   C) carbon dioxide.

1800. An accepted method of preventing frost formation on blast freezers is the use of
   A) hot gas.
   B) electric heaters.
   C) brine spray.

1801. An R-11 system generally utilizes a
   A) reciprocating compressor.
   B) rotary compressor.
   C) centrifugal compressor.

1802. The pH of brine may be increased by adding
   A) caustic soda.
   B) carbon dioxide gas.
   C) sodium chloride.

1803. Of the substances listed below, which is a chlorinated hydrocarbon?
   A) Carbon dioxide
   B) Ammonia
   C) Water
   D) Freon

1804. A liquid whose boiling point is above 32 °F could
A) not be used as a refrigerant.
B) be used as a refrigerant.
C) be used only with a CFC refrigerant.

1805. The addition of oil to R-12
A) raises its boiling point.
B) lowers its boiling point.
C) has no effect on its boiling point.

1806. Chlorine is used in water treatment to
A) kill micro-organisms.
B) reduce inorganic matter.
C) increase alkalinity.

1807. Shell-and-tube type condensers are set
A) vertically.
B) horizontally.
C) either vertically or horizontally.

1808. If an exact potential relay replacement is not available, what may be substituted?
A) Hot wire relay
B) One with a lower voltage rating
C) One with a higher voltage rating
D) Current relay

1809. An electrically grounded winding may be detected by
A) a shock.
B) the unit not running.
C) current flowing from one terminal to another.
D) current flowing from a terminal to the housing.

1810. In condenser service, the principle used for water and refrigerant circulation is the
A) up-flow principle.
B) parallel flow principle.
C) counterflow principle.

1811. The best remedy for a squeaky V belt (rubber composition) is to
A) check the belt alignment.
B) lubricate the belt with oil.
C) give the belt plenty of slack.

1812. Under most conditions, the best storage condition is when the relative humidity of the air
A) is well above the moisture content of the stored product.
B) is well below the moisture content of the stored product.
C) equals the moisture content of the stored product.

1813. Scale formation on heat transfer surfaces (such as the water side of condenser tubes)
A) may seriously hamper efficient operation.

B) should not be removed, as it retards corrosion.
C) allows an operator to use less refrigerant.

1814. Calcium chloride and lithium bromide are classified as
A) desaturators.
B) absorbents.
C) adsorbents.

1815. To cool one gallon of milk from 75 to 45 °F (specific heat of milk is 0.92, 1 gallon of milk weighs 8.6 lb) requires approximately
A) 130 Btu.
B) 240 Btu.
C) 360 Btu.

1816. What circuit is energized after the motor reaches operating speed?
A) Both starting and running windings
B) Only the running winding
C) Only the starting winding
D. Neither starting nor running windings

1817. What happens to the low-side pressure when the hot gas bypass valve opens?
A) Stays the same
B) Rises
C) Lowers
D) First it lowers, then it rises

1818. How many wires can be put under a screw terminal?
A) One
B) Two
C) Three
D) Four

1819. A two-stage or compound compressor has
A) cylinders of equal size.
B) a larger low-pressure cylinder.
C) a larger high-pressure cylinder.

1820. The primary function of a booster compressor is to
A) lower the discharge pressure.
B) lower the suction pressure.
C) increase the discharge pressure.

1821. Compressors equipped with valve unloaders are of the
A) positive displacement (piston) type.
B) centrifugal type.
C) hermetic type.

1822. The main function of the oil lantern in the compressor stuffing box is to
A) lubricate the rod.
B) prevent gas leakage.
C) separate impurities from the oil.

1823. A steam turbine is sometimes used to drive (prime mover) a
  A) centrifugal compressor.
  B) rotary compressor.
  C) reciprocating compressor.

1824. Condenser cooling water having a pH of 10 is considered
  A) alkaline.
  B) neutral.
  C) acid.

1825. The total heat removed from the refrigerant in the condenser is
  A) normally more than that absorbed in the evaporator.
  B) normally less than that absorbed in the evaporator.
  C) the latent heat of evaporation.

1826. Stationary bellows seals are used on
  A) motors.
  B) hermetic compressors.
  C) open-type compressors.
  D) serviceable hermetic compressors.

1827. Flywheels are usually used on
  A) hermetic compressors.
  B) shafts that are not tapered.
  C) open-type compressors.
  D) rotary-diaphragm compressors.

1828. Centrifugal compressors are used on
  A) small residential units.
  B) large air conditioning installations.
  C) commercial sized equipment.
  D) jobs requiring 50 tons or less.

1829. Of the following refrigerants, which is the most stable?
  A) R-12
  B) R-50
  C) R-22
  D) R-500

1830. Below-freezing temperature evaporators must
  A) have two fans.
  B) be controlled by suction pressure.
  C) have a defrost cycle.
  D) be made of steel.

1831. Before attempting to adjust the expansion valve in an open-type case,
  A) unload the case.
  B) load the case with product or a simulated load.
  C) neither A nor B.

1832. What is the name describing the plate evaporator where a solution for storage of cooling is provided?
  A) Finned tube

B) Eutectic plates
C) Shell and tube
D) Receiver plate

1833. What determines the final adjustment of the evaporator pressure regulator (EPR)?
A) Desired return air temperature of fixture
B) Desired outlet air temperature of the fixture
C) Neither A nor B

1834. What type of metering device is used on a flooded chiller?
A) Thermostatic expansion valve
B) Low-side float
C) Capillary tube
D) Hand expansion valve

1835. Which metering device will not shut off the flow of refrigerant when the compressor stops?
A) Thermostatic expansion valve
B) High-side float
C) Low-side float
D) Capillary tube

1836. Can the single model of thermostatic expansion valve be used for all refrigerants?
A) No
B) Sometimes
C) Occasionally
D) Yes

1837. Which pressure is the highest when employing an oil safety switch?
A) Oil
B) Back

1838. Which of the following metering devices cannot be used with a pressure control?
A) Expansion valve
B) High-side float
C) Low-side float
D) Capillary tube

1839. What prevents leakage past the stem of a packless valve?
A) A cap
B) Oil
C) Packing
D) A diaphragm

1840. What prevents leakage through the packing of a low-pressure valve?
A) A cap
B) Oil
C) Packing
D) Diaphragm

1841. What type of automatic closing valves are used on a receiver sight glass?
   A) Diaphragm
   B) Automatic ball-check valves
   C) Packing tube
   D) Solenoid

1842. How large should the receiver be?
   A) Tons of refrigeration times 5
   B) Equal to volume of the system
   C) Large enough to hold the full pump-down
   D) Full charge of the system should not occupy more than 85% of its volume

1843. Does an ammonia compressor use a plate or a screen baffle in the oil separator?
   A) Plate
   B) Screen

1844. In the case of a leak, water could possibly be of assistance with
   A) R-22.
   B) sulfur dioxide.
   C) ammonia.
   D) R-12.

1845. Under what conditions are two or more gas masks required?
   A) Group I exceeds 500 lb
   B) Group III exceeds 1,000 lb
   C) Group II exceeds 100 lb
   D) Any refrigerant exceeding 2,000 lb

1846. How long can a third class operator be absent from the post of duty in any one hour to inspect the equipment related to the system?
   A) 5 minutes
   B) 10 minutes
   C) 15 minutes
   D) 20 minutes

1847. If a piston weighing 100 lb exerts a force on an area of 20 in^2, what is the pressure exerted?
   A) 100 lb
   B) 5 lb/in^2
   C) 20 lb
   D) 2 lb/in^2

1848. If a refrigerant is condensed at 194 lb and 135 °F, the liquid is subcooled at a temperature of
   A) 140 °F.
   B) 120 °F.
   C) 180 °F.
   D) 160 °F.

1849. What is the maximum amount of refrigerant that can be stored in a machinery room in addition to that stored in the system (not to exceed 300 lb)?
    A) 20% of normal charge
    B) 30% of normal charge
    C) 40% of normal charge
    D) 50% of normal charge

1850. Are curves or tables best for finding the properties of refrigerant vapors?
    A) Curves
    B) Tables
    C) Neither curves nor tables

1851. How fast is the piston traveling in a modern high-speed compressor?
    A) 150 rpm
    B) 300 rpm
    C) 450 rpm
    D) 500 rpm

1852. What does VSA refer to in compressor design?
    A) Valves steel or aluminum
    B) Vertical single-acting
    C) Very slow acceleration
    D) Vefur sataci atture

1853. Which is a method of capacity modulation?
    A) Change in refrigerant charge
    B) Valve adjustment
    C) Two-staging
    D) Cylinder loading

1854. An application of variable speed capacity controls would be
    A) a low-temperature rotary compressor.
    B) aircraft air conditioning.
    C) automotive air conditioning.
    D) a centrifugal compressor.

1855. The disadvantage of strong compressor valve springs is
    A) sluggish operation.
    B) noise.
    C) difficult lubrication.
    D) short life.

1856. Which pressures have the highest volumetric efficiency?
    A) High back pressures
    B) Low back pressures

1857. The lowest capacity at which a centrifugal compressor can safely operate is what percent of full capacity?
A) 10%
B) 25%
C) 35%
D) 50%

1858. Safety heads on a compressor are used to
A) assist valve operation.
B) increase the compression ratio.
C) prevent damage due to liquid slugging.
D) prevent piston slapping.

1859. Ambient temperature below ____°F must require some sort of head pressure control?
A) 60
B) 50
C) 40
D) 30

1860. During summer operation, cooling tower water is available to the condenser at
A) 75 °F.
B) 80 °F.
C) 85 °F.
D) 90 °F.

1861. What type of condenser cannot be cleaned with a brush or router?
A) Double pipe
B) Shell and tube
C) Shell and coil

1862. Eliminator plates are used on
A) cooling towers.
B) evaporative condensers.
C) neither A nor B.

1863. What should be done before removing a valve head?
A) Add oil
B) Depressurize the compressor
C) Pressurize the compressor

1864. For a given R-12 unit, suction pressure at the compressor is 35 psig, there is an estimated suction line loss of 2 psig, and the temperature at the outlet of the evaporator is 51 °F. What is the superheat?
A) 9 °F.
B) 11 °F.
C) 14 °F.
D) 16 °F.

1865. The water level is maintained in the sump of a cooling tower by a
A) bypass valve.

B) hand valve.
C) solenoid valve.
D) float valve.

1866. A unit with low suction pressure and low superheat may
A) be overcharged.
B) have an oil-logged evaporator.
C) have the wrong superheat adjustment.
D) have an undersized compressor.

1867. Thermostatic expansion valve operation is determined by how many pressures?
A) One
B) Two
C) Three
D) Four

1868. On a TXV, if the inlet pressure is 27 psig and the spring pressure is 7 psig, what is the bulb pressure for the valve to be in equilibrium (valve has no equalizer connection)?
A) 27 psig
B) 7 psig
C) 34 psig
D) 20 psig

1869. On an evaporator used for a freezer room, fin spacing should be
A) 4 fins/inch.
B) 8 fins/inch.
C) 12 fins/inch.
D) 20 fins/inch.

1870. How many Btu are necessary to raise 1 gallon of milk from 45 to 80 °F? (Specific heat of milk is .92; one gallon of milk weighs 8.6 lb.)
A) 125 Btu
B) 177 Btu
C) 225 Btu
D) 277 Btu

1871. How many Btu are necessary to raise the temperature of 4 milk cartons from 45 to 80 °F? (Specific heat of milk carton = .32; one milk carton weighs .75 lb.)
A) 4.8 Btu
B) 33.6 Btu
C) 8.4 Btu
D) 36.3 Btu

1872. Why is the condenser always purged after charging?
A) To remove excess refrigerant
B) To help the unit pump-down
C) To remove noncondensable pressure
D) Purging is not necessary
E) To check the oil level

1873. In an operating water-cooled unit, head pressure temperature should correspond to
   A) 50 to 70 °F above outlet water temperature.
   B) 25 to 30 °F above outlet water temperature.
   C) 15 to 20 °F above outlet water temperature.
   D) 15 °F above room temperature.
   E) 25 °F above room temperature.

1874. What must be done to purge the condenser of a fuel gas operated water chiller?
   A) Attach gauges
   B) Isolate the condenser
   C) They have automatic purging devices
   D) Crack the line at the condenser header
   E) It is not necessary to purge

1875. What is wrong if the low-side pressure cannot be brought down to its normal setting?
   A) Too much refrigerant in the system
   B) Water in the system
   C) TXV is too small
   D) Dirt in the system
   E) Compressor is faulty

1876. What is one main advantage of a pressure-operated water valve?
   A) No cost of operation
   B) Turns the water on and off
   C) Easiest to install
   D) Varies the rate of water flow
   E) Operates satisfactorily under high water pressure

1877. Where is the refrigerant line to the water valve usually connected into the system?
   A) Suction line
   B) Compressor crankcase
   C) Receiver
   D) Compressor head
   E) Condenser

1878. What is the usual temperature difference between the water outlet on a unit with a water-cooled condenser without a cooling tower and the water inlet?
   A) 0 °F
   B) 10 °F
   C) 30 °F
   D) 40 °F
   E) 70 °F

1879. What happens to the water flow when the condensing unit stops?
   A) Keeps on running
   B) Continues running but at a lower rate of flow

C) Stops
D) Increases

1880. What is one of the advantages of an electric water valve?
A) Easy to install
B) Turns the water on and off
C) No cost of operation
D) Can operate under high pressure
E) Varies the rate of water flow

1881. How is ice cube making stopped automatically?
A) Bin-operated control
B) System produces a certain number of ice cubes and then stops
C) System stops when the water is used up
D) Timer
E) Pressure control

1882. What type of ice is produced when an auger is used?
A) Large sheets of ice
B) Solid cubes
C) Round, hollow cubes
D) Ice chips
E) Cone-shaped cubes

1883. On ice makers, the evaporator is usually made of
A) aluminum.
B) copper.
C) plastic.
D) brass.
E) stainless steel.

1884. What type of defrost system is used when solid, cubical ice cubes are made?
A) Hot gas
B) Electric
C) Thermal liquid
D) Electric and hot gas
E) Electric and thermal liquid

1885. The two-temperature valve is located in
A) the suction line.
B) the liquid line.
C) the condenser line.
D) place of the TXV.

1886. If the warmer evaporator is frosting back on a two-temperature system, the
A) two-temperature valve is leaking.
B) two-temperature valve is set too low.
C) system is overcharged.
D) TXV is leaking.
E) thermostat has points that are stuck closed.

1887. How is the warmer evaporator pressure checked?
A) A gauge opening on the two-temperature valve
B) It cannot be checked
C) Use the gauge manifold
D) Check the suction temperature
E) Remove, adjust and then re-install it

1888. Which evaporator has the greatest capacity?
A) Warm evaporator
B) Colder evaporator
C) Two capacities are the same
D) It does not make any difference
E) Only total capacity is important

1889. How far should external and internal pipe threads overlap or contact when correctly installed on evaporative condensers?
A) 3 threads
B) 5 threads
C) 7 threads
D) 9 threads
E) 11 threads

1890. Calibrating a float means
A) adjusting to provide the proper water level.
B) heating in order to test for leaks in the float.
C) dehydrating the float.
D) measuring the float size.
E) weighing the float.

1891. When using an evaporative condenser water must be added, as it is
A) cooler.
B) replacing that which has evaporated.
C) cleaner.
D) circulated faster.
E) under higher pressure.

1892. In a cooling tower system, static head is the
A) height of the unit.
B) length of the pipe to the condenser.
C) length of the pipe from the condenser.
D) vertical height of the piping.
E) water pressure when the pump is running.

1893. In a new system, a dehydrator is installed to
A) remove the copper chips.
B) remove any moisture that may be in the system.
C) keep moisture in the system from freezing.
D) remove the air.
E) keep the oil in the compressor.

1894. The suction line should be installed
    A) slanted upward toward the compressor.
    B) horizontally.
    C) with vertical loops.
    D) with no loops.
    E) horizontally and vertically only.

1895. The greatest decrease in unit capacity occurs with partial pinching at the
    A) suction line.
    B) liquid line.
    C) condenser line.
    D) evaporator coil tube.
    E) components equally in A, B, C, and D.

1896. The drier and sight glass are put in the system last, because
    A) they do not need evacuating.
    B) a vacuum cannot be pulled if they are in the system.
    C) a vacuum ruins the desiccant.
    D) a vacuum ruins the moisture indicator chemical.
    E) the drier and indicator must be free of impurities.

1897. During a joint brazing operation, protect nearby surfaces from being scorched by
    A) using a low-temperature flame.
    B) bending the pipe away from the surface.
    C) using a metal or asbestos plate shield.
    D) using flared connections.
    E) refinishing scorched surfaces.

1898. Reclaiming refrigerant means
    A) removing refrigerant in any condition from a system and storing it in an external container without necessarily testing or processing it in any way.
    B) cleaning refrigerant for reuse by oil separation and single or multiple passes through devices, such as replaceable core filter-driers, which reduce moisture, acidity and particular matter.
    C) reprocessing refrigerant to new product specifications by means which may include distillation. Will require chemical analysis of the refrigerant to determine that appropriate product specifications are met.
    D) none of the above.

1899. What is considered to be a good fast-freezing temperature?
    A) 32 °F
    B) 10 °F
    C) 5 °F
    D) -10 °F
    E) -20 °F

1900. A dry coil should be mounted
A) in a level position.
B) in the center of the cabinet.
C) 3" from the top of the cabinet.
D) at a slant toward the suction line.
E) as low as possible.

1901. Before starting an installed unit, the TXV adjustment should be
A) turned out.
B) turned in.
C) left alone.
D) turned to left.
E) turned to right.

1902. The average temperature difference between evaporator refrigerant and cabinet air is
A) 0 °F.
B) 10 °F.
C) 20 °F.
D) 30 °F.
E) 40 °F.

1903. On a reach-in cabinet, the thermostat should be located
A) on the coil.
B) in back of the coil.
C) on the cabinet wall.
D) just outside the door.
E) on the condensing unit.

1904. What is one of the advantages of a blower coil in a reach-in cabinet?
A) Requires little space
B) Does not dry up foods
C) Air movement is slow
D) Cheaper to operate
E) Easier to install

1905. How fast does the air flow over the typical blower coil in a reach-in cabinet?
A) 55 ft/min
B) 1,000 ft/min
C) 1,500 ft/min
D) 2,000 ft/min
E) 2,500 ft/min

1906. The lowest temperature at which refrigerant may be used in a direct water cooler is
A) 40 °F.
B) 35 °F.
C) 30 °F.
D) 32 °F.
E) 28 °F.

1907. In a restaurant, water should be cooled to
   A) 55 °F.
   B) 40 °F.
   C) 45 °F.
   D) 35 °F.
   E) 32 °F.

1908. Water coolers increase their efficiency with a heat exchanger by
   A) using a water-cooled condenser.
   B) cooling the incoming water with a heat transfer to the drain water.
   C) using the drain water to cool the condenser.
   D) using the incoming water to cool the condenser.
   E) running the liquid line through the cooled water.

1909. How many Btu must be removed to cool 20 gallons of water from 80 to 40 °F?
   A) 1,000
   B) 2,000
   C) 5,550
   D) 6,672
   E) 7,777

1910. In air conditioning, duct velocities are considered most during air
   A) heating.
   B) cooling.
   C) humidification.
   D) distribution.
   E) cleanliness.

1911. When the condensing unit is above the evaporator, a U-bend is put in the suction line to
   A) prevent frost back.
   B) assist the oil return.
   C) prevent back pressure in the evaporator.
   D) prevent high-/low-side pressures.
   E) act as a vibration control.

1912. Are CFCs harmful to the ozone layer?
   A) Yes
   B) No

1913. What temperature should the head pressure correspond to when the unit is running?
   A) 70 °F above the head pressure
   B) 10 °F above the room temperature
   C) 30 °F below the room temperature
   D) 30 °F above the room temperature
   E) It is independent of the temperature

1914. In sub-zero weather, normal head pressure is maintained by
A) an undersized condenser.
B) overcharging the system.
C) restricting the air flow over the condenser.
D) not changing anything.
E) shutting off the condenser fan.

1915. When there is a voltage drop at a terminal,
A) there is a loose, dirty connection.
B) the terminal is too large.
C) the electrical load is too small.
D) the voltmeter is connected incorrectly.
E) the terminal is grounded.

1916. Which type of piston crankshaft arrangement does not use a connecting rod?
A) Scotch yoke
B) Eccentric
C) Conventional
D) Hotchkiss
E) V type

1917. Which type of compressor has the cylinders parallel to the crankshaft?
A) Swash plate
B) Eccentric
C) Conventional
D) Scotch yoke
E) Hotchkiss

1918. If a 12" x 12" grille has 40 metal strips 1/32" thick mounted lengthwise, what is its net opening?
A) 144 in^2
B) 156 in^2
C) 129 in^2
D) 141-1/4 in^2

1919. What is the approximate setting for R-12 when the defrost is terminated by pressure?
A) 85 psig
B) 47 psig
C) 96 psig

1920. One horsepower equals how many foot pounds of work per minute?
A) 3,415
B) 31,415
C) 33,000
D) 946

1921. What are the possible results of operating a system when the compressor load is not properly balanced?
A) Short cycling, compressor may be damaged, and oil failure trips

B) Compressor will be too cold
C) Compressor will be too hot

1922. Refrigerant containers must be
A) steel.
B) 1/4" thick.
C) ASA approved.
D) ICC approved.

1923. The capacity of an evaporative condenser is based on what air condition?
A) Dry bulb temperature
B) Wet bulb temperature
C) Dew point temperature
D) Relative humidity

1924. Some sort of head pressure control is required when temperatures are below
A) 60 °F.
B) 50 °F.
C) 40 °F.
D) 30 °F.

1925. Eliminator plates are used on
A) air-cooled condensers.
B) evaporative condensers.
C) shell-and-coil condensers.

1926. The most efficient double-pipe condenser is designed for
A) parallel flow.
B) counter flow.
C) air flow.

1927. Most water-cooled condensers, using cooling tower water as applied to air conditioning installation at design conditions, raise the temperature of the water
A) 5 °F.
B) 10 °F.
C) 15 °F.
D) 20 °F.

1928. A bleed-off is used on
A) air-cooled condensers.
B) water-cooled condensers.
C) evaporative condensers or cooling towers.
D) air-cooled or water-cooled condensers.

1929. A water regulating valve is used on what kind of a system?
A) Waste water
B) Tower water
C) Air cooled
D) Evaporative

1930. For a florist cabinet that requires high humidity, the most desirable temperature difference between the refrigerant and surrounding air is
A) 7.5 °F.
B) 15 °F.
C) 20 °F.
D) 25 °F.

1931. If the inlet pressure regulator (IPR) valve is too high,
A) adjust the IPR until there is a slight increase in pressure.
B) adjust the PIR until there is a slight decrease in pressure.
C) neither A nor B.

1932. What are the steps in adjusting the differential pressure regulator (DPR)?
A) Fully close the valve; install a gauge on the liquid line; if receiver pressure is higher, purge off to the low side; turn adjustment until heat is felt on the line between the receiver and the DPR
B) Fully close the valve; install a gauge on the low side line; if receiver pressure is higher, purge off to the high side; turn adjustment until heat is felt on the line between the receiver and the DPR
C) Neither A nor B

1933. When evaporators must be used to remove large amounts of humidity, which would be the most desirable coil from the standpoint of number of tubes deep (rows deep)?
A) 2 tubes deep
B) 4 tubes deep
C) 6 tubes deep
D) 8 tubes deep

1934. In a direct expansion chiller, is the refrigerant in the tubes or in the shell?
A) Tubes
B) Shell
C) Tubes and the shell
D) None of the above

1935. The major factor in determining the filter design used on a particular installation is
A) degree of cleanliness required.
B) dust content of 8 grains per 1000 feet of air.
C) location of filter.
D) size of room where used.

1936. The power required to operate an ionizing-type electronic filter in a 2800 cfm system is approximately
A) 0.82 kilowatt.
B) 0.082 kilowatt.

C) 8.2 watts.
D) 820 watts.

1937. When the relative humidity of the incoming air and the outgoing air is the same, the dehumidifier has reached its
A) breakpoint capacity.
B) completion point.
C) equilibrium capacity.
D) regeneration time.

1938. Filters are used in air-conditioning systems where the dust content seldom exceeds
A) 0.4 grains/100 ft^3 of air.
B) 2 grains/1000 ft^3 of air.
C) 3 grains/1000 ft^3 of air.
D) 4 grains/1000 ft^3 of air.

1939. How many cfm of outside air is needed to remove odor from a gym that houses 275 people?
A) 82.50 cfm
B) 825 cfm
C) 8,250 cfm
D) 82,500 cfm

1940. How does the viscous impingement filter compare to the dry filter?
A) Dry filters have low lint-holding capacity
B) Viscous impingement filter holds less dust
C) Viscous impingement filter holds more lint
D) Efficiency of dry filter is higher

1941. The dielectric properties of the charged media electronic filter become impaired when
A) relative humidity exceeds 50%.
B) dry-bulb temperature exceeds 75 °F.
C) relative humidity exceeds 70%.
D) dew point temperature is 70 °F and dry-bulb temperature is 91 °F.

1942. If velocity pressure of a system is 30 inches water gauge and the static pressure is 2 inches water gauge, what is the dynamic pressure?
A) 15 inches water gauge
B) 28 inches water gauge
C) 32 inches water gauge
D) 60 inches water gauge

1943. To make measurements in a space that is not easily accessible, what is the best air measuring instruments to use?
A) Kata thermometer
B) Anemometer
C) Pitot tube
D) Velometer

1944. When calculating the heat load of an area that has a changeable load, such as day and night shifts of workers, it is necessary to consider
   A) time of day.
   B) occupant heat load.
   C) equipment heat load.
   D) minimum off-peak load.

1945. What is the minimum ambient temperature at which water-cooled equipment should be installed?
   A) 32 °F
   B) 35 °F
   C) 38 °F
   D) 40 °F

1946. The most probable cause of moist insulation is
   A) air passing through insulation.
   B) temperature differential.
   C) wrong type of insulation.
   D) high relative humidity.

1947. What type of insulation should be used on a 40 °F cold storage room?
   A) Rock wool
   B) Asbestos
   C) Cork
   D) Aluminum foil

1948. If the make-up water hardness is 125 ppm and the bleed-off hardness is 500 ppm, the cycle of concentration is
   A) 4 to 1.
   B) 1 to 4.
   C) 1.
   D) 4.

1949. To proportion the amount of outside and recirculated air returned to an air conditioner, the damper installed should be of which type?
   A) Volume
   B) Mixing
   C) Bypass
   D) Recirculating

1950. In the double-pipe condenser of floor-mounted units, the leaving refrigerant could be lowered to the same temperature as the incoming water; however, what degree of differential is acceptable?
   A) 15 °F
   B) 12 °F
   C) 10 °F
   D) 8 °F

1951. What occurs when the electric fire protection control is activated?
   A) Alarm is set off
   B) System fans shut off

C) Fans shut off and dampers close
D) Signal sent to fire department

1952. How many cfm will an evaporative cooler deliver if the air velocity is 60 fpm and the supply duct measures 4' long x 1.5' wide?
A) 3,600 cfm
B) 360 cfm
C) 36 cfm
D) 3.60 cfm

1953. A thermostat on an evaporative cooler usually controls the operation of the
A) blower and water pump.
B) automatic flush valve.
C) sump water heater.
D) blower.

1954. What capacity fan would be required for a room 25' x 20' x 10', requiring 12 air changes per hour?
A) 100 cfm
B) 1,000 cfm
C) 10,000 cfm
D) 11,000 cfm

1955. The ventilation requirement for a theater with a floor area of 150' x 75' is
A) 1,250 cfm.
B) 5,625 cfm.
C) 16,875 cfm.
D) 22,500 cfm.

1956. If the total air flow pressure is equal to 20 inches of water gauge and the static pressure equals 4 inches of water gauge, what is the velocity pressure?
A) 24 inches of water gauge
B) 20 inches of water gauge
C) 16 inches of water gauge
D) 14 inches of water gauge

1957. How much capacity will a condenser lose if the water supplied to it is 85 °F instead of 75 °F?
A) 25%
B) 15%
C) 10%
D) 5%

1958. The device used for moving large quantities of air against low pressure with free exhaust is a
A) multiblade fan with blades curved backward.
B) multiblade fan with blades curved forward.
C) multiblade fan with radical blades.
D) propeller fan.

1959. The average air velocity of a supply duct is 50 cfm. If the duct dimensions are 12" x 24", what is the quantity of airflow?
   A) 50 cfm
   B) 75 cfm
   C) 100 cfm
   D) 150 cfm

1960. In slinger-type evaporative units, the float actuated make-up valve controls the water
   A) valve.
   B) valve and bleed-off water.
   C) valve, water level, and bleed-off water.
   D) level.

1961. In installing an evaporative cooler, what diameter would be used for the water drain or waste system?
   A) 1.25 inches
   B) 1.5 inches
   C) 2 inches
   D) 2.25 inches

1962. Which of the following would be attached to the condenser to prevent water freeze-up during the heat cycle?
   A) Check valve
   B) Condenser pressure regulator
   C) Bypass valve
   D) Evaporator pressure regulator

1963. When setting the thermostat of a beverage cooler, it is necessary to consider
   A) anticipated heat load.
   B) freezing point of beverage.
   C) desired serving temperature.
   D) temperature range of the thermostat.

1964. The disadvantage of an expansion valve with a gas charge is that
   A) it may reach minimum operating pressure.
   B) it may cause flooding and searching if control is lost.
   C) control is lost if the diaphragm and case are colder than the bulb.
   D) control is lost if the diaphragm and case are warmer than the bulb.

1965. Which of these statements does not apply to a shell-and-tube condenser?
   A) It can be used both as condenser and receiver
   B) If water is scarce, a cooling tower may be used
   C) Exhaust water may be used to cool compressor heads
   D) Water normally circulates through the shell

1966. A compressor teardown is necessary if
   A) oil is low.
   B) metal particles are found in crankcase and strainer.
   C) oil pressure valve has failed.
   C) there is a defective oil pump.

1967. If an expansion valve has a cross charge, it has a
   A) temperature curve opposite refrigerant in system.
   B) temperature and pressure curve crossing refrigerant's curve.
   C) pressure curve opposite refrigerant in system.
   D) higher sensitivity to bulb temperature changes.

1968. A high-pressure control used in defrosting has a capillary tube to the high side. A system charged with R-12 will have the pressure range of the control set to open at
   A) 180 and close at 155 pounds.
   B) 155 and close at 180 pounds.
   C) 180 and close at 205 pounds.
   D) 205 and close at 180 pounds.

1969. Compressor off-time defrost is compatible with
   A) closed cabinets for dairy products.
   B) walk-in cabinets for bulk meats.
   C) display case for frozen vegetables.
   D) systems operating below 28 °F.

1970. Which one of the following is not often considered as contributing moisture to the system?
   A) Contaminated oil or refrigerant
   B) Leak in water-cooled condenser
   C) Open system
   D) New drier-strainer

1971. When installing a new set of rings, which of the following is used to position the rings correctly?
   A) Compression rings have a top
   B) Compression rings have no taper
   C) Oil rings have a taper
   D) Oil rings have a top

1972. Which defrost system has a high-pressure control to protect the unit?
   A) Hot water
   B) Hot gas
   C) Hot wire
   D) Secondary solution

1973. Which evaporator type is found in both cube and ice-flake machines?
   A) Tray
   B) Tube
   C) Cell
   D) Plate

1974. Anhydrous calcium sulfate cannot be used as a drier in a system charged with
- A) ammonia.
- B) SO_2.
- C) R-12.
- D) R-11.

1975. Good ice grade depends mostly on the
- A) specific gravity of the brine.
- B) use of a core sucker.
- C) evaporator temperature.
- D) use of fresh water.

1976. When mixing inhibited acid to clean scale from condenser tubing, what precautions should be taken for personal safety?
- A) Add acid to water never water to acid
- B) Use special pump for acids
- C) Mix solution in wooden barrel
- D) Do not use galvanized metals

1977. A cold storage plant made changes resulting in the system being unable to maintain freezer room temperature without a 50% increase in compressor operating time. What change caused the problem?
- A) Evaporative condenser was cleansed
- B) Bleed water flow at condenser was increased
- C) New set of belts was installed for low-temperature compressor
- D) New valves and valve plate installed in LT compressor

1978. In a cold storage system the pressure control used with a motor starter is set to cut out at
- A) desired suction pressure.
- B) 10 °F below coil temperature.
- C) desired high-side pressure.
- D) desired evaporative temperature.

1979. Personal safety in a cold storage plant may be affected by CO_2 gas in the
- A) machine room.
- B) meat freezer room.
- C) meat processing room.
- D) potato storage room.

1980. An ice plant is efficient when operating with a brine temperature of 15 °F. The specific gravity of the brine is adjusted so that its freezing point is
- A) 10 °F.
- B) 5 °F.
- C) 0 °F.
- D) -10 °F.

1981. When pumping down, the __________ suction-pressure regulating valves must be bypassed.
    A) bellows
    B) diaphragm
    C) snap-action
    D) check

1982. What is the refrigerant most commonly used for low temperature?
    A) Ammonia
    B) SO_2
    C) R-502
    D) Methyl chloride

1983. In a multiple-evaporator system, when the evaporator carrying more than half of the heat load is the one operated at the warmest temperature, the compressor will be
    A) too hot.
    B) at lower suction pressure.
    C) at abnormally high head pressure.
    D) at higher suction pressure.

1984. What is used for proper lubrication of multiple compressors?
    A) Oil and gas equalizer lines
    B) Equal length suction branches
    C) Manifold of sufficient size
    D) Parallel installation of compressors

1985. An ultralow temperature system with direct compounding
    A) uses intercoolers and subcoolers for increased efficiency.
    B) has two independent refrigerant systems.
    C) uses one evaporator to cool a condenser.
    D) uses R-12 at 24 inches of vacuum.

1986. What ultralow temperature system can use two kinds of refrigerants?
    A) Staging
    B) Cascade
    C) Compound
    D) Direct compounding

1987. How close may the piston of a reciprocating compressor be to the head?
    A) 2 inches
    B) 1 inch
    C) 0.001 inch
    D) 0.01 inch

1988. To ensure proper ice machine operation, a blowdown should be done
    A) once each defrost cycle.
    B) each 24-hour period.
    C) to cool condenser.
    D) as often as necessary.

1989. After installing a new set of rings, the compressor is pumping an excessive amount of oil from the crankcase. What most likely caused this trouble?
   A) Ring gaps staggered
   B) Oil rings upside down
   C) Compression rings upside down
   D) Compression rings substituted for oil rings

1990. When activated alumina is used as a drier, it can only be installed in the suction line when the system is charged with
   A) methyl chloride.
   B) SO_2.
   C) R-11.
   D) R-12.

1991. What are the operating advantages when four or more compressors are installed in a cold storage plant?
   A) Loss of two compressors will not halt operations
   B) Four can be used for high pressure operation
   C) Plant can be split into four symmetrical systems
   D) Plant can be split into two systems, with two compressors for high temperature and two for low

1992. In an ammonia system, what increase above normal in condenser temperature indicates scale formation?
   A) 5 °F
   B) 6 °F
   C) 8 °F
   D) 10 °F

1993. A valid step in preparing inhibited acid for cleaning scale from the tubes of a condenser is to
   A) dissolve 3-2/5 oz of inhibitor powder in 10 gallons of water.
   B) use solution cold for reducing cleaning time.
   C) use sulfuric acid in preference to hydrochloric acid.
   D) use commercial hydrochloric acid with specific gravity of 1.

1994. An installation requirement for satisfactory operation of compressors in parallel is
   A) connecting lines must be as long as possible.
   B) oil- and gas-equalizer lines are installed between crankcases.
   C) gas-equalizer connection is made below maximum oil level.
   D) oil and gas equalizers must have maximum of 5 tons capacity.

1995. An ordinary expansion valve is unsatisfactory at ultralow temperatures, because it
   A) has two power elements without thermal bulbs.
   B) cannot control flow of refrigerant.
   C) needs a solenoid valve behind it.
   D) necessitates excessive superheating at the bulb location in order to operate.

1996. How thick, in inches, must the insulation be for ultra-low temperature cabinets?
   A) 4 to 8
   B) 6 to 10
   C) 8 to 12
   D) 10 to 12

1997. Lubricating oil for low temperatures must not produce wax separation at or below the lowest expected operating temperature. To prevent this,
   A) use chlorinated refrigerants.
   B) increase percentage of oil in the mixture.
   C) use high grade oil processed for low temperatures and oil separators.
   D) keep oil-refrigerant mixture at low temperature.

1998. The sun is included in cooling load calculations by adding
   A) 5 °F.
   B) 10 °F.
   C) 15 °F.
   D) none of the above.

1999. If the outside temperature is 20 °F for 24 hours, how many degree days accumulate?
   A) 15
   B) 25
   C) 35
   D) 45

2000. What is the usual wind velocity design?
   A) 5 mph
   B) 10 mph
   C) 15 mph
   D) 20 mph

Heating

Level 1

2001. The leakage of air around windows, doors and cracks in a house is
    A) an air inlet.
    B) infiltration.
    C) a bypass.
    D) psia.
    E) none of the above.

2002. A limiting device installed on a forced air furnace is controlled by
    A) a vent.
    B) a bypass.
    C) temperature.
    D) pressure.
    E) none of the above.

2003. One cubic foot of natural gas would be
    A) 100 Btu.
    B) 2,500 Btu.
    C) 1,000 Btu.
    D) 140,000 Btu.
    E) 250 Btu.

2004. How much air is needed to burn 1,000 Btu?
    A) 100 ft^3
    B) 10 ft^3
    C) 1,000 ft^3
    D) 250 ft^3
    E) 1 ft^3

2005. The pressure from the gas meter to the appliance is
    A) 4 to 5 inches water column (wc).
    B) 3 to 4 inches wc.
    C) 8 to 11 inches wc.
    D) 6 to 7 inches wc.
    E) 9 to 11 inches wc.

2006. The manifold pressure of natural gas is
    A) 3 to 4 inches wc.
    B) 6 to 7 inches wc.
    C) 8 to 11 inches wc.
    D) 4 to 5 inches wc.
    E) 9 to 11 inches wc.

2007. What is the permissible pressure drop between the meter and the appliance?
A) 3.0" wc
B) 50.0" wc
C) 5.0" wc
D) 0.50" wc
E) 11.0" wc

2008. An oil tank of 220 gallons must be what gauge?
A) 14
B) 12
C) 16
D) 10
E) 8

2009. On a low-pressure, residential, forced-water boiler, the relief device opens up at
A) 15 lb.
B) 30 lb.
C) 125 lb.
D) 150 lb.

2010. On a residential, low-pressure, forced-water boiler, the pressure reducing valve is factory set at
A) 30 lb.
B) 15 lb.
C) 12 lb.
D) 125 lb.

2011. On a residential, low-pressure, steam boiler, the pressure reducing valve is set at
A) 30 lb.
B) 15 lb.
C) 12 lb.
D) none of the above.

2012. On residential, low-pressure, forced-water boilers, a gate valve must be at the
A) outlet of the boiler.
B) inlet of the boiler.
C) outlet and inlet of the boiler.
D) none of the above.

2013. On a residential, low-pressure, steam boiler, the device used for water expansion is a(n)
A) stack pipe.
B) compression tank.
C) airtrol.
D) none of the above.

2014. On a low-pressure, residential, steam boiler, the relief device opens at
    A) 30 lb.
    B) 12 lb.
    C) 15 lb.
    D) none of the above.

2015. The type of zone valve for White-Rodgers is
    A) diaphragm.
    B) heat-activated.
    C) motorized.
    D) none of the above.

2016. Minneapolis Honeywell zone valves are
    A) motorized.
    B) heat-activated.
    C) diaphragm.
    D) none of the above.

2017. Normal operating pressure for residential forced-water boilers is
    A) 30 lb.
    B) 12 to 15 lb.
    C) 15 to 30 lb.
    D) none of the above.

2018. The rate at which heat flows through the composite of several different materials used in a building is called the U or
    A) temperature factor.
    B) maintained temperature.
    C) heat transmission factor.
    D) none of the above.

2019. A pilot light has a safety device to
    A) relight the pilot light.
    B) stop the pilot light gas flow if there is no pilot flame.
    C) stop the main gas flow if there is no pilot flame.
    D) shut off the furnace if the stack temperature is too high.
    E) shut off the furnace if the furnace temperature is too high.

2020. The primary air adjustment is at the
    A) pilot light.
    B) gas spud.
    C) entrance to the burner.
    D) pressure regulator outlet.
    E) entrance to the gas manifold.

2021. In an intermittent, ignition primary control, the relay operates the
    A) motor and stack bimetal circuits.
    B) ignition and thermostat circuits.
    C) ignition and motor circuits.
    D) ignition transformer.
    E) overload manual reset circuits.

2022. The step-down transformer is located in the
 A) limit switch.
 B) thermostat.
 C) primary control.
 D) ignition transformer.
 E) manual switch box.

2023. The average gravity warm-air heating system needs
 A) a return for each room.
 B) usually two or three returns.
 C) no returns.
 D) five or more returns.

2024. The most difficult style of house to heat is probably the
 A) two story.
 B) ranch style.
 C) split level.
 D) six-room single level.

2025. The gas supply connection from the pilot shall be made
 A) between the main shut-off valve and the meter.
 B) between the main shut-off valve and the burner.
 C) by drilling and tapping the gas supply line.
 D) at any convenient location.

2026. The secondary air adjustment shutter should always be
 A) locked in the wide open position.
 B) properly adjusted by the installer.
 C) left in the position set by the factory.
 D) replaced by a barometric damper.

2027. Delayed gas ignition may be caused by a(n)
 A) defective thermostat.
 B) unvented pressure regulator.
 C) blown fuse.
 D) improperly located pilot.

2028. A separate electrical permit is required
 A) for all gas heating installations.
 B) for any connections to the 110-volt circuit other than the installation of a transformer on an existing outlet.
 C) when an electrician wires the installation.
 D) when a transformer is installed.

2029. In an electrical resistance heating system, a low-voltage thermostat can be used with a
 A) low-voltage power circuit.
 B) low amperage circuit.
 C) relay.
 D) parallel circuits.
 E) series circuits.

2030. How is a tubular electrical resistance heating unit insulated?
   A) It is not
   B) Plastic insulation
   C) Resistance wire is overlaid in an oxide powder
   D) Ceramic insulators
   E) Porcelain insulators

2031. How many grains of moisture will saturate one pound of dry air at 0 °F?
   A) 5.5 grains
   B) 10.5 grains
   C) 15.5 grains
   D) 20.5 grains
   E) 25.5 grains

2032. How many grains of moisture will saturate one pound of dry air 50% at 75 °F?
   A) 5.5 grains
   B) 36 grains
   C) 50 grains
   D) 66 grains
   E) 132 grains

2033. Which control devices may be used on the 24-volt circuit in a domestic, gas-fueled, heating hydronic system?
   A) Thermostat
   B) Thermostat, relay and safety pilot
   C) Thermostat, relay and gas solenoid
   D) Thermostat, relay, gas solenoid, limit control and safety pilot
   E) Thermostat, relay, gas solenoid, limit control, safety pilot and circulation pump

2034. The factor used to determine the percentage of a home's total heat loss by outside walls and windows is
   A) 60 to 80%.
   B) 100%.
   C) 50%.
   D) 25 to 50%.

2035. The flue pipe from a gas-designed furnace shall be
   A) at least 8 inches in diameter.
   B) the size indicated by the flue connection on the furnace.
   C) one size larger than the flue connection on the furnace.
   D) any convenient size.

2036. The flue pipe from a gas-fired hot water heater should
   A) be connected to the gas furnace flue pipe ahead of the diverter.
   B) be connected to the gas burner flue pipe behind the diverter.
   C) enter the chimney through a separate opening.
   D) be removed from the chimney.

2037. In case of a ruptured diaphragm in a pressure regulator, the escaping gas should
  A) discharge into the furnace room.
  B) cause the solenoid valve to close.
  C) be ignited by the pilot flame.
  D) cause an explosion.

2038. If the gas pilot flame goes out on equipment of 400,000 Btu or more, the safety pilot should shut off the gas to the main burner within
  A) 2 minutes.
  B) 3 minutes.
  C) 10 seconds.
  D) 60 seconds.

2039. If too much primary air is used, the flame will
  A) be too short.
  B) lift off the burner.
  C) be too hot.
  D) be irregular.

2040. Supply outlets (diffusers or registers) introduce conditioned air into the
  A) rooms.
  B) furnace plenum.
  C) wall stacks.
  D) attic.

2041. An area in which fuel or a gaseous substance is burned is called a
  A) combustion chamber.
  B) firebox liner.
  C) bonnet.
  D) radiator.

2042. To check the input of a gas burner on a residential, forced-water boiler,
  A) read the Btu input of the boiler.
  B) read the output of the Btu on the boiler.
  C) read the net Btu of the boiler.
  D) clock the revolutions of the gas meter dial.

2043. The purpose of the deflector plate in a draft diverter is to
  A) provide for adjustment of pressure level.
  B) prevent a backdraft from entering the combustion chamber.
  C) deflect the flue gases.
  D) none of the above.

2044. If the gas pilot goes out on equipment of less than 400,000 Btu, the safety pilot should shut off the gas to the main burner within
  A) 10 seconds.
  B) 90 seconds.
  C) 3 minutes.
  D) 2 minutes.

2045. The steam temperature of 5 inches of vacuum is
   A) 180 °F.
   B) 160 °F.
   C) 205 °F.
   D) 212 °F

2046. Oil tank vent pipes on tanks of 275 gallons must be at least
   A) 1".
   B) 1-1/4".
   C) 1-1/2".
   D) 2".
   E) 2-1/2".

2047. Oil tank fill pipes on tanks of 275 gallons must be how many feet above grade?
   A) One
   B) Two
   C) Three
   D) Four
   E) Five

2048. Oil tank vent pipes should terminate outside the building and be how many feet above grade?
   A) 1
   B) 2-1/2
   C) 3
   D) 3-1/2
   E) 4

2049. The system that provides a channel for air passage to and from a furnace is called the
   A) pipeline.
   B) duct system.
   C) passageway.
   D) tunnel.

2050. Heating system trunk ducts carry the air that is to be distributed to more than one
   A) supply inlet.
   B) furnace.
   C) residence.
   D) supply outlet.

2051. If part of a domestic heating system supply duct passes through an unexcavated space or a masonry wall, that portion should be covered with
   A) duct tape.
   B) insulation.
   C) vinyl film.
   D) aluminum foil.

2052. To combat added resistance of air flow (such as additional ductwork) in a heating system,
A) speed up a small blower motor.
B) slow down the blower motor.
C) install a smaller blower assembly.
D) install a larger blower assembly.

2053. The letter H on oil nozzles means
A) hollow.
B) solid.
C) special design.
D) semi-hollow.

2054. A correctly-adjusted natural gas flame is
A) orange.
B) red.
C) yellow.
D) blue.
E) green.

2055. If the air flow through a baseboard electric resistance heater is stopped,
A) the thermostat will open the circuit.
B) the relay will open the circuit.
C) the system will continue to operate.
D) a timer will shut off the unit.
E) a limit control will open the circuit.

2056. How much heat is created by one watt?
A) 760 Btu
B) 3,415 Btu
C) 3.42 Btu
D) 2545.6 Btu
E) 14.4 Btu

2057. What is the recommended maximum ampere flow in a gas furnace ?
A) 10 A
B) 15 A
C) 25 A
D) 30 A
E) 35 A

2058. What is the relative humidity at 75 °F db and 68 °F wb effective temperature?
A) 80%
B) 70%
C) 60%
D) 50%
E) 40%

2059. Some power humidifiers have electric motors to
   A) operate the valve.
   B) stir the water.
   C) expose more water surface to the air.
   D) pump water into the humidifier.
   E) scrape the scale out of the humidifier pan.

2060. What type of material varies in size as the humidity changes?
   A) Ceramic
   B) Metallic
   C) Hygroscopic
   D) Silicon
   E) Hydraulic

2061. Ducts located in attics, ventilated crawl spaces or other spaces exposed to outdoor weather conditions should be
   A) coated with paint to prevent corrosion.
   B) coated with asphalt paint to reduce heat loss.
   C) wrapped with at least three inches of insulation.
   D) wrapped with at least two inches of insulation.

2062. Return air grilles for perimeter heating systems may be located in the
   A) floor and baseboard.
   B) ceiling.
   C) higher part of inside walls.
   D) all of the above.

2063. The most common use of a clock in a heating system is to
   A) change from heating to cooling.
   B) turn the heating system on.
   C) reduce operating temperatures during the night.
   D) turn the cooling system on.
   E) operate the sequence controls.

2064. A domestic heating load must be determined on a
   A) room-by-room basis.
   B) wall-to-wall basis.
   C) floor-to-floor basis.
   D) floor-to-ceiling basis.

2065. What is the normal capacity of a 120-volt electric heating thermostat?
   A) 15 watts
   B) 20 watts
   C) 746 watts
   D) 2,000 watts
   E) 5,000 watts

2066. What voltage is used on the limit control of a gun-type oil burner system?
   A) 120 V
   B) 24 V
   C) 15 V
   D) Either 24 or 120 V
   E) 24-V solenoid operates a 120-V system

2067. A low voltage thermostat can operate a 120-V system by using a
   A) solenoid valve.
   B) transformer.
   C) relay.
   D) sequence control.
   E) resistor.

2068. A 24-V circuit is supplied by
   A) the power company.
   B) a solenoid.
   C) a thermostat.
   D) a step-up transformer.
   E) a step-down transformer.

2069. In a low voltage system, what voltage operates the heat anticipator?
   A) 24 V
   B) 6 V
   C) 120 V
   D) 12 V
   E) 240 V

2070. In the electrical circuits, a limit control is located in
   A) the 24-V circuit.
   B) the 120-V circuit.
   C) the 240-V circuit.
   D) both the 24 and 120-V circuits.
   E) either the 24 or 120-V circuit.

2071. How is velocity pressure measured with a pitot tube?
   A) Add the static pressure and total pressure
   B) Subtract the static pressure from the total pressure
   C) Add the static pressure and atmospheric pressure
   D) Subtract the atmospheric pressure from the static pressure
   E) Subtract the total pressure from the static pressure

2072. What is the minimum number of pitot tube readings one should take in a rectangular duct?
   A) 4
   B) 8
   C) 10
   D) 12
   E) 16

2073. The pressure level in an installed gas conversion burner should be
   - A) below the fire door.
   - B) under 5 pounds/in^2.
   - C) approximately at the center of the fire door.
   - D) limited by a pressuretrol.

2074. To measure outlet grille air velocity, use a(n)
   - A) anemometer.
   - B) barometer.
   - C) manometer.
   - D) micrometer.
   - E) draft gauge.

2075. What is the average effective area of an outlet grille?
   - A) 50%
   - B) 60%
   - C) 70%
   - D) 80%
   - E) 90%

2076. Why must an air duct elbow be considered as having an equivalent length?
   - A) The elbow is smaller
   - B) The air flow friction is lower
   - C) The air flow friction is higher
   - D) To know its curved length
   - E) Because all ducts have an equivalent length

2077. To clean thermostat contact points, use
   - A) sandpaper.
   - B) a file.
   - C) paper.
   - D) stone.
   - E) an emery cloth.

2078. Which burner control series indicates a three-wire 24-volt system?
   - A) Series 10
   - B) Series 20
   - C) Series 40
   - D) Series 80
   - E) Series 100

2079. To determine the entire distribution system layout and size, make a
   - A) blueprint survey.
   - B) heat loss survey.
   - C) structure estimate survey.
   - D) transmitting surface survey.

2080. The sensible heat of dry air is
A) 0.24 Btu/ft^3.
B) 0.24 Btu/lb.
C) 0.32 Btu/ft^3.
D) 1.43 Btu/lb.
E) 0.043 Btu/lb.

2081. How much pressure drop is acceptable across a domestic filter?
A) 1/2" wc
B) 1" wc
C) 2" wc
D) 3" wc
E) 6" wc

2082. What control is used to operate the fan motor?
A) Room thermostat
B) Primary control
C) Stack thermostat
D) Bonnet thermostat
E) Outdoor thermostat

2083. What is the return air volume compared to the warm air volume?
A) Less
B) More
C) Same
D) 10% more
E) 20% more

2084. The average main duct temperature of a warm-air system is
A) 100 °F.
B) 120 °F.
C) 140 °F.
D) 160 °F.
E) 180 °F.

2085. The check opening in the fire door of a coal to gas converted furnace should be
A) locked partly open.
B) left as is for adjustment.
C) locked fully open.
D) sealed closed.

2086. Which of the following does not affect heat loss from a duct?
A) Temperature in the furnace bonnet
B) Rate of heat loss per foot of pipe
C) Type of fuel used in combustion to produce heat
D) Distance from the furnace to warm air registers
E) How fast the air travels through the pipe

2087. In a hydronic system, how is hot water heated for domestic use?
A) Separate hot water heater
B) Heat exchanger
C) System water is used
D) Pipe within a pipe is used
E) Reserve water in the compression tank is used

2088. What is the most probable cause of oil around the base of the gun-type oil burner?
A) Improper combustion
B) Loose filter cover
C) Excessive pressure
D) Leaking pump shaft seal
E) Dirty nozzle

2089. The gas supply line to the gas burner must be
A) under 10 feet in length.
B) an independent line from the meter.
C) under 30 feet in length.
D) installed by a plumber.

2090. The proper method for checking a suspected gas leak is
A) to depend upon the sense of smell.
B) with a match or small flame.
C) by sense of touch.
D) with soap and water.

2091. What keeps the oil pressure constant at the nozzle?
A) Oil level in the tank
B) Pressure relief valve
C) Oil pump
D) Size of the nozzle orifice
E) Pressure regulator

2092. What is the usual oil pressure at the nozzle in a high-pressure, gun-type oil burner?
A) 15 psig
B) 30 psig
C) 50 psig
D) 100 psig
E) 200 psig

2093. To start combustion, the atomized oil must be heated to
A) 72 °F.
B) 100 °F.
C) 212 °F.
D) 400 °F.
E) 700 °F.

2094. It is important to keep water out of the oil burner passages because it
- A) causes rust.
- B) causes corrosion.
- C) is heavier than oil.
- D) will overload the motor.
- E) may cause a flame out.

2095. Gas piping that is to be concealed must be
- A) extra heavy pipe.
- B) installed by a plumber.
- C) provided with unions.
- D) inspected before being concealed.

2096. What is the efficiency of a gun-type oil burner installation if the stack temperature is 400 °F and the CO_2 is 8%?
- A) 59.5%
- B) 69.5%
- C) 78.5%
- D) 83.5%
- E) 85.5%

2097. What is the efficiency of a gun-type oil burner installation if the stack temperature is 500 °F and the CO_2 is 5%?
- A) 59.5%
- B) 69.5%
- C) 78.5%
- D) 83.5%
- E) 85.5%

2098. What is the efficiency of a gun-type oil burner installation if the stack temperature is 600 °F and the CO_2 is 8%?
- A) 66.5%
- B) 68.5%
- C) 72.5%
- D) 78.5%
- E) 83.5%

2099. What is the efficiency of a gun-type oil burner installation if the stack temperature is 700 °F and the CO_2 is 10%?
- A) 68.5%
- B) 71.5%
- C) 73%
- D) 74%
- E) 76.5%

2100. How much excess air is needed to produce 8% CO_2 in an oil burner system?
- A) 30%
- B) 50%
- C) 80%

D) 125%
E) 150%

2101. How is a smoke test made using the Ringelmann Scale?
A) Binoculars are used to observe stack fumes
B) A sample is collected in distilled water
C) A sample is forced through a white filter
D) A filter is put in the stack for a definite time
E) The soot deposit on stack internal wall is inspected

2102. The amount of excess air can be determined by
A) measuring the amount of air fed to the furnace.
B) using a draft gauge.
C) using a pitot tube.
D) using a stack CO_2 indicator.
E) using a stack oxygen indicator.

2103. A hydronic hot water system has a compression tank to
A) create a pressure.
B) allow for expansion of the water.
C) fill the system when the water level is low.
D) collect the sediment in the system.
E) provide reserve heat.

2104. What type of pump is used on a hydronic system?
A) Centrifugal
B) Piston
C) Rotary
D) Diaphragm
E) Gear

2105. An instrument used to determine specific gravity is a
A) hygrometer.
B) hydrometer.
C) viscosimeter.
D) none of the above.

2106. Basement heat loss is usually determined by heat loss measurements of the
A) wall above the grade or outside ground level.
B) wall below the grade and the floor.
C) basement floor.
D) both A and B.

2107. A publication that contains information about heat transfer factors, heat transfer multipliers and U factors is
A) MHAW's Encyclopedia of Heat.
B) Webster's New Collegiate Dictionary.
C) Manual J of the ACCA.
D) The Heating Specialist's Guide.

2108. Supply ducts receive warmed air from this part of the casing.
A) Tight casing
B) Combustion chamber
C) Firebox
D) Bonnet

2109. The average temperature of air discharged from the furnace outlet duct (or ducts) is called the
A) outlet air temperature.
B) inlet duct temperature.
C) stack loss.
D) duct temperature.

2110. The average temperature of air entering the forced-air furnace is called the
A) flue gas temperature.
B) inlet duct temperature.
C) bonnet efficiency.
D) fuel heat output.

2111. The amount of heat input that escapes in flue gases is called
A) inlet duct temperature.
B) outlet air temperature.
C) stack loss.
D) chimney vapor.

2112. To keep moisture from seeping into the heating system of houses built over crawl spaces, the ground should first be covered with
A) sawdust.
B) concrete.
C) vapor barrier.
D) gravel.

2113. How many electrode installation dimensions should be checked?
A) One
B) Two
C) Three
D) Four
E) Five

2114. On residential, low-pressure, steam boilers, normal operating pressure is
A) 30 lb.
B) 12 to 15 lb.
C) 15 to 30 lb.
D) none of the above.

2115. On Taco circulating pumps the bearing assembly is lubricated with
A) No. 20 non-detergent motor oil.
B) grease.
C) water.
D) petroleum jelly.

2116. Residential, low-pressure, forced-water boilers are sized
    A) to the load of the building.
    B) 10% greater than the load of the building.
    C) 10% less than the load of the building.
    D) none of the above.

2117. On residential, low-pressure, forced-water boilers, the most important rating in dealing with load calculation is
    A) Btu input.
    B) Btu output.
    C) Btu net.
    D) none of the above.

2118. On residential, low-pressure, steam boilers, low water cut-offs are required.
    A) True
    B) False

2119. On residential, low-pressure, forced-water boilers, low water cut-offs are not required.
    A) True
    B) False

2120. The amount of 3/4" copper pipe equaling 1 gallon of water is approximately
    A) 13 ft.
    B) 23 ft.
    C) 33 ft.
    D) 43 ft.

2121. Residential, low-pressure, forced-water fin tubing (baseboard radiation) is sized by
    A) Btu input of the boiler.
    B) Btu output of the boiler.
    C) Btu net.
    D) none of the above.

2122. The maximum allowable temperature for residential, low-pressure, forced-water boilers is
    A) 212 °F.
    B) 210 °F.
    C) 250 °F.
    D) 200 °F.

2123. On residential, low-pressure, forced-water boilers, the pilot safety could be
    A) hydraulic.
    B) diaphragm.
    C) motorized.
    D) none of the above.

2124. What effect does the fan have on the air?
 A) Cooling
 B) Heating
 C) Cleaning
 D) Drying
 E) Nothing

2125. What type of air pressure is measured by a differential manometer?
 A) Velocity
 B) Total pressure
 C) Static
 D) Vacuum
 E) Partial vacuum

2126. When an adhesive filter needs to be changed, the pressure difference across it will be
 A) 0.10" wc.
 B) 0.20" wc.
 C) 0.30" wc.
 D) 0.40" wc.
 E) 0.50" wc.

2127. If the filter is on the fan intake side, what is the pressure on both sides of the filter?
 A) Positive
 B) Negative
 C) Both positive and negative
 D) Either positive or negative
 E) Negative on one side of filter only

2128. When the air flow into a duct is reduced, the pressure drop across the filter
 A) stays the same.
 B) increases.
 C) decreases.
 D) may either increase or decrease.
 E) reduces to zero.

2129. What does one inch of water column equal?
 A) 29.9 psi
 B) 14.7 psi
 C) 0.432 psi
 D) 0.036 psi
 E) 2.31 psi

2130. When a combination heating and cooling unit has an electronic air cleaner, the electronic air cleaner should operate
 A) in the heating cycle only.
 B) only when the indoor fan is operating.
 C) in the cooling cycle only.

D) only by itself without heat or cooling.
E) none of the above

2131. In a forced-air system, an electronic air cleaner would be installed
A) in the plenum chamber.
B) in the main trunk line.
C) on the return air side of the furnace fan.
D) at the return registers.

3132. Wet bulb depression means
A) a cavity in the bulb.
B) the wet bulb is located below the dry bulb.
C) the difference between the wet and dry bulb thermometers.
D) the thermometer is inaccurate.
E) there is mercury in the thermometer.

2133. Mean temperature indicates the
A) average temperature.
B) lowest temperature.
C) highest temperature.
D) design temperature.
E) dry bulb temperature.

2134. Ambient temperature indicates the
A) evaporator temperature.
B) condenser temperature.
C) surrounding temperature.
D) desired temperature.
E) dew point.

2135. How many heat sources must be considered when calculating the heat load?
A) One
B) Two
C) Three
D) Four
E) Five

2136. The most efficient type of duct is
A) round.
B) square.
C) rectangular.
D) oblong.
E) any kind.

2137. A duct fitting that attaches the duct to a wall stack, register or grille is a
A) plenum.
B) duct sleeve.
C) boot.
D) take-off.
E) none of the above.

2138. An anemometer is not used for measuring drafts because it
   A) is too accurate.
   B) won't operate on slow air velocities.
   C) always indicates too high draft velocities.
   D) measures temperature.
   E) only operates inside ducts.

2139. When do air velocities out exceed the air velocities in?
   A) Never
   B) Always
   C) When the inlet area is larger than the outlet area totals
   D) When the inlet area is smaller than the outlet area totals
   E) When a high velocity fan is used

2140. The installation of a gas burner in a warm air furnace must include
   A) low water cut-off.
   B) hot air limit control.
   C) "stack switch" combustion control.
   D) circulating fan.

2141. The purpose of a pressure regulator is to
   A) increase the gas pressure.
   B) maintain a constant gas pressure regardless of variations in gas supply.
   C) prevent overheating the furnace.
   D) provide a proper mixture of gas and air.

2142. The oil pump coupling is attached to the motor shaft with a(n)
   A) woodruff key.
   B) flat key.
   C) Allen setscrew.
   D) capscrew.
   E) machine screw.

2143. How many controls does a gun-type oil burner have when used on a warm air system?
   A) One
   B) Two
   C) Three
   D) Four
   E) Five

2144. What voltages are found in most domestic gun-type oil burner systems?
   A) 24 and 120 volt
   B) 120 and 240 volt
   C) 24 and 230 volt
   D) 208 and 440 volt
   E) 24 and 220 volt

2145. In a warm-air system, the primary control (bimetal type) is mounted on the
 A) plenum chamber.
 B) oil burner housing.
 C) furnace stack.
 D) cold air return.
 E) thermostat.

2146. In a gun-type oil burner system, the relay is located in the
 A) thermostat.
 B) limit control.
 C) primary control.
 D) motor terminal box.
 E) line switch.

2147. To find out more information about ductwork design and installation, contact
 A) ACCA.
 B) US Army Engineer Corps.
 C) Encyclopedia of Heating Science.
 D) Manual B of the Better Heating Bureau.

2148. Closing the fingers on the diverter adjusts the
 A) primary air.
 B) safety pilot.
 C) burner input.
 D) pressure level.

2149. The pour point of oil is the
 A) congealing point.
 B) point of slowest oil flow.
 C) point where oil stops flowing.
 D) none of the above.

2150. 0 °C is equal to
 A) 32 °F.
 B) 100 °F.
 C) 212 °F.
 D) 0 °F.

2151. Using the conversion formula, -40 °F is equal to
 A) 60 °C.
 B) 10 °C.
 C) -40 °C.
 D) -60 °C.

2152. An example of a change in state is
 A) heating water from 32 to 212 °F.
 B) changing water from 32 °F ice to 32 °F liquid.
 C) heating steam at 212 °F above its boiling point at atmospheric pressure.
 D) changing ice from 0 °F to ice at 32 °F.

2153. The type of heat added when changing water at 32 °F to water at 212 °F is
   A) critical heat.
   B) latent heat.
   C) sensible heat.
   D) heat of fusion.

2154. The amount of heat required to raise 10 lb of water from 40 to 100 °F is
   A) 40 Btu.
   B) 400 Btu.
   C) 60 Btu.
   D) 600 Btu.

2155. An example of sublimation is the evaporation of
   A) water.
   B) dry ice.
   C) milk.
   D) steam.

2156. Water has a specific heat of 1.0 Btu/lb. The amount of heat required to raise the temperature of 5 lb of water from 10 to 20 °F is
   A) 10 Btu.
   B) 4.8 Btu.
   C) 50 Btu.
   D) 100 Btu.

2157. The specific heat of water vapor is
   A) 0.50.
   B) 1.00.
   C) 1.15.
   D) 2.00.

2158. Indoor velocities are usually measured in
   A) miles per minute.
   B) feet per minute.
   C) meters per hour.
   D) centimeters per second.

2159. When heat flows upward from a register, heat is transferred by
   A) vaporization.
   B) convection.
   C) conduction.
   D) sublimation.

2160. Comfort is
   A) the feeling of contentment with the environment.
   B) an atmosphere of 86 db, 40% rh.
   C) air conditioning, like an ocean breeze.
   D) warm, moist, clean air.

2161. The body temperature is normally maintained at
   A) 97 to 100 °F.
   B) 35 °C.
   C) 97 to 100 °C.
   D) 40 to 45 °C.

2162. To be comfortable, one must
   A) wear warm clothing.
   B) lose heat at the proper rate to the surrounding air.
   C) be fairly active.
   D) use only forced warm air heat.

2163. The body transfers heat
   A) by air that blows past it.
   B) mainly by conduction.
   C) through normal body functions.
   D) by radiation, convection and evaporation.

2164. The amount of heat required to evaporate 1 lb of water at 212 °F is approximately
   A) 970 Btu.
   B) 144 Btu.
   C) 300 kcal.
   D) 500 kcal.

2165. Thermal comfort conditions include
   A) freedom from noise.
   B) freedom from disagreeable odors.
   C) clean air.
   D) temperature of air.

2166. The ASHRAE Comfort Standard recommends
   A) 75 °F, 75% relative humidity.
   B) 76 °F, 40% relative humidity.
   C) 80 °F, 40% relative humidity.
   D) 27 °C, 50% relative humidity.

2167. A psychrometer is
   A) a chart measuring relative humidity.
   B) a metric gauge for wire.
   C) a device with two thermometers for measuring wet and dry bulb temperatures.
   D) the name of a heating unit.

2168. A psychrometric chart is a
   A) device for measuring humidity.
   B) graph plotting air motion and temperature.
   C) chart where the various properties of air are plotted.
   D) chart showing how to assemble a heating unit.

2169. An electronic filter can remove dust particles ranging in size from
  A) 0.01 to 10 mm.
  B) 10 to 100 mm.
  C) 0.001 to 0.0001 mm.
  D) 100 mm or more.

2170. Combustion is what type of process?
  A) Mechanical
  B) Chemical
  C) Electrical
  D) Pneumatic

2171. What are the three conditions necessary for combustion?
  A) Fuel, air and moisture
  B) Carbon, air and pressure
  C) Fuel, heat and oxygen
  D) Carbon, hydrogen and oxygen

2172. Complete combustion occurs when all carbon and hydrogen in the fuel are
  A) oxidized.
  B) heated.
  C) combined.
  D) dissipated.

2173. Incomplete combustion may be caused by
  A) a fire that is too hot.
  B) too much draft.
  C) insufficient fuel supply.
  D) insufficient air supply.

2174. What is the percentage of oxygen in the air?
  A) 79%
  B) 21%
  C) 69%
  D) 31%

2175. Expressed in percentage, efficiency equals the useful heat divided by the
  A) heat loss of the building.
  B) capacity of the furnace.
  C) heating value of the fuel.
  D) temperature of the flue gases.

2176. The amount of excess air required for complete combustion is
  A) 5 to 30%.
  B) 25 to 60%.
  C) 40 to 75%.
  D) 5 to 50%.

2177. The percent of CO_2 in the flue gas is not affected by the amount of excess air used by the combustion equipment.
    A) True
    B) False

2178. Natural gas is chiefly composed of
    A) methane.
    B) propane.
    C) butane.
    D) ethylene.

2179. Is it necessary to preheat Oil No. 6?
    A) Yes
    B) No
    C) Only on units of less than 100,000 Btu

2180. Which of the following items is not included in a gravity-type furnace?
    A) Heat exchanger
    B) Cabinet
    C) Air filter
    D) Humidifier

2181. The location of the fan on a counterflow furnace is
    A) above the heat exchanger.
    B) below the heat exchanger.
    C) at the side of the heat exchanger.
    D) in a separate unit.

2182. The material used to construct the heat exchanger is
    A) galvanized iron.
    B) copper.
    C) aluminum.
    D) cold-rolled, low-carbon steel.

2183. An individual-section type of heat exchanger is used on which kind of furnace?
    A) Gas
    B) Oil
    C) Coal
    D) Electric

2184. The type of fuel that requires the greatest chimney draft is
    A) gas.
    B) oil.
    C) coal.
    D) electric.

2185. The fan blades on forced-air furnaces are
A) curved backward.
B) curved forward.
C) flat.
D) propeller-shaped.

2186. If a motor has a 3" pulley, the fan a 6" pulley, and the motor runs at 1200 rpm, the speed of the fan is
A) 600 rpm.
B) 1,200 rpm.
C) 1,800 rpm.
D) 2,400 rpm.

2187. If the fan is mounted on the shaft of the motor, the type of drive is called
A) unified.
B) V-belt.
C) indirect.
D) direct.

2188. A downflow furnace is also called a
A) high-boy.
B) low-boy
C) horizontal.
D) counterflow.

2189. As the outside temperature becomes colder, the need for humidification
A) increases.
B) decreases.
C) neither A nor B.

2190. Which of the following fuel oil(s) should generally be used with a high-pressure atomizing burner?
A) No. 1 only
B) No. 2 only
C) No. 1 or 2
D) None of the above

2191. A hand shut-off valve is used to turn gas on and off to the
A) burner.
B) pilot.
C) burner and pilot.
D) none of the above.

2192. The proper gas manifold pressure for a residential gas unit is
A) 2 1/2" wc.
B) 3 1/2" wc.
C) 7" wc.
D) 11" wc.

2193. A thermocouple produces how many millivolts of electric power?
   A) 25 to 30
   B) 35 to 50
   C) 500 to 600
   D) 750 to 800

2194. CGV stands for
   A) continuous gas valve.
   B) correct gas volume.
   C) cast gas vent.
   D) combination gas valve.

2195. The maximum safety drop-out time for units with thermocouples is
   A) 1 minute.
   B) 2 minutes.
   C) 3 minutes.
   D) 4 minutes.

2196. When gas is mixed with air before it enters the burner, the type of pilot is
   A) primary aerated.
   B) non-primary aerated.
   C) neither A nor B.

2197. What type of fuel requires 100% shutoff at the gas valve?
   A) Natural gas
   B) LP gas
   C) Neither A nor B

2198. Air that is mixed with gas before burning is called
   A) outside air.
   B) total combustion air.
   C) secondary air.
   D) primary air.

2199. A draft diverter
   A) creates a draft of 0.02" wc over fire.
   B) connects directly to the gas burner.
   C) is opened at the bottom.
   D) is really not necessary on a modern furnace.

2200. Oil is vaporized by
   A) pressurization.
   B) heat.
   C) centrifugal force.
   D) atomization.

2201. The most popular oil burner is the
   A) pot type.
   B) rotary type.
   C) low-pressure type.
   D) high-pressure type.

2202. A high-pressure burner mixes oil and air
 A) before entering the nozzle.
 B) after oil passes through the nozzle.
 C) neither A nor B.

2203. A flexible connector should be made of what material?
 A) Rigid
 B) Flexible
 C) Approved
 D) None of the above

2204. The type of pump used with an outside tank located below the oil burner is called
 A) two-stage.
 B) single-stage.
 C) neither A nor B.

2205. The type of motor used on an oil burner is a
 A) capacitor start.
 B) split-phase.
 C) capacitor run.
 D) shaded pole.

2206. The purpose of the nozzle on a high-pressure burner is to
 A) blow air.
 B) mix the oil and air.
 C) atomize the oil.
 D) burn the oil.

2207. The shape of the oil/air spray must
 A) correspond to the type of combustion chamber.
 B) fit the burner supplied.
 C) make a good-looking fire.
 D) fit the heating load.

2208. The oil capacity (in gallons) of a single inside tank is usually
 A) 275.
 B) 550.
 C) 1,000.
 D) 1,500.

2209. What size copper tubing is usually used for inside tank oil lines?
 A) 1/4" OD
 B) 3/8" OD
 C) 1/2" OD
 D) 5/8" OD

2210. The products of combustion are
 A) C, SO_2, H_2O, N_2, and H.
 B) CO_2, NaCe, and $AgNO_3$.
 C) CO_2, O_2, N_2 and H_2O.
 D) AgCl, KCl, C_6 and H_5.

2211. In practice, a gas burner should be adjusted to produce how much CO_2 in flue gas when the unit is 80% effective?
   A) 10 to 12%
   B) 8 to 9%
   C) 4 to 7%
   D) 20 to 25%

2212. An instrument used to check CO_2 is a
   A) flue gas analyzer.
   B) true spot tester.
   C) micrometer.
   D) psychrometer.

2213. A burner should run for a minimum of how many minutes before combustion tests are made?
   A) 2
   B) 10
   C) 20
   D) 30

2214. What is the maximum number of holes that should be in the flue for combustion tests?
   A) One
   B) Two
   C) Three
   D) Four

2215. Draft at the burner (operating) should be
   A) negative.
   B) positive.
   C) neutral.
   D) none of the above.

2216. In a new oil furnace, the proper reading for the smoke test should be
   A) No. 1 or 2.
   B) No. 3 or 4.
   C) No. 5 or 6.
   D) No. 7 or 8.

2217. In making the CO_2 test on an oil furnace, the bulb should be collapsed how many times?
   A) Six
   B) Eight
   C) Twelve
   D) Eighteen

2218. Combustion efficiency should be
   A) 50% or better.
   B) 60% or better.
   C) 75% or better.
   D) 85% or better.

2219. If the combustion efficiency is 80%, the stack loss should be
- A) 80%.
- B) 20%.
- C) 70%.
- D) 30%.

2220. Which of the following is not a load device?
- A) Fan
- B) Humidifier
- C) Heater
- D) Fan control

2221. Which basic element of the furnace control system includes the sensor?
- A) Power supply
- B) Gas valve
- C) Limit devices
- D) Fan motor

2222. People can sense a variation in temperature of
- A) 1/2 °F.
- B) 1-1/2 °F.
- C) 2-1/2 °F.
- D) 4 °F.

2223. A low-voltage control system operates on how many volts?
- A) 24 V
- B) 30 V
- C) 120 V
- D) 240 V

2224. On a bimetallic thermostat, the metal used in addition to copper is
- A) zinc.
- B) chromium.
- C) invar.
- D) iron.

2225. What is the largest number of switches commonly found in a single mercury tube element?
- A) One
- B) Two
- C) Three
- D) Four

2226. A self-generating system usually uses its own control circuit of
- A) 30 mV.
- B) 750 mV.
- C) 24 V.
- D) 120 V.

2227. If a thermostat cuts in at 70 °F and cuts out at 72 °F, the differential is
 A) 0 °F.
 B) 2 °F.
 C) 70 °F.
 D) 72 °F.

2228. A heating system should cycle how many times per hour?
 A) 1 to 3
 B) 4 to 7
 C) 8 to 10
 D) 11 to 15

2229. The color code terminals of a thermostat are
 A) Q, H, C, F.
 B) R, G, B, V.
 C) T, A, E, H.
 D) R, W, Y, G.

2230. A timed fan start delays the operation of the fan approximately how many seconds after the heater is turned on?
 A) 30
 B) 60
 C) 90
 D) 120

2231. A limit control is usually set to cut out the source of heat at
 A) 180 °F.
 B) 200 °F.
 C) 220 °F.
 D) 240 °F.

2232. For a heating furnace, the code-recommended branch circuit is
 A) 5 A.
 B) 10 A.
 C) 15 A.
 D) 25 A.

2233. How many circuits are required to operate a two-speed fan?
 A) One
 B) Two
 C) Three
 D) Four

2234. On a call for heating, which two terminals on the thermostat close?
 A) R and W
 B) R and Y
 C) R and B
 D) G and W

2235. With a cut-out temperature of 200 °F for the limit control, the normal cut-in is
   A) 150°F.
   B) 175 °F.
   C) 200 °F.
   D) 225 °F.

2236. What is the usual cut-out temperature for the secondary limit control?
   A) 145 °F
   B) 160 °F
   C) 175 °F
   D) 200 °F

2237. When the thermostat calls for cooling, what two terminals on the thermostat close?
   A) R and W
   B) R and Y
   C) R and G
   D) G and W

2238. What control device activates a low-voltage, timed fan start?
   A) Gas valve
   B) Limit control
   C) Thermostat
   D) Fan control

2239. What accessory can be added that requires a second power supply?
   A) Humidifier
   B) Air conditioner
   C) Electrostatic air cleaner
   D) Stoker

2240. What new thermostat terminals are required when a two-stage gas valve arrangement is used?
   A) R1 and R2
   B) W1 and W2
   C) B1 and B2
   D) Y1 and Y2

2241. In testing a fan relay, what voltage is usually applied to the relay coil?
   A) 24 V
   B) 120 V
   C) 240 V
   D) 480 V

2242. On the stack relay type of oil primary control, are the cold contacts normally open or normally closed?
   A) Normally open
   B) Normally closed
   C) There are no cold contacts

2243. The light-sensitive material used on a cad cell is
   A) cadmium nitrate.
   B) copper sulfate.
   C) sodium chloride.
   D) cadmium sulfate.

2244. What is the resistance of a cad cell in the absence of light?
   A) 5,000 ohms
   B) 50,000 ohms
   C) 100,000 ohms
   D) 150,000 ohms

2245. What is the minimum size for a boiler pressure relief valve?
   A) 1-1/2 inches
   B) 1 inch
   C) 3/4 inch
   D) 1/2 inch

2246. How many low-voltage terminals are on a cad cell primary control?
   A) Two
   B) Four
   C) Six
   D) Eight

2247. All boilers shall have the potable water protected with
   A) a check valve.
   B) backflow preventer.
   C) neither A nor B.

2248. On the oil burner stack relay, the power supply is wired to which terminals?
   A) 0 and 1
   B) 1 and 2
   C) 2 and 3
   D) 3 and 4

2249. In checking a switch with a voltmeter, a reading of zero indicates a(n)
   A) closed switch.
   B) open switch.
   C) short.
   D) ground.

2250. The specific gravity of natural gas is
   A) 0.65.
   B) 1.00.
   C) 1.50.
   D) none of the above.

2251. If the pilot goes out on gas burners having an hourly input of less than 400,000 Btu, the main gas shall shut off within
   A) 10 seconds.
   B) 50 seconds.
   C) 3 minutes.
   D) none of the above.

2252. Combustion efficiency is determined from
   A) stack temperature and percent CO.
   B) stack temperature and inches of stack draft.
   C) stack temperature and smoke spot.
   D) none of the above.

2253. The heat value of natural gas is
   A) 500 Btu/ft^3.
   B) 1,000 Btu/ft^3.
   C) 2,500 Btu/ft^3.
   D) none of the above.

2254. The minimum size of a hand hole plate is
   A) 3" x 4".
   B) 2-3/4" x 3-1/2".
   C) 2" x 4".
   D) none of the above.

2255. The limit setting on a standard efficiency forced warm air furnace is
   A) 140 °F.
   B) 200 °F.
   C) 300 °F.
   D) none of the above.

2256. When the heating system does not work with the thermostat jumpered, there is
   A) a bad thermostat.
   B) a problem other than the thermostat.
   C) a definite electrical problem.
   D) none of the above.

2257. A good thermocouple will read open circuit
   A) 4 to 7 mV.
   B) 12 to 17 mV.
   C) 17 to 30 mV.
   D) none of the above.

2258. A good pilotstat power unit will operate in the range of
   A) 4 to 7 mV.
   B) 12 to 17 mV.
   C) 17 to 30 mV.
   D) none of the above.

2259. A good thermocouple will read in a loaded circuit
    A) 4 to 7 mV.
    B) 12 to 17 mV.
    C) 17 to 30 mV.
    D) none of the above.

2260. Continuous blower operation can be used for
    A) cooling only.
    B) both heating and cooling.
    C) heating only.
    D) none of the above.

2261. A DPDT blower relay (fan center) protects the system from
    A) inadequate speed control.
    B) backfeed through multispeed motor windings.
    C) overloading the multispeed motor.
    D) none of the above.

2262. What is the design temperature of PVC pipe?
    A) 63 °F
    B) 70 °F
    C) 73 °F
    D) 83 °F

2263. Compared to natural gas with the same Btu rating, the orifice opening in an LP gas-fired furnace is
    A) smaller.
    B) larger.
    C) the same size.
    D) none of the above.

2264. How many times the working pressure is the pressure test for low pressure residential boilers?
    A) 1-1/2 times
    B) 2 times
    C) 3 times
    D) None of the above

2265. Allowable pressure loss due to pipe friction is
    A) 0.3" wc.
    B) 0.5" wc.
    C) 0.65" wc.
    D) none of the above.

2266. When checking for gas leaks, use
    A) soapy water.
    B) safety matches.
    C) electronic leak detector.
    D) none of the above.

2267. Concealed joints in gas piping must be
   A) painted yellow where they come out of the wall.
   B) inspected before covering.
   C) covered with anti-leak compound.
   D) none of the above.

2268. If units at the highest points of a hot-water heating system are cool, these units are most likely
   A) undersized.
   B) air-bound.
   C) on a vacuum.
   D) none of the above.

2269. When operating a fuel oil furnace with a one-pipe system on the fuel pump, the
   A) bypass plug should be installed.
   B) single pipe should be teed into both suction parts.
   C) bypass plug should be removed.
   D) none of the above.

2270. A gas drip leg must
   A) be installed in humid or wet areas.
   B) have the capacity to catch 150% of the drippings.
   C) not be smaller in diameter than the pipe to which it connects.
   D) none of the above.

2271. Gas piping cannot be run through
   A) unvented basements.
   B) other apartments.
   C) return ducts and clothes chutes.
   D) none of the above.

2272. A thermostat that cycles too often or not enough indicates
   A) a thermostat that is not level.
   B) a defective or burnt heat anticipator.
   C) improper heat anticipator setting.
   D) none of the above.

2273. Recuperative furnaces have an efficiency of
   A) 70%.
   B) 80%.
   C) 90%.
   D) none of the above.

2274. Cycle pilot furnaces have an efficiency of more than
   A) 70%.
   B) 80%.
   C) 90%.
   D) none of the above.

2275. Condensing furnaces have an efficiency of more than
   A) 70%.

B) 80%.
C) 90%.
D) none of the above.

2276. The main difference between recuperative and condensing furnaces is in the
A) secondary heat exchanger.
B) control system.
C) vent system.
D) none of the above.

2277. Furnace flue gas condensate is
A) basic.
B) neutral.
C) acidic.
D) none of the above.

2278. The maximum flue gas temperature for furnaces utilizing PVC pipe is
A) 100 °F.
B) 140 °F.
C) 180 °F.
D) none of the above.

2279. On a call for heat, the first step on a mechanical vent furnace is
A) burner operation.
B) blower operation.
C) combustion chamber purge cycle.
D) none of the above.

2280. Most condensing furnaces may not be connected to the existing chimney unless
A) stainless steel connector pipe is used.
B) PVC connector pipe is used.
C) the furnace manufacturer permits this application.
D) none of the above.

2281. Redundant gas valves utilizing flame rectification to prove a pilot flame must be grounded to
A) earth ground.
B) the neutral ground bar in the fuse box.
C) any ground available.
D) none of the above.

2282. To prove a pilot flame, most White Rodgers redundant gas valves use a
A) bimetal sensor.
B) mercury vapor sensor.
C) flame rod (rectification) sensor.
D) none of the above.

2283. To prove a pilot flame, most Bryant and Carrier redundant gas valves use a
    A) bimetal sensor.
    B) mercury vapor sensor.
    C) flame rod (rectification) sensor.
    D) none of the above.

2284. To prove a pilot flame, most Honeywell redundant gas valves use a
    A) bimetal sensor.
    B) mercury vapor sensor.
    C) flame rod (rectification) sensor.
    D) none of the above.

2285. The redundant gas valve with the quickest response in sensing a pilot flame and energizing the main gas valve utilizes the
    A) bimetal sensor.
    B) mercury vapor sensor.
    C) flame rod (rectification) sensor.
    D) none of the above.

2286. During a call for heat, the redundant gas valves with pick and hold coils continuously energize
    A) the pick coil.
    B) the hold coil.
    C) neither A nor B.

2287. The 100% lockout switch is opened when the
    A) pilot flame is not established within a defined time limit.
    B) gas valve is not energized within a defined time limit.
    C) combustion vent fan is not proven.
    D) none of the above.

2288. The combustion vent fan proving switch operates between
    A) -0.03 and -0.06" wc.
    B) -0.3 and -0.6" wc.
    C) -3.0 and -6.0" wc.
    D) none of the above.

2289. On redundant gas valves, the thermostat heat anticipator setting is
    A) higher than with standard gas valves.
    B) lower than with standard gas valves.
    C) the same as with standard gas valves.
    D) none of the above.

2290. Excessive sooting in the heat exchangers of a condensing furnace may be caused by
    A) high gas pressure.
    B) partial blocking by condensate.
    C) an inoperative flue gas vent fan.
    D) none of the above.

2291. To check for dirty exterior fin tubing on a secondary heat exchanger, look at
    A) high flue gas temperatures.
    B) high temperature rise through the furnace.
    C) low gas pressure drop.
    D) none of the above.

2292. To check for plugged interior tubing on a secondary heat exchanger, look at
    A) high flue gas temperatures.
    B) high temperature rise through the furnace.
    C) low gas pressure drop.
    D) none of the above.

2293. Two basic types of gas burner ignition systems are
    A) thermal and mercury.
    B) direct and indirect.
    C) lockout and non-lockout.
    D) none of the above.

2294. Lennox pulse furnaces have controlled explosions at the rate of
    A) 50 to 70 cycles/second.
    B) 50 to 70 cycles/minute.
    C) 50 to 70 cycles/hour.
    D) none of the above.

2295. When reducing the input of a gas furnace to more closely represent the house heat loss,
    A) reduce the orifice size by no more than 20%.
    B) reduce the gas pressure by no more than 20%.
    C) either A or B.
    D) neither A nor B.

2296. Natural gas burner orifices are
    A) larger in diameter than propane gas burner orifices.
    B) smaller in diameter than propane gas burner orifices.
    C) the same diameter as propane gas burner orifices.
    D) none of the above.

2297. A smoke spot of 5 will produce soot that
    A) is extremely light if there is any at all.
    B) will require cleaning once per year.
    C) will collect heavily and rapidly.
    D) none of the above.

2298. Excessive draft can
    A) increase the stack temperature.
    B) reduce the percentage of CO_2 in the flue gases.
    C) both A and B.
    D) neither A nor B.

2299. The hole for the stack thermometer should be
- A) 1/4" in diameter.
- B) 6 inches from the furnace side of the draft regulator.
- C) both A and B.
- D) neither A nor B.

2300. The maximum operation voltage for three-wire convenience outlets is
- A) 240 V.
- B) 125 V.
- C) 100 V.
- D) 50 V.

2301. Does the pilot burn continuously when using a spark igniter?
- A) Yes
- B) No

2302. The purpose of the current type standing pilot is to
- A) heat the thermocouple.
- B) ignite the main burner.
- C) generate a millivoltage.
- D) all of the above.

2303. The output voltage of a thermocouple may be as high as
- A) 30 mV.
- B) 250 mV.
- C) 500 mV.
- D) 750 mV.

2304. The pressure regulator on an LP system is set at
- A) 3.5" wc.
- B) 7.0" wc.
- C) 11.0" wc.
- D) 14.7" wc.

2305. The output voltage of a thermopile may be as high as
- A) 30 mV.
- B) 250 mV.
- C) 500 mV.
- D) 750 mV.

2306. The pilot flame should envelop the end of the thermocouple
- A) 1/8 to 3/8 inch.
- B) 3/8 to 1/2 inch.
- C) 1/2 to 3/4 inch.
- D) 1/8 to 3/4 inch.

2307. Primary air is mixed with the gas within the
- A) manifold.
- B) orifice.
- C) venturi.
- D) burner head.

2308. On flanged joints, bolts should be tightened
   A) using the crossover method.
   B) by rotation.
   C) by hand
   D) through welding.

2309. The Pittsburgh lock is used on
   A) square ducts.
   B) round pipe.
   C) galvanized metal only.
   D) expanded metal.

2310. A yellowish flame at the burner is usually caused by
   A) too little primary air.
   B) too much primary air.
   C) the secondary air ratio factor.
   D) the primary air dilution source.

2311. Manifold gas pressure that is too low could cause the flame to
   A) flash back through the burner.
   B) lift off the burner.
   C) reduce burner efficiency.
   D) dilute the secondary air ratio factor.

2312. Excessive primary air could cause the flame to
   A) flash back through the burner.
   B) lift off the burner.
   C) reduce burner efficiency.
   D) dilute the secondary air ratio factor.

2313. What is the minimum size of steam main off a boiler?
   A) 1 inch
   B) 2 inches
   C) 2-1/2 inches
   D) 3 inches

2314. Combustion efficiency is established through
   A) percent CO_2.
   B) flue gas temperatures.
   C) both A and B.
   D) neither A nor B.

2315. Average gas burning combustion efficiencies are
   A) 40 to 60%.
   B) 50 to 65%.
   C) 77 to 80%.
   D) 77 to 95%.

2316. The function of the gas burner assembly is to
   A) produce proper fire at the base of the heat exchanger.
   B) heat the air that goes to the rooms.
   C) mix gas and air.
   D) provide safe lighting of the burner.

2317. The type of gas valve operator with a delayed action feature is a
   A) solenoid.
   B) diaphragm.
   C) bimetal.
   D) bulb.

2318. A fast acting type of gas valve is the
   A) solenoid.
   B) diaphragm.
   C) bimetal.
   D) bulb.

2319. Which moves gas to several burner units in a low pressure system?
   A) Multiple feeder
   B) Thermopile
   C) Spud
   D) Manifold

2320. Gas burners that are horizontal firing through spuds are called
   A) radiant.
   B) Bunsen.
   C) upshot.
   D) inshot.

2321. Upshot burners are
   A) vertical.
   B) horizontal.
   C) easily converted to oil.
   D) uniform in surface design.

2322. An oil-fired unit using No. 2 fuel oil and burning 105,000 Btu has a nozzle size of
   A) 0.65 gph.
   B) 0.75 gph.
   C) 0.85 gph.
   D) 0.90 gph.

2323. To which terminals on the primary control is the cad cell connected?
   A) CR
   B) RW
   C) TT
   D) FF

2324. What is the usual cut-out temperature for the secondary limit control on a horizontal furnace?
    A) 145 °F
    B) 160 °F
    C) 175 °F
    D) 200 °F

2325. Which type of oil burner mixes the oil and air before it enters the nozzle?
    A) Low-pressure
    B) Rotary
    C) High-pressure
    D) Pot type

2326. What is the minimum diameter opening of a blowdown valve on a water-gauge glass?
    A) 1 inch
    B) 3/4 inch
    C) 1/2 inch
    D) 1/4 inch

2327. What is the approximate resistance of a cad cell in the presence of light?
    A) 100 ohm
    B) 1,000 ohm
    C) 10,000 ohm
    D) 100,000 ohm

2328. On a cad cell primary control, the hot line is connected to the
    A) orange lead.
    B) black lead.
    C) brown lead.
    D) white lead.

2329. On the stack primary control, line power is connected to terminals
    A) 1 and 2.
    B) 2 and 3.
    C) 3 and 4.
    D) T and T.

2330. If the flame fails, the cad cell relay will lock out on safety in about
    A) 30 seconds.
    B) 60 seconds.
    C) 90 seconds.
    D) 120 seconds.

2331. If the flame fails, the stack relay will lock out on safety in about
    A) 30 seconds.
    B) 60 seconds.
    C) 90 seconds.
    D) 120 seconds.

2332. Oil primary relays are usually wired directly to the
A) line power.
B) fan switch.
C) limit switch.
D) transformer.

2333. On an oil primary relay (the constant ignition type), the oil burner motor and ignition transformer are wired in
A) series.
B) parallel.
C) combination series and parallel.
D) either series or parallel.

2334. The secondary voltage of an oil burner ignition transformer is
A) 1,000 V.
B) 5,000 V.
C) 10,000 V.
D) 100,000 V.

2335. The stack primary control should be installed in the flue pipe between
A) the furnace and the barometric damper.
B) the barometric damper and the chimney.
C) heat exchanger sections.
D) the burner housing and the flame.

2336. The cad cell sensor is located in the
A) burner housing.
B) cad cell relay.
C) combustion chamber.
D) stack relay.

2337. On a counterflow type oil furnace, the auxiliary limit switch and the normal limit switch are wired in
A) series.
B) parallel.
C) combination series and parallel.
D) either series or parallel.

2338. If oil in suspension with air is allowed to touch a cold surface, it will
A) heat that cold surface.
B) condense into a liquid.
C) produce a flash fire.
D) burn with a cracking sound.

2339. Assuming perfect combustion, the amount of CO_2 found in the flue gas would be
A) 4 to 6%.
B) 66 to 10%.
C) 10 to 12%.
D) 15%.

2340. Assuming normal good combustion, approximately how much CO_2 should be found in the flue gas of an oil unit?
    A) 4% to 6%
    B) 6% to 10%
    C) 8% to 9%
    D) 15%

2341. When the flame is being well utilized, there should be
    A) high stack temperature.
    B) low stack temperature.
    C) cold stack temperature.
    D) diluted stack temperature.

2342. Excessive draft will result in
    A) high stack temperature.
    B) low stack temperature.
    C) cold stack temperature.
    D) diluted stack temperature.

2343. The transfer of heat through mediums such as water, air and steam is called
    A) radiation.
    B) fusion.
    C) convection.
    D) conduction.

2344. What is needed to provide better lift from the tank to the burner?
    A) Single-stage pump
    B) Two-stage pump
    C) Single-pipe system
    D) Two-pipe system

2345. What is required to provide better removal of air in the oil supply?
    A) Single-stage pump
    B) Two-stage pump
    C) Single pipe system
    D) Two pipe system

2346. The electrodes on a gun-type oil burner are located
    A) within the air/oil spray.
    B) outside of the air/oil spray.
    C) on the ignition transformer.
    D) in front of the nozzle.

2347. A boiler having a self-supporting, water-cooled shell or furnace bottom is called a
    A) wet bottom boiler.
    B) wet back boiler.
    C) water-cooled boiler.
    D) water-tube boiler.

2348. On the bimetal stack relay, which terminal is for the burner motor?
- A) No. 4
- B) No. 2
- C) No. 1
- D) No. 3

2349. A time start fan control is usually used on
- A) counterflow furnaces.
- B) downflow furnaces.
- C) horizontal furnaces.
- D) any of the above.

2350. High-pressure steam starts at
- A) 30 lb.
- B) 125 lb.
- C) 3 to 5 lb.
- D) none of the above.

2351. Fuel oil storage that is attached to the oil-fired room heater by the manufacturer is not to exceed
- A) 20 gallons.
- B) 10 gallons.
- C) 5 gallons.
- D) 15 gallons.

2352. A permit is not required on gas burners with an input less than
- A) 50,000 Btu.
- B) 40,000 Btu.
- C) 30,000 Btu.
- D) 20,000 Btu.

2353. Roof-mounted equipment not exceeding 400 lb should be supported by how many wood joists?
- A) One
- B) Two
- C) Three
- D) Four

2354. Roof-mounted equipment not exceeding 400 lb should be supported by how many trusses?
- A) One
- B) Two
- C) Three
- D) Four

2355. What is the maximum allowable percentage above the rated input Btu at which a gas furnace can be safely operated?
- A) 20%
- B) 10%
- C) 0%
- D) None of the above

2356. For a typical municipality, the clearance in front of an oil furnace is
    A) 6 inches.
    B) 24 inches.
    C) 18 inches.
    D) none of the above.

2357. Inside a building, the air for combustion must be at least one square inch per
    A) 1,000 Btu.
    B) 2,000 Btu.
    C) 3,000 Btu.
    D) 4,000 Btu.

2358. Through horizontal ducts, the outdoor air for combustion must be at least one square inch per
    A) 1,000 Btu.
    B) 2,000 Btu.
    C) 3,000 Btu.
    D) 4,000 Btu.

2359. Through vertical ducts, the outdoor air for combustion must be at least one square inch per
    A) 1,000 Btu.
    B) 2,000 Btu.
    C) 3,000 Btu.
    D) 4,000 Btu.

2360. The setting of the limit control on a mechanical warm air furnace shall not exceed
    A) 100 °F.
    B) 200 °F.
    C) 300 °F.
    D) 212 °F.

2361. Minimum oil pipe size for domestic-type burners is
    A) 1/2".
    B) 3/8".
    C) 3/4".
    D) 1".

2362. Oil tank vent pipes are
    A) 1-1/2" black iron pipe.
    B) 1-1/4" black iron pipe.
    C) 1-3/4" black iron pipe.
    D) 1-7/8" black iron pipe.

2363. Oil tank fill pipes are
    A) 1-1/2" black iron pipe.
    B) 1-1/4" black iron pipe.
    C) 1-3/4" black iron pipe.
    D) 1-7/8" black iron pipe.

2364. The thickness of an oil tank installed indoors must be
 A) 10 ga.
 B) 12 ga.
 C) 14 ga.
 D) 16 ga.

2365. How far away from a building opening must an oil tank vent pipe terminate?
 A) 1 ft
 B) 1-1/2 ft
 C) 2 ft
 D) 3 ft

2366. The difference between the opening and closing pressures of a safety or relief valve is known as the
 A) differential pressure.
 B) pressure drop.
 C) blow down.
 D) none of the above.

2367. On oil-fired units with No. 2 oil and less than 5 gallons, the burner will shut off within
 A) 60 seconds.
 B) 10 seconds.
 C) 2 minutes.
 D) 3 minutes.

2368. The number of Btu for one cubic foot of natural gas is
 A) 2,500 Btu.
 B) 1,000 Btu.
 C) 84,000 Btu.
 D) 100,000 Btu.

2369. The number of Btu for one gallon of No. 2 oil is
 A) 100,000 Btu.
 B) 137,000 Btu.
 C) 84,000 Btu.
 D) 2,500 Btu.

2370. Which valve makes no noise when opening or closing?
 A) Diaphragm
 B) Pilot
 C) Gas metering
 D) Recycling solenoid

2371. Thermocouples and thermopiles are used to operate
 A) thermostats.
 B) venturi.
 C) safety devices.
 D) transformers.

2372. The usual cause of an inoperative gas burner is the
   A) thermostat.
   B) pilot light.
   C) transformer.
   D) pressure regulator.

2373. The number of Btu in one gallon of No. 6 fuel oil is about
   A) 135,000.
   B) 165,000.
   C) 190,000.
   D) none of the above.

2374. What is the permissible pressure drop between the meter and the appliance?
   A) 3.0" wc
   B) 0.05" wc
   C) 5.0" wc
   D) None of the above

2375. Underground tanks storing flammable liquids must have a vent pipe draining to the tank. The top of this vent pipe must be how far away from the building?
   A) 1 ft
   B) 3 ft
   C) 5 ft
   D) 10 ft

2376. Duct systems conveying exhaust from restaurant hoods shall operate at a velocity no less than
   A) 1,000 fpm.
   B) 1,500 fpm.
   C) 1,700 fpm.
   D) 1,850 fpm.

2377. Metal smoke stacks 12 to 16 inches in diameter shall be not less than
   A) 24 ga.
   B) 22 ga.
   C) 12 ga.
   D) 10 ga.

2378. The flue pipe from a gas designed furnace shall have a diameter of at least
   A) 8".
   B) 6".
   C) 5".
   D) none of the above.

2379. A high velocity round duct 23 inches in diameter with a longitudinal seam construction must be at least
- A) 14 gauge sheet steel.
- B) 16 gauge sheet steel.
- C) 18 gauge sheet steel.
- D) 20 gauge sheet steel.

2380. What is the normal wattage capacity of a 120-volt electric heating thermostat?
- A) 100 watts
- B) 1,000 watts
- C) 3,000 watts
- D) 5,000 watts

2381. What voltage is used on the limit control of a gun-type oil burner system?
- A) 24 V
- B) 115 V
- C) 30 mV
- D) 750 mV

2382. Which pipe has the thickest wall?
- A) Schedule 20
- B) Schedule 40
- C) Schedule 60
- D) Schedule 80

2383. Clean-out openings shall be provided in every metal smoke stack at
- A) the furnace or heat-producing apparatus.
- B) the base.
- C) 6 ft from the floor.
- D) none of the above.

2384. Chimneys for oil-fired warm air furnaces shall be
- A) Class A or B.
- B) Class A or C.
- C) Class B or C.
- D) Class B.

2385. One boiler horsepower is equal to
- A) 144 Btu.
- B) 970 Btu.
- C) 33,475 Btu.
- D) 134,475 Btu.

2386. An oil tank of 500 gallons must be
- A) 10 ga.
- B) 12 ga.
- C) 14 ga.
- D) 16 ga.

2387. What is the minimum allowable percentage of Btu at which a gas furnace can safely be fired?
    A) 80%
    B) 90%
    C) 100%
    D) None of the above

2388. The effective height of a chimney is the vertical distance from the chimney point of entry to the
    A) ceiling.
    B) roof.
    C) peak of the roof.
    D) top of the chimney.

2389. A limiting device installed on a steam boiler is controlled by
    A) a vent.
    B) temperature.
    C) pressure.
    D) hydraulic.

2390. A limiting device installed on a hot water boiler is controlled by
    A) vent.
    B) temperature.
    C) pressure.
    D) hydraulic.

2391. Warm air is circulated in the gravity warm-air system by
    A) direct conduction.
    B) indirect radiation.
    C) natural convection.
    D) indirect gravitation.

2392. Gas burners having an hourly input of 400,000 Btu will have a trial for ignition period not exceeding
    A) 2 minutes.
    B) 3 minutes.
    C) 60 seconds.
    D) 10 seconds.

2393. Vent and chimney connectors shall be at least
    A) 26 ga.
    B) 28 ga.
    C) 30 ga.
    D) 14 ga.

2394. The flame failure timing period on furnaces with firing rates of 5 gph is
    A) 2 minutes.
    B) 3 minutes.
    C) 10 seconds.
    D) 60 seconds.

2395. Ducts shall be supported at intervals of
A) 5 ft.
B) 8 ft.
C) 10 ft.
D) 12 ft.

2396. One inch gas pipe shall be supported at intervals of
A) 5 ft.
B) 8 ft.
C) 10 ft.
D) 12 ft.

2397. The minimum clearance in front of a gas furnace is
A) 6".
B) 12".
C) 18".
D) 24".

2398. The minimum clearance in front of an oil furnace is
A) 6".
B) 12".
C) 18".
D) 24".

2399. The minimum clearance at the top of an oil furnace is
A) 6".
B) 12".
C) 18".
D) 24".

2400. The minimum clearance at the side of a gas furnace is
A) 6".
B) 12".
C) 18".
D) 24".

2401. When equipment is installed in a place such as a roof, supplying safe access and approach to this equipment is the responsibility of the
A) building owner.
B) heating contractor.
C) building contractor.
D) none of the above.

2402. In building construction, equipment openings shall be at least
A) 24" x 36".
B) 18" x 36".
C) 24" x 24".
D) 36" x 36".

2403. Natural gas pressure at the street is usually
A) 15 to 30" wc.
B) 15 to 60 lb.

C) 5 to 30 lb.
D) 15 to 60" wc.

2404. Propane manifold pressure is
A) 3 to 5" wc.
B) 6 to 7" wc.
C) 5 to 7" wc.
D) none of the above.

2405. Propane tank pressure is
A) 15 to 30 lb.
B) 50 to 100 lb.
C) 100 to 200 lb.
D) 200 to 300 lb.

2406. The three types of pilot safety devices are
A) current, hydraulic, bimetal.
B) current, motorized, diaphragm.
C) current, potential, bimetal.
D) voltage, hydraulic, bimetal.

2407. The three types of gas valves are
A) current, motorized, potential.
B) current, motorized, diaphragm.
C) current, hydraulic, bimetal.
D) motorized, current, bimetal.

2408. Chimney heights for No. 2 oil or gas-fired equipment of 500,000 Btu or less are ____ ft above flat roofs, and at least ____ ft above ridges, peaks, etc., within ____ ft.
A) 3, 2, 10
B) 1, 2, 3
C) 2, 3, 3
D) 2, 3, 10

2409. The maximum length of chimney connector is
A) 3/4 of the distance from the chimney entrance to the chimney top.
B) 2/3 that of the effective chimney height.
C) effective chimney height.
D) 15 ft.

2410. Cold air returns from bathrooms should be
A) the same size as the hot air inlet.
B) 1" up from the floor.
C) omitted.
D) supplied with a damper.

2411. Smoke pipe, Class B vent, must have a rise of
A) 1/4" per inch.
B) 1/2" per inch.
C) 1/4" per foot.
D) 1/2" per foot.

2412. For smoke pipe, Class B vent, the gauge is
A) 26 ga.
B) 28 ga.
C) 30 ga.
D) none of the above.

2413. Portable ladders cannot be over
A) 15 ft.
B) 20 ft.
C) 25 ft.
D) 30 ft.

2414. Portable ladders must have how many rungs projecting above the roof?
A) One
B) Two
C) Three
D) Four

2415. On a gravity warm air furnace, the limit control setting shall not exceed
A) 160 °F.
B) 200 °F.
C) 250 °F.
D) 300 °F.

2416. According to BOCA regulations for low heat appliances, products of combustion at the point of entrance to the flue must be
A) 500 to 1,100 °F.
B) 1,000 °F or less.
C) 1,001 to 2,000 °F.
D) 2,000 to 3,000 °F.

2417. According to BOCA regulations for medium heat appliances, products of combustion at the point of entrance to the flue must be
A) 500 to 1,000 °F.
B) 1,001 to 2,000 °F.
C) 2,001 to 3,000 °F.
D) 3,000 to 4,000 °F.

2418. According to BOCA regulations for high heat appliances, products of combustion at the point of entrance to the flue must be
A) 100 to 500 °F.
B) 501 to 1,000 °F.
C) 1,001 to 2,000 °F.
D) 2,001 °F and up.

2419. The location of outside air exhaust and intake openings shall be located a minimum of _____ feet from lot lines or buildings on the same lot.
A) 5
B) 10

C) 15
D) 20

2420. Duct supports shall be fitted with approved hangers at intervals not exceeding
A) 8 ft.
B) 10 ft.
C) 12 ft.
D) 14 ft.

2421. Installations exhausting more than _____ cfm shall be provided with make-up air.
A) 50
B) 100
C) 200
D) 450

2422. In gas heating, standing pilots burn
A) sporadically.
B) only when called for.
C) continuously.

2423. A 125-volt ac grounding-type outlet shall be available for all appliances. The outlet shall be located on the same level, within _____ feet of the appliance.
A) 25
B) 50
C) 75
D) 100

2424. The quantity of gas to be provided at each outlet shall be determined by the
A) size of the pipe at the gas valve.
B) Btu output of the unit.
C) Btu input of the unit.
D) none of the above.

2425. The approximate gas input for a free-standing domestic range is
A) 4,000 Btu.
B) 12,500 Btu.
C) 25,000 Btu.
D) 65,000 Btu.

2426. The approximate gas input for a domestic water heater (30 to 40 gallons) is
A) 15,000 Btu.
B) 20,000 Btu.
C) 45,000 Btu.
D) 85,000 Btu.

2427. The approximate gas input for a domestic clothes dryer is
A) 10,000 Btu.
B) 15,000 Btu.
C) 35,000 Btu.
D) 65,000 Btu.

2428. Gas pipe of 1/2 to 1 inch will have threading of
A) 8 threads.
B) 9 threads.
C) 10 threads.
D) 11 threads.

2429. Gas pipe of 1-1/4 to 2 inches will have threading of
A) 11 threads.
B) 12 threads.
C) 13 threads.
D) 14 threads.

2430. Gas pipe of 2-1/2 to 3 inches will have threading of
A) 11 threads.
B) 12 threads.
C) 13 threads.
D) 14 threads.

2431. Iron gas pipe smaller than ______ may not be used in any concealed location.
A) 1/4"
B) 3/8"
C) 1/2"
D) 3/4"

2432. Gas piping outlets shall extend at least _______ through finished ceilings and walls.
A) 1"
B) 2"
C) 3"
D) 6"

2433. Gas piping outlets shall extend at least _______ through floors.
A) 1"
B) 2"
C) 3"
D) 6"

2434. When testing piping for tightness, use
A) oxygen.
B) oil.
C) air.
D) cool gas.

2435. When bending steel gas pipe,
   A) heat the pipe.
   B) use a pipe bender.
   C) both A and B.
   D) neither A nor B.

2436. Gas pipe may be installed in an air duct if the gas pipe is
   A) hard rubber gas line.
   B) copper gas line.
   C) steel gas line.
   D) none of the above.

2437. An oil furnace hanging from a 15-foot ceiling in a service garage requires the use of a
   A) two-pipe, single-stage oil pump.
   B) two-pipe, two-stage oil pump.
   C) one-pipe, single-stage oil pump.
   D) one-pipe, two-stage oil pump.

2438. Flexible gas connectors shall not be more than
   A) 3 ft.
   B) 4 ft.
   C) 6 ft.
   D) 8 ft.

2439. Floor registers shall not be allowed in any bathrooms.
   A) True
   B) False

2440. Floor registers shall not be allowed in any washrooms.
   A) True
   B) False

2441. Floor registers shall not be allowed in any toilet rooms.
   A) True
   B) False

2442. Floor registers shall not be allowed in any laundry rooms.
   A) True
   B) False

2443. Floor registers shall not be allowed in any utility rooms.
   A) True
   B) False

2444. Floor registers shall not be allowed in any kitchens.
   A) True
   B) False

2445. Floor registers shall not be allowed in any basements.
   A) True
   B) False

2446. Access openings to attics shall be so located that a clearance of ___ exists between the top of the ceiling joists and the bottom of the rafter at the point of entrance.
A) 1 ft
B) 2 ft
C) 3 ft
D) 4 ft

2447. Gas pipe testing for tightness must be
A) 3.5" wc for 3 minutes.
B) 6.7" wc for 10 minutes.
C) 10 psig for 3 minutes.
D) 3 psig for 10 minutes.

2448. An approved gas log lighter for a wood burning fireplace shall have a gas input of less than ____ Btu per hour.
A) 5,000
B) 10,000
C) 20,000
D) 30,000

2449. An approved gas log lighter for a wood burning fireplace shall have a timing device that permits gas flow for a period of not more than
A) 10 minutes.
B) 15 minutes.
C) 30 minutes.
D) 1 hour.

2450. The method of heat transfer employed by a forced air heating system is
A) radiation.
B) forced conduction.
C) forced radiation.
D) forced convection.

2451. Maximum current flows through an ohmmeter circuit when
A) there is minimum amount of resistance between ohmmeter terminals.
B) scale indicates 6000 ohms.
C) there is maximum resistance to the flow.
D) scale indicates INF.

2452. Universal motors may be used with dc or
A) ac.
B) single-phase ac.
C) two-phase ac.
D) three-phase ac.

2453. At what temperature will all molecular movement in a substance stop?
   A) -100 °F
   B) -200 °F
   C) -300 °F
   D) Absolute zero

2454. What is 1 Btu equal to in foot-pounds?
   A) 678
   B) 700
   C) 760
   D) 778

2455. What is entropy?
   A) Total heat in 10 lb of a substance
   B) Pressure at which a liquid will remain liquid.
   C) Critical temperatures
   D) Mathematical constant for calculating energy in a system

2456. What is sensible heat?
   A) Hidden heat present in a substance
   B) Movement of molecules within a substance
   C) Heat that changes a substance from liquid to vapor
   D) Heat that can be added or subtracted without changing state of a substance

2457. What is the specific heat of water?
   A) 1.1
   B) 1.0
   C) 0.6
   D) 0.1

2458. Most heating, air-conditioning, and ventilating ducts are made from
   A) galvanized sheet iron.
   B) aluminum.
   C) stainless steel.
   D) black iron.

2459. Heating vents, vertical flues, and chimneys are sometimes called
   A) collars.
   B) stacks.
   C) thimbles.
   D) roof jacks.

2460. The type of heat with which the heating specialist deals is produced by
   A) friction.
   B) chemical action.
   C) electrical resistance.
   D) burning fuel.

2461. On the Fahrenheit thermometer, the range between the freezing and boiling point of water is
   A) 100.
   B) 150.
   C) 170.
   D) 180.

2462. Heat that can either be measured by a thermometer or sensed by touch is referred to as
   A) sensible.
   B) specific.
   C) latent.
   D) total.

2463. Which of the following differences in temperature will cause the fastest heat flow?
   A) Room 90 °F, outside 32 °F
   B) Room 80 °F, outside 90 °F
   C) Room 70 °F, outside 60 °F
   D) Room -30 °F, outside 40 °F

2464. The name given to the point at which a substance ignites is
   A) kindling.
   B) pressure.
   C) static.
   D) vaporizing.

2465. Incomplete combustion is usually caused by
   A) increased carbon dioxide.
   B) too much oxygen.
   C) insufficient carbon monoxide.
   D) lack of oxygen.

2466. Carbon monoxide (CO) and carbon dioxide are gases given off during the combustion process. The percentage of these gases depends upon the
   A) type of coal and amount of smoke.
   B) retention of heat and increase of air.
   C) caloric dissipation and density of flue gas.
   D) amount of air (oxygen) mixed with the fuel.

2467. The greatest heat loss in a heating unit occurs in the
   A) grates.
   B) peephole.
   C) chimney.
   D) radiation.

2468. What percent of the heat from a burning fireplace is lost in the gases and smoke that go up the chimney?
   A) 50 to 70%
   B) 60 to 70%

C) 70 to 80%
D) 85 to 90%

2469. An efficient, modern space heater utilizes what percent of the heat from its fuel?
A) 50%
B) 65%
C) 75%
D) 85%

2470. One reason for using copper tubing when installing an oil-fired space heater is
A) it eliminates reaming of cut ends.
B) that it cannot be easily bent.
C) its high resistance to corrosion.
D) it requires more threading than iron piping.

2471. The gas pipe used to move natural gas to a space heater is usually made of
A) steel.
B) black iron.
C) copper.
D) lead.

2472. The purpose of the draft diverter in a gas-fired space heater is to
A) increase downdraft.
B) divert flue gas from chimney.
C) prevent excessive updraft or downdraft.
D) divert heater room air away from flue pipe.

2473. Insufficient amounts of primary air cause the flame to burn
A) white.
B) blue.
C) yellow.
D) red.

2474. Filters are installed in warm-air furnaces to
A) clean the circulating air.
B) filter out mechanical noise.
C) preheat the cold air.
D) decrease air temperature.

2475. Most steel furnaces are lined with firebrick to prevent the flames from burning through the thin metal during operation. This barrier is called a
A) fire plate.
B) refractory.
C) damper.
D) humidifier.

2476. Replace the air filters in a warm-air furnace at least
A) monthly.
B) quarterly.
C) yearly.
D) twice a year.

2477. When installing an 80-ft, 1/2" gas line along the ceiling of a building, what is the fewest number of hangers that could hold this pipe?
A) 8
B) 10
C) 12
D) 14

2478. The nozzle type selected for a gun-type, high-pressure burner furnace depends upon the
A) speed of compressors.
B) number of electrodes.
C) size of furnace or boiler.
D) number of insulators.

2479. When installing an oil burner that has a single-stage fuel unit, the fuel tank should be
A) underground.
B) outside boiler room.
C) alongside burner.
D) elevated.

2480. To clean the nozzle in a domestic oil burner, use
A) a safety solvent.
B) soapy water.
C) wire.
D) a carbon mixture.

2481. The pressure reading on a manometer is expressed in inches of
A) oil.
B) water.
C) gas.
D) air.

2482. A gas burner having 100% cut-off safety pilots uses
A) LPG.
B) natural gas.
C) propane.
D) butane.

2483. A yellow flame in a gas burner indicates
A) absence of carbon.
B) poor air-fuel mixture.
C) correct air-fuel mixture.
D) complete combustion.

2484. What happens to the gas solenoid valve when its supply of electric current fails?
   A) It opens
   B) It closes
   C) The valve disc is unseated
   D) The coil is energized

2485. Which of the following gas valves has a device that allows manual operation of the valve during failure of electric current?
   A) Recycling solenoid
   B) Diaphragm
   C) Thermocouple
   D) Thermopile

2486. Total heat is
   A) sensible heat plus latent heat.
   B) superheat less specific heat.
   C) sensible heat plus specific heat.
   D) specific heat plus sensible heat.

2487. The correct flow of air to vaporizing-type oil burners depends primarily on
   A) chimney draft.
   B) oil grade.
   C) weight of damper.
   D) fuel supply tubing.

2488. The continuous action of hot air rising in the chimney and being replaced by cold air drawn through the space heater is called
   A) draft.
   B) conversion.
   C) combustion.
   D) conduction.

2489. Which of the following burners would be used to extract the maximum amount of heat from natural gas?
   A) Combination
   B) Premix
   C) Upshot
   D) Inshot

2490. To seal (gastight) the sections of a cast iron furnace, use
   A) graphite rope.
   B) tar paper.
   C) manila fibertape.
   D) furnace cement.

2491. The primary difference between the forced warm-air heating system and the gravity heating system is in the
A) placement of the cold-air returns.
B) method of circulating the air.
C) number of registers.
D) number of warm-air pipes.

2492. A firebrick lining is installed in the combustion chamber of a steel furnace to
A) increase flame.
B) reduce fuel consumption.
C) act as heat exchanger.
D) protect its wall.

2493. Which of the following accounts for the circulation of warm air in a gravity warm-air furnace?
A) The blower fan
B) Back pressure
C) Warm air is lighter than cool air
D) Setting

2494. In the past, what was the most common type of insulation used to cover old heating equipment?
A) Block
B) Sheet
C) Rock wool
D) Magnesium asbestos

2495. When placing a 20-ft ladder against a wall, how many feet from the wall should the foot of the ladder be placed?
A) 2
B) 4
C) 5
D) 8

2496. When working on a boiler installation that requires measuring the amount of pull used in tightening some nuts, what type of wrench is used for this measurement?
A) Adjustable
B) Box end
C) Open end
D) Torque

2497. In the past, metal pipes on old furnaces were covered with asbestos paper to
A) lower furnace temperature.
B) prevent leaks.
C) increase radiation.
D) reduce heat loss.

2498. In gravity heating, clean the air ducts with a vacuum cleaner once each
   A) day.
   B) week.
   C) month.
   D) season.

2499. In maintaining a forced, warm-air furnace with air conditioning, inspect the air filters
   A) once a month.
   B) twice a month.
   C) once a year.
   D) twice a year.

2500. How much play is allowed in the blower belt of a forced, warm-air furnace?
   A) 1/4 inch
   B) 1/2 inch
   C) 1 inch
   D) 2 inches

2501. In the forced, warm-air heating system, the air ducts are usually
   A) larger than gravity air ducts.
   B) attached to the walls of the heated rooms.
   C) sloped up 0.50 inch per 6 feet.
   D) hung from the ceiling.

2502. Humidifiers in furnaces usually consist of a pan and float-operated needle control valve. This valve controls the
   A) temperature of room.
   B) air flow.
   C) water level in the pan.
   D) back pressure of warm air.

2503. In maintaining gravity warm-air furnaces, bear in mind that soot deposits on the heat exchanger will
   A) increase heat transmission.
   B) reduce furnace draft.
   C) insulate it.
   D) create air leaks.

2504. In maintaining a forced warm-air furnace, what is the minimum number of times the blower and electric motor bearings should be oiled?
   A) Monthly during heating season
   B) Once each quarter
   C) Once each heating season
   D) Twice each heating season

2505. The unit of pressure commonly used in air ducts is measured in
A) inches of water.
B) inches of air.
C) velocity.
D) inches of mercury.

2506. Which of the following electrical wires will carry the most current?
A) 14 ga
B) 10 ga
C) 12 ga
D) 24 ga

2507. An electrical network having 120 volts to a neutral from all legs is called
A) delta.
B) star or wye.
C) polyphase.
D) two-phase.

2508. A squirrel cage fan is another name for a(n)
A) backward inclined fan.
B) forward inclined fan.
C) radial flow fan.
D) exhaust fan.

2509. All filters must
A) be coated.
B) be constructed of fiberglass.
C) be 100% fireproof.
D) have a pressure drop.

2510. The thermostat range is the
A) open and close settings.
B) temperature difference.
C) capacity of the thermostat.
D) variety of the models available.

2511. If a gas-fired furnace has a 6" flue stack, what other exhaust may be connected to this stack?
A) Bathroom vent
B) Kitchen range hood
C) Laundry vent
D) No other exhaust may be connected

2512. Flues for oil-fired warm air furnaces shall be
A) Class A.
B) Class D.
C) Class E.
D) Class W.

2513. To check the rpm of a belt-driven fan, use a(n)
A) ammeter.
B) tachometer.

C) velometer.
D) manometer.

2514. What kind of threads does a street ell have?
A) External threads only
B) Internal threads only
C) Both external and internal threads
D) NF and NP threads

2515. Which of the following is a device for removing dust from the air by means of electric charges induced on the dust particles?
A) Electric air cleaner
B) Electric precipitator
C) Electric magnet
D) Electric ejector

2516. A U-shaped tube partially filled with a liquid (usually water, mercury or a light oil) and constructed so that the amount of displacement of the liquid indicates the pressure being exerted on the instrument is a(n)
A) potentiometer.
B) velometer.
C) manometer.
D) anemometer.

2517. On a oil-fired furnace, the primary control
A) opens oil solenoid valve.
B) stops operation of unit in case of fire.
C) stops burner operation if flame is not established or if flame is extinguished after being in operation.
D) controls oil pressure at burner nozzle.

2518. Air in the oil line to burner may cause
A) pulsation.
B) excessive fuel consumption.
C) blue flame.
D) dirty nozzle.

2519. The specific gravity of most gases varies from
A) 0.45 to 0.65.
B) 0.65 to 1.0.
C) 1.0 to 1.4.
D) none of the above.

2520. Mechanical ventilation for paint spray booths must have a(n)
A) explosion-proof motor.
B) two-speed motor.
C) backdraft damper.
D) fusestat.

2521. To check motor performance on a mechanical ventilation system use a(n)
A) pressure gauge.
B) ammeter.
C) pitot tube.
D) velometer.

2522. To check air volume at a side wall supply grille, use a
A) tachometer.
B) velometer.
C) ammeter.
D) mercury gauge.

2523. The heating element required for hot water heaters is the
A) spiral.
B) cast-in.
C) tubular.
D) immersion.

2524. On a hydrotherm boiler, the maximum length of air supply and exhaust piping is
A) 30 feet.
B) 40 feet.
C) 50 feet.
D) 60 feet.

2525. The maximum temperature for CPVC piping is
A) 250 °F.
B) 180 °F.
C) 140 °F.
D) 100 °F.

2526. On a hydrotherm boiler, the maximum number of elbows in the air supply or exhaust piping is
A) two.
B) three.
C) four.
D) six.

2527. On a hydrotherm boiler, the air supply piping is
A) 28 ga.
B) 26 ga.
C) PVC.
D) CPVC.

2528. On a hydrotherm boiler, the exhaust piping is
A) 28 ga.
B) 26 ga.
C) PVC.
D) CPVC.

2529. The pressure relief valve on a hydrotherm boiler opens at
   A) 10 lb.
   B) 15 lb.
   C) 30 lb
   D) 45 lb.

2530. When installing a forced air system in a building, the duct distribution system must carry the required flow of air with the least amount of
   A) resistance.
   B) velocity.
   C) heat.
   D) dynamic loss.

2531. The pressure loss through friction in a straight duct
   A) decreases when inside of duct is rough.
   B) increases when inside of duct is rough.
   C) decreases when duct is lengthened.
   D) increases when larger duct is used.

2532. To reduce turbulence and pressure loss in the elbows of ducts, use
   A) louvers.
   B) registers.
   C) turning vanes.
   D) diffusers.

2533. When a hot-water system is first filled with water, it is normally necessary to
   A) close air vents on the radiators.
   B) keep water temperature below 150 °F.
   C) increase intake of air.
   D) bleed air out of system.

2534. The radiators in a forced circulation hot-water heating system will not heat. There is sufficient water in the system, radiator valves are open, no corrosion is present, and the air has been bled from the system. The next step is to
   A) check the circulation pump.
   B) clean boiler flues.
   C) change to larger boiler.
   D) increase draft.

2535. Boiler water is not used for domestic purposes because
   A) it is too hot.
   B) of chemicals added.
   C) quantity is limited.
   D) it has been aerated.

2536. A pipe fitting shaped like an ell but with one female end and one male end is called a
   A) male union L.
   B) female union L.
   C) street L.
   D) female-to-male L.

2537. Threaded and tapered fittings that are screwed into the ends of other fittings or valves to reduce the size of the end openings are known as
   A) reducers.
   B) nipples.
   C) bushings.
   D) increasers.

2538. When joining two pieces of copper pipe, the melted solder is drawn into the joint by
   A) capillary attraction.
   B) gravity.
   C) flow rate.
   D) brushing.

2539. The acetylene connection hose on the oxyacetylene outfit has
   A) a quick-disconnect fitting.
   B) black and white stripes.
   C) left-hand threads.
   D) green and white stripes.

2540. The neutral or balanced flame that is used in welding is produced when the mixed gases consist of approximately
   A) 1 volume oxygen, 1 volume acetylene, 1-1/2 volumes outside air.
   B) 3 volumes oxygen, 1 volume acetylene, 1 volume atmosphere.
   C) 2 volumes oxygen, 1 volume acetylene, 1/2 volume atmosphere.
   D) 1-1/2 volumes oxygen, 1 volume acetylene, no outside air.

2541. When using oxyacetylene welding equipment, first
   A) remove protective caps.
   B) crack cylinder valves slightly.
   C) secure the cylinders.
   D) connect hose to torch.

2542. The color of the pure acetylene flame is
   A) yellowish.
   B) blue.
   C) red.
   D) white.

2543. Normal operating pressure for residential steam boilers is
   A) 3 to 5 lb.
   B) 12 to 15 lb.
   C) 30 lb.
   D) none of the above.

2544. On a low-pressure hot water boiler, the maximum temperature is not to exceed
    A) 210 °F.
    B) 160 °F.
    C) 250 °F.
    D) 212 °F.

2545. HVAC systems handling over 2000 cfm must have
    A) fire suppression.
    B) smoke detectors.
    C) neither A nor B.

2546. The limit control setting on a low-pressure, hot water, space-heating boiler shall not exceed
    A) 125 °F.
    B) 212 °F.
    C) 210 °F.
    D) 250 °F.

2547. Inside an extrol tank is a(n)
    A) air pocket.
    B) hard rubber ball.
    C) vacuum.
    D) rubber diaphragm full of air.

2548. The installation of a gas burner in a hot-water boiler must include a(n)
    A) low water cut-off.
    B) closed expansion tank.
    C) water temperature limit control.
    D) automatic water feeder.

2549. On most circulators, how many oil cups are there?
    A) One
    B) Two
    C) Three
    D) Four

2550. On a B & G circulator, the term B & G stands for
    A) Ball and Gossett.
    B) Bell and Gossett.
    C) Ball and Green.
    D) Bell and Green.

2551. On residential circulating pumps, the device that moves the water is called a(n)
    A) coupler.
    B) bearing assembly.
    C) impeller.
    D) flange.

2552. On a residential circulating pump, the device connected between the motor and bearing assembly is a(n)
   A) mechanical seal.
   B) bronze bearing.
   C) coupler.
   D) impeller.

2553. During normal operation in a low pressure F.W. boiler, B & G flow control valves should be kept in the
   A) open position.
   B) closed position.
   C) halfway open position.
   D) none of the above.

2554. B & G zone control valves are
   A) motorized.
   B) heat-activated.
   C) water activated.
   D) none of the above.

2555. How much heat does one pound of steam release as it is condenses to water at 212 °F?
   A) 144 Btu
   B) 212 Btu
   C) 1,000 Btu
   D) 970 Btu

2556. In the pipe of a single-pipe, residential, low-pressure steam system,
   A) water travels to the boiler.
   B) steam travels to the radiator.
   C) steam travels to the radiator while water drains back to the boiler.
   D) steam travels during heating, water returns during off cycle.

2557. The limit control setting on a hot water supply heater or boiler must not exceed
   A) 200 °F.
   B) 250 °F.
   C) 210 °F.
   D) 212 °F.

2558. Room heating units, baseboard radiation, convectors, radiators, ,etc., are sized to the heat loss of
   A) the room.
   B) the boiler.
   C) inside rooms.
   D) none of the above.

2559. A steam trap is a device that allows steam, but not condensation, to pass through it.
   A) True
   B) False

2560. Soot deposits in a boiler cause stack temperatures to
A) remain the same.
B) increase.
C) decrease.
D) none of the above.

2561. In hydronics, the term convector refers to a unit that emits the greater portion of its heat by
A) radiation.
B) conduction.
C) convection.
D) none of the above.

2562. One boiler horsepower is equal to supplying how many (EDR) square feet of steam radiation per hour?
A) 140
B) 200
C) 180
D) 233

2563. The load piping and pickup according to MCA standards is
A) 13%.
B) 25%.
C) 20%.
D) 30%.

2564. How many square feet of heat transfer surface is equal to 1 boiler horsepower?
A) 3
B) 5
C) 7
D) 15

2565. If the input is 100,000 Btu, the efficiency 80% and the 25% allowance is made for piping and pickup, what is the net capacity of the boiler?
A) 80,000 Btu
B) 40,000 Btu
C) 60,000 Btu
D) 20,000 Btu

2566. What type of burner is used with a Scotch boiler?
A) Vaporizing
B) Horizontal in-shot
C) Low pressure
D) Rotary

2567. The area of water surface at the water line is called
A) disengaging area.
B) heating surface.
C) steaming area.
D) blow down area.

2568. The ratio of heating surface to boiler horsepower is
A) 3 to 1.
B) 4 to 1.
C) 5 to 1.
D) 6 to 1.

2569. All fire-tube boilers can produce what percent dry steam?
A) 90%
B) 94%
C) 96%
D) 98%

2570. A low-water cutoff is a de-energizer set to trip when the
A) low water reaches 180 °F.
B) low water rises.
C) boiler water drops.
D) none of the above.

2571. For a boiler operating at more than 100 psi, the blow-off piping thickness shall not be less than
A) standard black iron pipe.
B) heavy copper pipe.
C) Schedule 80 pipe.
D) Schedule 40 pipe.

2572. An ASME pressure relief valve on a boiler is usually provided with a hand lever, which
A) closes the valve once it has opened.
B) tests the valve and relief of air on initial fill.
C) changes operating pressure.
D) provides manual safety.

2573. A steam or hot water boiler must have a(n)
A) drain valve.
B) individually controlled make-up valve.
C) both A and B.
D) neither A nor B.

2574. Which valve is the most suitable for throttling flow?
A) Gate
B) Check
C) Globe
D) Swing

2575. What type of cleaning compound is used in starting up a boiler?
A) Trisodium phosphate
B) Sodium chlorate
C) Sulfur oxide
D) Lithium bromide

2576. Does UL require a motorized safety shut-off valve?
A) Yes
B) No

2577. Which approval group requires a pre-purge cycle?
A) FIA
B) FM
C) UL
D) All of the above

2578. What initiates the boiler control sequence?
A) Scanner
B) Timer
C) Limit control
D) Operating control

2579. Do industrial boilers always start on low fire?
A) Yes
B) No

2580. The three main types of boilers are
A) water-tube, HRT, locomotive.
B) cast iron, horizontal, vertical.
C) fire-tube, HRT, vertical.
D) water-tube, cast iron, fire-tube.

2581. Scotch marine boilers are used to develop pressures
A) below 250 psig.
B) above 250 psig.
C) neither A nor B.

2582. Water-tube boilers operate with efficiencies up to
A) 70%.
B) 80%.
C) 90%.
D) 95%.

2583. The temperature of steam at 5 psig pressure is
A) 212 °F.
B) 180 °F.
C) 230 °F.
D) 205 °F.

2584. When checking the primary safety on a gas hot water heater, check the
A) T and P valve.
B) burner.
C) gas valve.
D) none of the above.

2585. Boiler horsepower is based on an evaporation temperature of
   A) 230 °F.
   B) 220 °F.
   C) 224 °F.
   D) 212 °F.

2586. A boiler horsepower is equal to an evaporation rate of how many pounds of steam per hour?
   A) 43.5 lb
   B) 34.5 lb
   C) 35.4 lb
   D) 53.4 lb

2587. What is the typical width of a ligament?
   A) 1/2"
   B) 3/4"
   C) 1"
   D) 1-1/4"

2588. What is the thickness of the tube sheet on low-pressure boilers?
   A) 3/16"
   B) 5/16"
   C) 7/16"
   D) 9/16"

2589. Which boilers use the least refractory materials?
   A) 2 and 4 pass
   B) 3 pass
   C) 5 and 7 pass
   D) 6 pass

2590. How much insulation is applied to the shell?
   A) 1"
   B) 2 to 3"
   C) 3 to 4"
   D) 5"

2591. The ASME code for low-pressure boilers requires how many relief valves?
   A) 1 or more
   B) 2 or more
   C) 3 or more
   D) 4 or more

2592. How many low-pressure boilers are used in the commercial market?
   A) 50%
   B) 60%
   C) 75%
   D) 90%

2593. What certificate is necessary to obtain insurance?
A) UL
B) FM
C) National Board Inspector
D) FIA

2594. A column of water that exerts a pressure of one pound is
A) 0.433 ft high.
B) 3.95 ft high.
C) 2.31 ft high.
D) none of the above.

2595. For threading pipe in close quarters, use
A) combination pipe and stock dies.
B) power pipe threading machine.
C) ratchet-type pipe dies.
D) none of the above.

2596. A fluctuating water level is usually caused by
A) a load that is too high.
B) foaming.
C) a firing rate that is too high.
D) wrong orifice.

2597. How often should boiler tubes be cleaned?
A) Once a year
B) Once a month
C) Every two months
D) As required

2598. How often should the low water cut-off be cleaned?
A) Annually
B) Monthly
C) Weekly
D) Every two years

2599. Boiler tubes are cleaned with
A) trisodium phosphate.
B) sodium chloride.
C) acid.
D) wire brush.

2600. The highest operating pressure used on packaged Scotch boilers is
A) 30 lb.
B) 50 lb.
C) 100 lb.
D) 150 lb.

Heating
Level 2

2601. On a high-pressure boiler, tube diameter is
  A) 1 to 2 inches.
  B) 2 to 4 inches.
  C) 4 to 6 inches.
  D) none of the above.

2602. An ac motor rated at 4,950 watts has a horsepower of
  A) 5.9.
  B) 6.6.
  C) 7.5.
  D) 8.2.

2603. In an ac circuit, apparent power will be equal to the true power when
  A) apparent power is greater than true power.
  B) true power is greater than apparent power.
  C) the ac circuit is made up of pure resistance.
  D) reactive power and apparent power are equal.

2604. The critical temperature of water is
  A) 600 °F.
  B) 650 °F.
  C) 675 °F.
  D) 689 °F.

2605. When making a double-wall stack, how much air space is allowed between the inside and the outside stacks?
  A) 1/4 inch
  B) 1/2 inch
  C) 3/4 inch
  D) 1 inch

2606. A safety feature for the stack assembly is
  A) roof jacks.
  B) chimney flashings.
  C) wall thimbles.
  D) draft diverters.

2607. Which of the following is used to prevent leaks around the top of a roof jack?
  A) Draft diverter
  B) Storm collar
  C) Wall thimble
  D) Vent cap

2608. In order to exhaust air by gravity flow or forced air, install
 A) diverters.
 B) collars.
 C) thimbles.
 D) ventilators.

2609. Which would be installed in a ventilator to regulate the air flow?
 A) Collar
 B) Thimble
 C) Damper
 D) Roof jack

2610. After installing a draft ventilator and nailing the flashing to the roof deck, make the flashing watertight by covering it with
 A) asbestos.
 B) asphalt tar.
 C) felt.
 D) rubber.

2611. Transferring electrical energy from one coil of wire to another coil by means of a varying magnetic field is called
 A) repulsion.
 B) magnetic induction.
 C) mutual induction.
 D) electromagnetic induction.

2612. The primary winding of a transformer is connected to a 240-V power source. What is the secondary voltage if the transformer has a ratio of 1 to 2?
 A) 480
 B) 240
 C) 120
 D) 100

2613. An ac motor uses 500 watts of power. If the motor is operated for 3 hours, how much power in kilowatt-hours is used?
 A) 1
 B) 1.5
 C) 166
 D) 1,500

2614. To control an individual circuit from two locations, it is necessary to use
 A) single-pole switches.
 B) double-pole switches.
 C) three-way switches.
 D) four-way switches.

2615. What must be placed under a space heater that is installed on a wood floor?
 A) Sheet metal

B) Waterproof canvas
C) Rubber mat
D) Plywood

2616. When installing a 5-foot horizontal smoke pipe from the space heater to the chimney, how many inches of upward pitch is the smoke pipe given?
A) 1 inch
B) 1-1/4 inch
C) 1-1/2 inch
D) None

2617. When installing an oil-fired space heater on a building whose roof peak is 30 feet from the ground, how many feet should the top of the chimney be above the ground?
A) 20
B) 25
C) 30
D) 33

2618. When the burner goes out in a oil-fired space heater, the probable cause is
A) low oil supply.
B) stuck needle valve.
C) dirty float valve.
D) slag in heat exchanger.

2619. When an oil-fired space heater shows high fuel consumption, the probable cause is
A) improper fuel.
B) dirt in oil supply line.
C) clogged oil strainer.
D) dirty float valve.

2620. Approved flexible metal tubing having a nominal diameter not less than the inlet connection to the appliance as provided by the manufacturer of the appliance and which is not more than ________ feet in length may be used to connect overhead mounted unit heaters.
A) 6
B) 5
C) 4
D) 3

2621. For proper burner control maintenance, the contact points of the individual control units should be cleaned and all electrical contacts tightened
A) once each year.
B) semi-annually.
C) monthly.
D) weekly.

2622. The smoke pipe between the space heater and chimney should have as few bends and joints as possible in order to
   A) increase resistance to flow of smoke.
   B) reduce resistance to flow of smoke.
   C) increase convection.
   D) reduce conduction.

2623. When a gun-type oil burner is in operation and a buzzing sound is heard in the oil tank,
   A) check antihum diaphragm.
   B) adjust intake port.
   C) tighten pressure regulating screw cap.
   D) clean bypass plug.

2624. After installing a downdraft diverter in a gas-fired furnace, make sure that it
   A) has been calibrated for this unit.
   B) is the adjustable type.
   C) has large vent-pipe sheet.
   D) is the correct size.

2625. A horizontal smoke pipe 6 feet long requires a pitch of at least
   A) 1/2 inch.
   B) 1 inch.
   C) 1-1/2 inch.
   D) 6 inches.

2626. When the roof peak is 25 feet, the minimum height of its chimney should be
   A) 25 feet.
   B) 28 feet.
   C) 31 feet.
   D) 34 feet.

2627. If a warm-air heating system is keeping some rooms at 65 °F and other rooms at 79 °F,
   A) remake furnace fire.
   B) decrease furnace fire.
   C) adjust the dampers.
   D) balance the butterfly valves.

2628. In a forced warm-air heating system, air from the register should measure
   A) 35 fpm.
   B) 40 fpm.
   C) 45 fpm.
   D) 50 fpm.

2629. When installing a furnace, use a base of
   A) sand.
   B) masonry.
   C) asbestos.
   D) rubber.

2630. When installing a furnace, make sure it is level by using a
   A) spirit level.
   B) steel tape.
   C) T-square.
   D) mason's square.

2631. A square duct symbol that has two diagonal lines represents what kind of duct?
   A) Exhaust
   B) Supply
   C) Recirculation
   D) Flow in

2632. In sheet metal work, a register is a(n)
   A) air speed recorder.
   B) covering grille.
   C) air intake regulator.
   D) supply outlet.

2633. Standard practice requires a qualified boiler inspector to inspect a steam boiler at least once a
   A) day.
   B) week.
   C) month.
   D) year.

2634. All boilers are constructed to incorporate a furnace or firebox for
   A) storing the fuel.
   B) burning the fuel.
   C) holding the gases.
   D) preventing boiler-surface heat.

2635. In most cases, baffles are provided in the boiler to
   A) guide the gases.
   B) prevent exit of gases.
   C) prevent formation of gases.
   D) reduce heat absorption by boiler water.

2636. The boiler access door permits the cleaning of tubes and boiler sides. Between cleanings, seal this door with
   A) asbestos rope.
   B) rivets.
   C) cement powder.
   D) bolts.

2637. How are sections of low-pressure sectional steel boilers usually joined together?
A) Bolted
B) Riveted
C) Welded
D) Hooked

2638. An example of a complete, self-contained, boiler unit is the
A) Scotch marine.
B) HRT.
C) locomotive.
D) bent tube.

2639. When a radiator fails to heat or when water-hammer occurs, the probable cause is
A) failure of air vent to function.
B) a fluctuating boiler-water line.
C) a completely open valve.
D) a completely closed valve.

2640. What component of a two-pipe vapor system would be found at the bottom of the opposite end of the radiator?
A) Air vent
B) Radiator valve
C) Thermostatic trap
D) Flow-control valve

2641. How are sections of high-pressure steel sectional boilers usually joined together?
A) Bolted
B) Welded
C) Hooked
D) Riveted

2642. For low-pressure, low-capacity purposes, install a
A) box-header cross drum boiler.
B) box-header longitudinal boiler.
C) bent-tube boiler.
D) fire-tube boiler.

2643. Which boiler has its firebox constructed separately of firebrick?
A) HRT
B) Firebox
C) Water tube
D) Scotch marine

2644. In the storage-type hot water tank, the stored water temperature should not exceed
A) 170 °F.
B) 180 °F.

C) 185 °F.
D) 200 °F.

2645. What type valve is used for throttling?
A) Nonreturn
B) Gate
C) Check
D) Globe

2646. It is necessary to be a certified welder before being allowed to weld a
A) boiler shell.
B) firebox door.
C) water line.
D) pressure vessel.

2647. The wire solder used on copper water pipe is composed of
A) 100% lead.
B) 50% lead and 50% tin.
C) 40% tin and 60% lead.
D) 10% copper, 10% spelter, 80% lead.

2648. The pure acetylene flame is
A) unsuitable for welding.
B) short and bushy.
C) a bluish color.
D) colorless.

2649. The type of insulation used to cover the outside surface of a boiler is
A) rockwool.
B) fiberglass.
C) sheet.
D) blanket.

2650. What type of insulation is referred to as asbestos paper?
A) Roll
B) Sheet
C) Magnesium asbestos
D) Blanket

2651. The refractory lining in a firebox protects the
A) boiler lining.
B) fire tubes.
C) crown sheet.
D) insulation.

2652. Which valve must always be kept either fully open or fully closed?
A) Quick opening
B) Gate
C) Glove
D) Check

2653. How many pipe wrenches should be used to ensure tight joints when installing wrought iron or steel fittings?
A) One
B) Two
C) Three
D) Four

2654. After installing a refractory lining in the firebox of a boiler, how many hours must the firebox dry out before starting a fire in it?
A) Two
B) Four
C) Eight
D) Twelve

2655. When installing a conversion oil burner in a steam boiler, there are 15 radiators with 100 square feet of EDR (equivalent direct radiation) each. What is the size of the nozzle, in gallons per hour, needed?
A) 5.0
B) 7.5
C) 10.0
D) 15.0

2656. In the two-pipe, open-type gravity system, the amount of temperature drop between the beginning and end of the line depends upon
A) radiator size and atmosphere vent.
B) location of expansion tank.
C) size of pneumatic compression tank.
D) length of main and heating load.

2657. In a gravity, open-tank system with an average boiler temperature of 170 °F, the radiator emission rate of Btu is
A) 100.
B) 125.
C) 135.
D) 150.

2658. In the one-pipe, closed-tank, forced circulating system, what must be installed to improve the circulation through individual radiators?
A) Pressure valves
B) Connecting tees
C) Elbows
D) Branches

2659. A badly cracked section in a cast iron radiator should be
A) riveted.
B) welded.
C) replaced.
D) inverted.

2660. The boiler in a hot water system smokes through the feed doors, but there are no chimney leaks. What should be done next?
    A) Blow down the boiler
    B) Increase furnace draft
    C) Change kind of fuel
    D) Clean flues and flue pipes

2661. The distribution piping in a hot-water system fails to transfer water to the upper radiators. There is no stoppage in the lines, the circulation pump runs, and air has been bled from the system. The cause of the problem is
    A) insufficient water.
    B) decreased furnace draft.
    C) wrong type of fuel.
    D) chimney leaks.

2662. For water at 60 psig and 308 °F, what is the total Btu/ft^3?
    A) 13,800
    B) 14,000
    C) 15,000
    D) 15,840

2663. If water and steam were each subjected to 70 psig at 316 °F, the water would contain how many times more Btu/ft^3 than the steam?
    A) 50
    B) 58
    C) 63
    D) 71

2664. When the temperature of the water in an HTHW system is 100 °F, the density is about 63 lb/ft^3. What is the density of water in lb/ft^3 when the temperature rises to 200 °F?
    A) 60
    B) 70
    C) 90
    D) 120

2665. A 1 inch steam line transfers 9,000 Btuh. If this heating system is changed to HTHW, how many Btu will the line to transmit?
    A) 9,000
    B) 27,000
    C) 270,000
    D) 900,000

2666. For a given volume, steam contains how many times more heat than air?
    A) 38 to 65
    B) 38 to 70
    C) 40 to 80
    D) 40 to 90

2667. The high Fahrenheit temperature range for most HTHW heating plants is
A) 100 to 200 °F.
B) 200 to 300 °F.
C) 300 to 400 °F.
D) 350 to 450 °F.

2668. The second pump in the two pump HTHW system is used to
A) pump water to the distribution system.
B) circulate water through generator.
C) circulate water throughout system.
D) pump water to expansion tank.

2669. Which of the following could cause an explosion in a steam boiler?
A) Steam pressure equaling atmospheric pressure
B) Excessive steam pressure inside steam drum
C) Sudden lowering of water level
D) Collapse of generator tube

2670. After the loss of water inside of the tubes, tube failure in a forced circulation HTHW system can take place in
A) 2 seconds.
B) 30 seconds.
C) 1 minute.
D) 2 minutes.

2671. All piping in an HTHW system should be
A) riveted.
B) soldered.
C) cemented.
D) welded.

2672. In a one-pipe, open tank gravity system, the larger radiators are located at the end of the system in order to
A) equalize heat radiation.
B) raise water temperature.
C) lower water temperature.
D) decrease rate of water circulation.

2673. The open gravity, hot-water system is designed to operate at the maximum boiler temperature of 180 °F. This gives an average radiator temperature of
A) 150 °F.
B) 160 °F.
C) 170 °F.
D) 180 °F.

2674. At what temperature does the HTHW system maintain its water?
A) At 210 °F
B) Below 210 °F
C) At 212 °F
D) Above 212 °F

2675. The heat in the HTHW system is about how many times greater than the heat in the steam system?
    A) 20
    B) 25
    C) 30
    D) 35

2676. The specific fuel type used in firing the boilers of HTHW systems depends on the
    A) type of firing equipment.
    B) location of the heating plant.
    C) size of fuel storage area.
    D) amount of impurities in the water.

2677. A significant advantage of the HTHW system is its
    A) high energy content.
    B) relative safety.
    C) generator tubes.
    D) slow rate of corrosion.

2678. How many inches must the water be above the top row of tubes when a fire-tube boiler is ready to light off?
    A) One
    B) Two
    C) Three
    D) Four

2679. Before starting a new pump or one that has been idle for a long period of time, turn the pump shaft by hand in order to verify the
    A) impeller is locked.
    B) impeller is free.
    C) valves are closed.
    D) air discharge tube is clear.

2680. A high-pressure, steam steel boiler is rated at how many horsepower?
    A) 85 to 90
    B) 90
    C) 95 to 100
    D) 100 or more

2681. In the cascade heater method, water is heated by
    A) direct contact with steam.
    B) inert gas pressure.
    C) a steam boiler.
    D) a hot-water generator.

2682. When pressurizing the HTHW system with the saturated steam cushion, it is necessary to generate an excess amount of heat in order to
A) allow for expansion in drum.
B) increase flow of hot water.
C) provide saturated steam.
D) offset radiant heat loss.

2683. A characteristic of the mechanical gas cushion design for pressurizing the HTHW system is the
A) extra large steam drum.
B) expansion tank being part of generator.
C) expansion tank being independent of generator.
D) frequent flashing of steam.

2684. What is the minimum number of gallons of sodium sulfite used to treat five million gallons of water?
A) 50
B) 100
C) 130
D) 150

2685. A low-pressure steam line is at 5 psi and 230 °F. A break in this line would cause the steam to discharge into the atmosphere at a velocity of how many feet per second?
A) 1,000
B) 1,200
C) 1,500
D) 1,600

2686. A high-pressure steam line is at 125 psi and 300 °F. A break in this line would cause the steam to discharge at a velocity of how many feet per second?
A) 1,200
B) 1,400
C) 1,600
D) 1,800

2687. A break in a low-pressure water line at 15 psi and 200 °F would cause discharge of the water at a velocity of how many feet per second?
A) 175
B) 500
C) 1,000
D) 1,750

2688. Round, cast-iron boilers are built in sizes that can supply a maximum of how many square feet of radiation?
A) 1,200
B) 1,500
C) 1,700
D) 1,850

2689. When water is at a temperature of 300 °F, what is its approximate density in lb/ft^3?
 A) 49
 B) 52
 C) 55.5
 D) 57.5

2690. Water at 100 °F is increased in temperature to 300 °F. What is the approximate increase in water volume?
 A) 3%
 B) 4%
 C) 5%
 D) 7%

2691. In a HTHW system, internal corrosion is practically eliminated by
 A) high pH alkaline water.
 B) low pH alkaline water.
 C) pressure-reducing valves.
 D) rapid absorption of additional oxygen.

2692. Water is how many times as heavy as steam?
 A) 3
 B) 5
 C) 7
 D) 10

2693. The water used in the HTHW heating system is drawn from the lower part of the expansion tank. It is mixed with the system's return water, and circulated throughout the system. This mixing is necessary in order to
 A) facilitate cavitation.
 B) reduce gas pressure.
 C) prevent cavitation.
 D) reduce corrosion.

2694. In starting up the HTHW system, the boiler is fired at what percent of its rated capacity?
 A) 10
 B) 15
 C) 20
 D) 25

2695. Boilers that must be provided with at least two ways of feeding water are
 A) those that have 300 ft^2 of heating surface.
 B) those that have 500 ft^2 of heating surface.
 C) those operating below 15 psig .
 D) low-pressure units.

2696. A duplex steam reciprocating pump having a normal speed of 40 strokes per minute suddenly increases to 80 strokes per minute. The most probable cause is
   A) loss of water pressure.
   B) feed water is too hot.
   C) steam pressure doubled.
   D) feed water is too cold.

2697. Leaks around the packing glands of a steam pump cause
   A) pump to increase its speed.
   B) loss of condensate.
   C) pump to stop.
   D) increase of condensate.

2698. The regulating valve of a thermohydraulic feed-water regulator is opened by
   A) steam pressure in closed system.
   B) electrical contacts.
   C) float-operated switch.
   D) expansion of thermostat.

2699. A turbine type centrifugal pump is suitable as a feed-water pump because of its
   A) high-discharge pressure.
   B) low-discharge pressure.
   C) high-temperature balance.
   D) high-suction force.

2700. The stop valves on the suction and discharge sides of a boiler with an electric motor permit repair or replacement of the pump without interrupting operation of the
   A) boiler.
   B) motor.
   C) impeller.
   D) condensate receiver.

2701. A check valve is installed between the boiler and the pump discharge in order to prevent hot water or steam from traveling from the
   A) pump to packing glands.
   B) strainer to pump.
   C) boiler to pump.
   D) pump to injector.

2702. An injector is correctly installed on a boiler with enough water and steam pressure available, but the injector will not feed water. The most probable cause is
   A) injector is not primed.
   B) mineral coatings in injector.
   C) injector holes are plugged.
   D) water is very hot.

2703. For the injector to operate, the water should have a maximum temperature of
   A) 100 °F.
   B) 110 °F.
   C) 115 °F.
   D) 120 °F.

2704. Feed water is preheated to
   A) eliminate oxygen from water.
   B) change boiler water chemically.
   C) increase efficiency of boiler operation.
   D) reduce temperature of make-up water.

2705. When using exhaust steam to heat a feed-water temperature from 60 to 160 °F, the efficiency of the boiler plant is increased
   A) 6%.
   B) 10%.
   C) 12%.
   D) 15%.

2706. If steam is generated in a downward-flow economizer-type heater, the cause is
   A) improper circulation through the unit.
   B) the water is too cold.
   C) the tubes are incorrectly pitched.
   D) soot has accumulated in tubes.

2707. To precipitate the hardness salts of calcium and magnesium from the water, use
   A) sodium zeolite.
   B) hydrogen zeolite.
   C) soda ash.
   D) lime soda.

2708. What function does a float assembly perform in a tray-type de-aerator?
   A) Decreases nozzle pressure
   B) Opens spray valves
   C) Increases steam condensation
   D) Controls water level

2709. How many pounds of disodium phosphate are used when boiling out a boiler that holds 10,000 pounds of water?
   A) 3
   B) 15
   C) 30
   D) 45

2710. The chemically-treated water used to cleanse a scaled and corroded boiler is raised to 200 °F. For how many hours is this temperature maintained?
A) 5 to 10
B) 15 to 20
C) 20 to 30
D) 24 to 48

2711. When using the dry method of storing a 3000 gallon boiler, how many pounds of quick lime are used?
A) 2
B) 3
C) 5
D) 6

2712. When using the wet method to store a boiler that contains 5000 pounds of water, how many pounds of sodium sulfite should be used?
A) 5
B) 6.25
C) 25
D) 50

2713. CO_2 is causing corrosion in condensate return lines of a water boiler. Partially neutralize this by treating the condensate with
A) bicarbonate.
B) carbon.
C) amines.
D) quick lime.

2714. The makeup rate of feed water for a central heating plant is equal to what percent of the steam produced?
A) 5 to 10
B) 5 to 15
C) 10 to 15
D) 15 to 20

2715. What is the size of grille or register to pass 800 cfm at 600 ft velocity (assume free area at 80%)?
A) 133 in^2
B) 197 in^2
C) 239 in^2
D) 314 in^2

2716. By increasing a vent stack from 4" to 8", capacity will increase
A) 2-1/2 times.
B) 4 times.
C) 2 times.
D) 3 times.

2717. Which is a desirable velocity for a low-velocity system?
A) 500 fpm
B) 1,000 fpm
C) 1,500 fpm
D) 2,000 fpm

2718. Which is a desirable friction loss per 100 ft of duct?
A) 0.08 fpm
B) 0.12 fpm
C) 0.02 fpm
D) 0.05 fpm

2719. Duct design criteria is not based on
A) friction loss.
B) velocity.
C) Btu capacity.
D) static pressure.

2720. The weight of 26 ga galvanized sheet metal per square foot is
A) 0.906 lb.
B) 0.708 lb.
C) 1.010 lb.
D) 0.560 lb.

2721. In connecting ductwork to a unit with two fan outlets, use a
A) Pittsburgh lock.
B) mixing Y.
C) splitter H.
D) none of the above.

2722. Triangulation is a term used in duct
A) layout.
B) insulation.
C) sizing.

2723. A duct transition is a fitting for
A) changing duct size or shape.
B) the splitter damper.
C) the branch take-off.
D) the fire damper.

2724. A true grooved seam allowance is
A) 1-1/2 times the width of seam.
B) 2 times the width of seam.
C) 3 times the width of seam.
D) 3-1/2 times the width of seam.

2725. If a rectangular duct is 32" x 18", the equivalent round duct would be
A) 20".
B) 26".
C) 30".
D) 22".

2726. In taking off a duct to determine weight, elbows are always measured
 A) on center line.
 B) twice.
 C) on the radius.
 D) on the perimeter.

2727. Galvanized sheet metal is made by applying a protective coating of
 A) lead.
 B) zinc.
 C) tin.
 D) nickel.

2728. To anneal copper,
 A) heat and cool slowly.
 B) heat and quench.
 C) apply electricity.
 D) none of the above.

2729. To cut out the cheeks of a 20 ga elbow, use
 A) a groover.
 B) snips.
 C) duck bills.
 D) dividers.

2730. When setting a seam by hand on a round pipe, use a
 A) duck bill.
 B) rivet set.
 C) setting tool.
 D) punch.

2731. Ductwork is bent up on a
 A) ductulator.
 B) slitter.
 C) lock former.
 D) brake.

2732. Drive cleats can be made on a
 A) seamer.
 B) bar folder.
 C) anvil or rail.
 D) none of the above.

2733. When drawing rivets, use a(n)
 A) rivet set.
 B) Whitney punch.
 C) awl.
 D) divider.

2734. A duct system is considered high velocity when air travels in excess of
 A) 12 miles/hour.
 B) 15,000 cfm.

C) 800 gpm.
D) 2,500 fpm.

2735. How many degrees are there in a semi-circle?
A) 45°
B) 90°
C) 180°
D) 360°

2736. In low-velocity air distribution systems, air flow to the branch take-offs is regulated by a(n)
A) amprobe.
B) volume controller.
C) splitter damper.
D) draft indicator.

2737. One of the three fundamental blower laws states that the power varies at the ______ of speed.
A) rate
B) cube
C) square
D) circumference

2738. The following choices show relationships between the height and width of an elbow. Which one indicates the aspect ratio of the elbow?
A) H + W
B) H x W
C) W ÷ H
D) H ÷ W

2739. For best air flow, the turning vanes in an elbow should be spaced to have an aspect ratio of
A) one.
B) two.
C) four.
D) five.

2740. To vary the volume of air flowing through a duct system, install a
A) damper.
B) diffuser.
C) grille.
D) flashing.

2741. After a balanced air flow is achieved using a deflecting damper, it is
A) locked in position.
B) riveted to duct.
C) changed weekly.
D) removed.

2742. When making a duct from 24 gauge sheet metal, a butterfly damper is required. What gauge sheet metal is used for the blade of the damper?
A) 26
B) 24
C) 22
D) 20

2743. On some working drawings, louver dampers may be called
A) ceiling diffusers.
B) turning vanes.
C) deflecting dampers.
D) multiblade dampers.

2744. Which are usually purchased as factory-made assemblies?
A) Splitters
B) Duct elbows
C) Turning vanes
D) Louver dampers

2745. Which is used as a fire damper in a duct that passes through a fire wall?
A) Diffuser
B) Control quadrant
C) Automatic louver damper
D) Register

2746. Which is installed at the supply outlets of ducts to control volume and direction of the airflow?
A) Registers
B) Elbows
C) Splitters
D) Grilles

2747. If a register is located close to a duct elbow, what should be installed to ensure even distribution of air flow?
A) Butterfly damper
B) Turning vanes
C) Automatic louver
D) Ceiling diffuser

2748. When installing a register in a duct, what is the approximate clearance given the blade frames?
A) 1/8"
B) 1/4"
C) 1/2"
D) 3/4"

2749. When installing a grille in a door, how much clearance is given the cutout hole for the grille?
A) 3/4"
B) 1/2"

C) 1/4"
D) 1/8"

2750. Use the cross brake to brace a rectangular duct if the duct width is
A) 30".
B) more than 30".
C) less than 24".
D) 24".

2751. When using galvanized sheet iron to make a circular duct with a diameter of 23 inches, the galvanized sheet iron should be what gauge?
A) 20
B) 22
C) 24
D) 26

2752. After completing cutting and notching patterns, the next step in fabricating sheet metal components is to form the patterns into shape. Which of the following is the correct forming sequence?
A) Cross braking, seam allowance bending, corner bends, joint connection allowance bending
B) Cross braking, seam allowance bending, joint connection allowance bending, corner bends
C) Seam allowance bending, corner bends, joint connection allowance bending, cross braking
D) Cross braking seam allowance bending, corner bends, joint connection allowance bending

2753. Machines that can be used to form and assemble a round duct joint are
A) Pittsburgh, slip roll, crimping, beading.
B) crimping, slip roll, beading.
C) beading, Pittsburgh.
D) Pittsburgh, crimping, slip roll.

2754. Hangers are used to support and level heating and air-conditioning ducts. Which is used to make holes in the hangers for connecting the ducts?
A) Machine screw
B) Electric drill
C) Whitney punch
D) Screw nail

2755. The final steps of a duct installation job are to install the
A) insulation, volume dampers, turning vanes.
B) louvers, transitions, seams.
C) drop joints, hangers, plenum.
D) registers, grilles, diffusers.

2756. When installing a diffuser, attach it to the
   A) grille.
   B) hanger.
   C) damper assembly.
   D) return air register.

2757. When patching a hole in a duct joint, how many inches apart are the holes spaced when attaching the patch with rivets?
   A) 1/2
   B) 1
   C) 1.5
   D) 2

2758. The stack or ventilator opening should be above the highest part of the building roof to take advantage of the
   A) wind from any direction.
   B) low pressure area.
   C) high pressure area.
   D) area of least suction.

2759. The hot-water boiler that is somewhat portable and has a self-contained firebox is
   A) self-contained.
   B) skid-mounted.
   C) circulating.
   D) package.

2760. What type of boiler is constructed in several sections?
   A) Cast iron
   B) Steel
   C) Wrought iron
   D) Brick

2761. The number of intermediate sections in a square cast iron boiler is determined by the size of the
   A) push nipples.
   B) header.
   C) boiler.
   D) firebox.

2762. One of the two sections of a steel, hot-water boiler consists of the base and either the grates or burner. It is constructed according to the
   A) number of water jackets.
   B) type of fuel used.
   C) size of smoke passages.
   D) location of combustion chamber.

2763. The pressure relief valve on a hot-water boiler may corrode or stick and should be forced to operate once each
   A) hour.
   B) day.

C) week.
D) month.

2764. In the event of an induced or forced draft failure in a boiler, what would shut down the firing equipment?
A) Air flow switch
B) Pressure gauge
C) Pressure relief gauge
D) Water level valve

2765. The baffles in a hot-water boiler
A) reduce water evaporation.
B) hold the hot gases.
C) mix air and fuel.
D) clean boiler tubes.

2766. If a steel boiler develops a very large hole in the boiler flue,
A) weld the hole.
B) cover the hole with asbestos paper.
C) replace the flue.
D) replace a boiler section.

2767. The expansion tanks in gravity and forced circulation systems
A) allow water in distribution system to expand.
B) hold the water extracted from the steam lines.
C) hold the steam extracted from the water lines.
D) force water in distribution system to condense.

2768. How many gallons do domestic hot-water heaters hold?
A) 5 to 10
B) 5 to 20
C) 10 to 25
D) 20 to 50

2769. The HTHW system uses very little make-up water, because it
A) is a closed system.
B) requires frequent blow downs.
C) has considerable leakage.
D) operates at low thermal level.

2770. A HTHW system has
A) constant water density
B) no pipe radiation
C) simple construction.
D) no pump leakage.

2771. In pressurizing the HTHW system, an expansion tank is required because
A) water expands when heated.
B) it reduces saturation temperatures.
C) it prevents vaporization when temperature falls.
D) it keeps water below 212 °F.

2772. How many valves are in the water circulating pump of a forced hot water heating system?
A) None
B) One
C) Two
D) Three

2773. How often should the flow control valve in a forced hot water circulating system be checked?
A) Daily
B) Weekly
C) Monthly
D) Bimonthly

2774. Hot-water heaters are glass lined to
A) strengthen tank wall.
B) prevent heat loss.
C) resist corrosion.
D) maintain water temperature.

2775. Make-up water is needed for an operating boiler because of
A) incoming-water pressure.
B) change in boiler temperature.
C) too much condensate.
D) insufficient condensate.

2776. An automatic water-level control is used in connection with an electric-driven feed-water pump. When the float rises sufficiently, the pump
A) starts.
B) doubles its speed.
C) runs slowly.
D) shuts off.

2777. The standard automatic feed-water control is a(n)
A) injector.
B) valve.
C) float.
D) trap.

2778. What returns the condensate to the boiler by means of gravity?
A) Automatic feed
B) Centrifugal pump
C) Hartford loop
D) Manual feeder

2779. The symbol CO_2 stands for
A) one part calcium and two parts oxygen.
B) one part coal and two parts oil.
C) carbon dioxide.
D) carbonic oxygen.

2780. For an ordinary installation, space pipe hangers or pipe supports at intervals of
   A) 5 ft.
   B) 20 ft.
   C) 10 ft.
   D) 30 ft.

2781. A method of joining metal using fusible alloys having a melting point under 700 °F is
   A) acetylene welding.
   B) brazing.
   C) soldering.
   D) arc welding.

2782. A straight line drawn from the center to the extreme edge of a circle is known as
   A) circumference of a circle.
   B) area of a circle.
   C) diameter of a circle.
   D) radius of a circle.

2783. What will the readings of wet and dry hygrometers be if exposed to air that is completely saturated?
   A) They will both read alike
   B) Wet bulb will be lower
   C) Dry bulb will be lower
   D) 90% relative humidity

2784. What type of electric motor has a set of field coils and a rotating armature?
   A) Induction motor
   B) Capacitor split-phase motor
   C) Repulsion-start induction-run motor
   D) Dual-winding, repulsion-induction motor

2785. An anemometer measures
   A) air speed.
   B) relative humidity.
   C) water temperature.
   D) air pressure.

2786. The term *induced draft* could refer to a type of
   A) control.
   B) compressor.
   C) cooling tower.
   D) diagram.

2787. The purpose of a Hartford loop is to
   A) prevent water from backing out of a boiler.
   B) remove air from the return line.
   C) allow for pipe expansion.
   D) provide a balance for the steam header.

2788. On a steam system, the operation of an inverted bucket trap is based on the
- A) rise and fall of a float.
- B) combination of steam pressure and weight of the condensate.
- C) steam pressure drop.
- D) weight of the water in the trap.

2789. Where oil fuel tanks are lower than the burner on fuel-burning equipment, it is recommended that they have a
- A) plastic pipe system.
- B) 3-pipe system.
- C) 2-pipe system.
- D) 1-pipe system.

2790. A pop safety valve on a boiler is usually provided with a hand lever, for
- A) closing the valve once it pops open.
- B) relieving air.
- C) manual safety.
- D) testing the valve.

2791. Steam mains and returns should be pitched not less than one inch per ______ feet in the direction of the steam flow.
- A) 10
- B) 20
- C) 30
- D) 40

2792. Each steam boiler shall have at least how many water glasses?
- A) None
- B) One
- C) Two
- D) Three

2793. The steam pressure gauge must be connected to the
- A) steam space of the boiler.
- B) main steam line.
- C) water space of the boiler.

2794. A conventional, float-operated, low water, fuel cut-off will turn off a(n) ______ but will not mechanically feed water to a boiler.
- A) thermostatic fuel burner
- B) automatic fuel burner
- C) automatic water feeder
- D) electric water feeder

2795. A steam trap is an automatic valve used in steam systems to permit passage of
- A) condensate.
- B) air.
- C) condensate, air and noncondensable gases.
- D) steam and air.

2796. In a hot water heating system using a closed expansion tank, vents are required at
   A) expansion tank only.
   B) discharge side of pump.
   C) last coil in system.
   D) all high points.

2797. In an automatically-fired steam boiler, the lowest safe water line, with reference to the water glass, is
   A) the top of water glass.
   B) midway in water glass.
   C) no lower than lowest visible part of the water glass.
   D) 3/4 up from bottom of water glass.

2798. A warm air furnace is sized for a space requiring 112,000 Btuh. The temperature rise of supply air over return air is 100 °F. How many cfm are required?
   A) 1,020
   B) 870
   C) 1,100
   D) 1,210

2799. A noisy heating system may be caused by
   A) piping pockets.
   B) boiling water in an open system.
   C) floor holes too small for risers.
   D) all of the above.

2800. Correct safety control requires an ASME relief valve on all
   A) high-pressure steam systems.
   B) low-pressure steam systems.
   C) ASME rated systems.
   D) hot-water and steam-heating boilers.

2801. An ASME relief valve should be connected
   A) to the top of the boiler.
   B) to the supply main at the boiler.
   C) ahead of the fill-valve.
   D) on the return main downstream of the pump.

2802. What type of metal is not used on a safety valve seat?
   A) Brass-copper and zinc
   B) Carbon steel
   C) Monel-nickel and copper

2803. A pipe conducting condensation from the supply side to the return side of a steam heating system is a
   A) drip.
   B) riser.
   C) runout.
   D) Hartford connection.

2804. An HRT boiler and a Wickes boiler are
   A) both fired internally.
   B) both fired externally.
   C) not fired the same way.

2805. A siphon on a steam gauge is used to
   A) reduce fluctuation of the gauge.
   B) reduce excessive pressure.
   C) eliminate false readings caused by high temperature.

2806. A 300 psi spring on a safety valve can be set to operate at
   A) 290 to 310 psi.
   B) 285 to 315 psi.
   C) 270 to 330 psi.

2807. The oil burner requiring the highest oil pressure is the
   A) steam atomizing.
   B) air atomizing.
   C) mechanical atomizing.
   D) rotary cup.

2808. Carbonate hardness is
   A) permanent.
   B) temporary.
   C) a chloride.

2809. If the belt driving a throttling flyball governor is slipping, the
   A) engine speeds up.
   B) cut-off will be later.
   C) engine slows down.

2810. A locomotive boiler
   A) is not pitched.
   B) pitches to the rear.
   C) pitches to the front.

2811. An HRT boiler expands
   A) to the front.
   B) to the rear.
   C) downward.

2812. Scale on the inside of waterwall tubes causes
   A) an interruption in water circulation.
   B) high tube metal temperatures.
   C) an increase in steam generation in the tubes.

2813. Steam purification is accomplished by
   A) line separators.
   B) filters.
   C) drum internals.

2814. On a water strainer, the area of holes in the strainer is
A) the same as the pipe area.
B) greater than the pipe area.
C) less than the pipe area.

2815. Steam in a boiler is washed with incoming feed water to reduce the
A) moisture in the steam.
B) solids in the water.
C) solids in the steam.

2816. Turbines used for boiler feed pumps are usually
A) single-stage impulse.
B) single-stage reaction.
C) multistage reaction.

2817. The difference between ac and dc is
A) dc changes direction.
B) ac doesn't change direction.
C) ac changes direction and magnitude.

2818. Valves between water column and boiler
A) are not permitted.
B) have non-rising stems.
C) are of CS & Y construction.

2819. A pump pumps against
A) pressure.
B) a head.
C) velocity.

2820. The discharge temperature of water from a blow-down tank should not exceed
A) 140 °F.
B) 180 °F.
C) 210 °F.

2821. On large HRT boilers, the heads are stayed with
A) through stays.
B) diagonal stays.
C) radial stays.

2822. Engine knocks due to bearing wear can be quieted down by using
A) fuel oil.
B) light engine oil.
C) heavy engine oil.

2823. Before soot blowing on an automatic combustion control system,
A) have a high water level.
B) speed up the fuel supply.
C) place control system on manual or hand operation.

2824. When loss of fire occurs on an automatic combustion control system, the dampers
A) open.
B) close.
C) stay the same.

2825. The pump that is started with its discharge closed is the
A) rotary.
B) centrifugal.
C) reciprocating.

2826. A firebox on a vertical fire-tube boiler is stayed with
A) radial stays.
B) one Adamson ring.
C) stay bolts.

2827. The temperature of a Copes feed water regulator is
A) higher at the top.
B) lower at the top.
C) constant throughout.

2828. A non-return valve acts as a(n)
A) globe valve.
B) angle valve.
C) check valve.

2829. Hydrostatic lubricators are found on
A) steam inlet sides of engines.
B) steam inlet sides of turbines.
C) inlet sides of centrifugal pumps.

2830. Wet atomizing steam to an oil burner
A) causes sparking.
B) increases amount of steam needed.
C) carbons up the burner tips.

2831. A gauge on a steam boiler should have one or more turns of tubing placed between it and the boiler. This is known as a
A) series loop.
B) pigtail.
C) hartford loop.
D) none of the above.

2832. A boiler rated at 150 hp and an SWP of 200 psi would have a hard time maintaining steady flow at a pressure of
A) 50 psi.
B) 100 psi.
C) 180 psi.
D) 200 psi.

2833. A conventional Sterling boiler has
A) three steam drums with a mud drum.

B) two steam drums with a mud drum.
C) two steam drums, one water drum and one mud drum.

2834. The Detroit Roto-Stoker is
A) overfeed.
B) underfeed.
C) spreader.
D) pulverized fuel.

2835. The Detroit Roto-Stoker needs
A) a windbox.
B) no windbox.
C) a large ash pit.

2836. An 8" x 6" x 7" pump has a water piston with a diameter of
A) 8".
B) 7".
C) 6".

2837. The pressure against which a pump can work is in proportion to the diameter of the
A) piston.
B) piston and length of the stroke.
C) piston, length of the stroke, and pump factor.

2838. A chain grate stoker is best used with
A) coking coal.
B) coal at 3% ash content.
C) coal at 6% to 15% ash content.

2839. In the economic boiler, the short tubes are
A) smaller in diameter than the long tubes.
B) larger in diameter than the long tubes.
C) the same diameter as the long tubes.

2840. In a closed heater, the
A) pressure of the water is higher than the steam.
B) pressure of the steam is higher than the water.
C) water and steam pressures are the same.

2841. An open heater is located on
A) the suction side of the boiler feed pump.
B) the discharge side of the boiler feed pump.
C) either side of the feed pump.

2842. If an area of 1 in^2 is at 10 psi, to find the pressure on 1 ft^2 multiply by
A) 12.
B) 144.
C) 1,728.

2843. An increase in engine speed will cause boiler water level to
A) rise.
B) lower.
C) stay the same.

2844. The steam valve on a simplex pump is thrown by
A) mechanical linkage.
B) steam.
C) pilot valve.

2845. A single Bourdon tube
A) may be spiral wound.
B) must be positioned so it can drain.
C) may not be spiral wound.

2846. A gear pump has
A) no valves.
B) one suction and one discharge valve.
C) two suction and two discharge valves.

2847. A globe valve used for boiler blow down
A) may be used if of straight-away type.
B) may not be used.
C) none of the above.

2848. Chemically pure water has
A) high resistance.
B) low (reluctance) resistance.
C) none of the above.

2849. Chemically pure water has
A) high conductivity.
B) low conductivity.
C) none of the above.

2850. Low water level on an HRT boiler would first overheat
A) top tubes.
B) end of diagonal stays.
C) dry sheet.

2851. Fuses are connected in
A) series with the load.
B) parallel with the load.
C) series-parallel with the load.

2852. A centrifugal pump has
A) one moving part.
B) two moving parts.
C) three moving parts.

2853. 23 °F is equivalent to
A) -5 °C.
B) 15 °C.
C) 0 °C.

2854. A device used to measure high temperatures is called a
A) pyrometer.
B) manometer.
C) none of the above.

2855. When joining threaded pipe, put the compound on
A) the female thread.
B) the male thread.
C) both male and female threads.

2856. Grease is composed of
A) vegetable oil and soap.
B) mineral oil and soap.
C) mineral oil and lye.

2857. The majority of water tube ends in a Sterling boiler are
A) beaded.
B) flared.
C) welded.

2858. The load on an engine is the
A) brake horsepower.
B) indicated horsepower.
C) actual horsepower.

2859. 1,000 watts is equal to
A) 34.5 Btu/min.
B) 3,413 Btuh
C) 777.5 Btuh

2860. The inverted bucket trap has a
A) hole in the bucket to vent out air.
B) ball float.
C) none of the above.

2861. More sulfur in fuel oil makes
A) more heat.
B) less heat.
C) more corrosion.

2862. A simplex pump is usually used as a(n)
A) vacuum pump or to pump air.
B) oil pump.
C) water pump.

2863. A pressure gauge may have
A) a Bourdon tube.
B) a diaphragm.
C) either A or B.

2864. Pressure gauges may be installed
A) in vertical position.
B) in horizontal position.
C) tilted forward for better visibility.

2865. Grease fittings on modern machinery are
A) Alemite fittings.
B) Zerk fittings.
C) none of the above.

2866. The bursting strength of steel is the
   A) elasticity.
   B) ultimate strength.
   C) yield point.

2867. Which burner requires the lowest oil pressure?
   A) Steam atomized
   B) Air atomized
   C) Rotary cup

2868. The amount of CO_2 in an efficiently operating oil burner boiler stack discharge is
   A) 5 to 8 percent.
   B) 10 to 14 percent.
   C) 12 to 20 percent.

2869. A fire in an operating pulverizer may best be controlled by
   A) spraying water into mill.
   B) maintaining high fuel to air ratio.
   C) shutting off coal feed.

2870. The type of coal that should be pulverized the finest is
   A) anthracite.
   B) subbituminous.
   C) bituminous.

2871. A generator with 4 poles is turning
   A) 1,200 rpm.
   B) 1,800 rpm.
   C) 2,400 rpm.

2872. A generator with 6 poles turning at 1,800 rpm produces current with
   A) 60 cycles frequency.
   B) 90 cycles frequency.
   C) 120 cycles frequency.

2873. The thickness of a manhole gasket is
   A) 1/4".
   B) 1/2".
   C) 3/4".

2874. Which is the most volatile matter?
   A) Coke
   B) Bituminous coal
   C) Anthracite coal

2875. To prevent water hammer,
   A) use superheated steam.
   B) make sure the right person has the hammer.
   C) drain the lines.

2876. Coal is of what origin?
   A) Mineral

B) Vegetable
C) Animal

2877. For complete combustion, excess air should be kept at
A) a maximum.
B) a minimum.
C) zero.

2878. Power is measured by
A) Btu.
B) horsepower.
C) foot-pound.

2879. Spontaneous combustion in stored coal is caused by coal high in
A) moisture and sulfur.
B) hydrocarbons.
C) fixed carbons.

2880. The water at the bottom of an HRT boiler water glass should be
A) 3 inches above the top row of tubes.
B) 4 inches above the top row of tubes.
C) 6 inches above the top row of tubes.

2881. The frequency used most often is
A) 0 cycles.
B) 50 cycles.
C) 60 cycles.

2882. Soda ash is used to
A) remove scale from drum.
B) alkalize water.
C) remove temporary hardness.

2883. When temperature and pressure increase, latent heat
A) increases.
B) decreases.
C) stays the same.

2884. An evaporator is used to
A) distill make-up water.
B) take impurities out of oil.
C) leave air out of open heater.

2885. Boiler feed water can be
A) above 212 °F.
B) below 212 °F.
C) at 212 °F.

2886. Which is not a positive pump?
A) Reciprocating
B) Rotary
C) Centrifugal

2887. Pump pressure in a globe valve
- A) is above seat.
- B) is below seat.
- C) has no pressure.

2888. Oil burners serve to
- A) vaporize oil.
- B) atomize oil.
- C) burn the oil.

2889. Which of these coals would be most difficult to keep from smoking?
- A) Coking
- B) Anthracite
- C) Bituminous

2890. Gun-type burners (such as Troy, Peabody, etc.) should be
- A) left in burner.
- B) removed, cleaned and hung on torch rack.
- C) cleaned and left in burner.

2891. Transformers are used to
- A) increase or decrease voltage.
- B) lessen transmission losses.
- C) change ac to dc.

2892. In starting a large motor, use a
- A) rheostat.
- B) transformer.
- C) exciter.

2893. In cleaning electrical parts, use
- A) gasoline.
- B) emery cloth.
- C) sandpaper.

2894. Which of these motors uses an exciter?
- A) Synchronous
- B) Induction
- C) Squirrel cage

2895. A synchronous motor uses
- A) ac.
- B) dc.
- C) ac and dc.

2896. Frequency most often used in ac is
- A) 0 cycles per second (cps).
- B) 60 cps.
- C) 50 cps.

2897. The entrance to a blowdown tank points
- A) downward.
- B) upward.
- C) straight in.

2898. All fire-tube HRTs are
    A) externally fired.
    B) internally and externally fired.
    C) internally fired.

2899. Locomotive boilers are fired
    A) externally.
    B) internally.
    C) both externally and internally.

2900. Blowdown tanks are vented to
    A) atmosphere.
    B) sewer.
    C) nowhere, as they are not vented.

2901. Finned tubes on waterwalls are to
    A) protect refractory.
    B) extract more radiant heat.
    C) put maintenance cost on furnace settings.

2902. On the water end, there are at least
    A) 8 valves on a duplex pump.
    B) 10 valves on a duplex pump.
    C) 12 valves on a duplex pump.

2903. Gas piping installed to operate at or above 7 inches of water column pressure shall be tested at
    A) 1-1/2 times the proposed column pressure.
    B) 2 times the proposed column pressure.
    C) 2-1/2 times the proposed column pressure.
    D) 3 times the proposed column pressure.

2904. The chart showing the properties of air is
    A) psychrometric chart.
    B) pyrometer chart.
    C) psychrometer chart.

2905. The heat added to water at 32 °F to bring temperature to 212 °F is
    A) sensible heat.
    B) latent heat.
    C) specific heat.

2906. The heat added to water at 212 °F to make steam at 212 °F is
    A) sensible heat.
    B) latent heat.
    C) specific heat.

2907. The minimum pipe size on column to water glass is
    A) 1".
    B) 3/4".
    C) 1-1/2".

2908. The minimum drain size on water column is
- A) 1".
- B) 3/4".
- C) 1/2".

2909. Most large factories use
- A) one-phase motors.
- B) two-phase motors.
- C) three-phase motors.

2910. Hot water heating systems have a
- A) relief valve.
- B) spring-loaded pressure safety valve.
- C) spring-loaded relief valve.

2911. If an electric motor fails to start, first look at the
- A) fuses.
- B) rheostat.
- C) switch.

2912. A pH of 7 in boiler water indicates
- A) alkalinity.
- B) acidity.
- C) a condition of neutral.

2913. Fusible plugs on standard Sterling boilers are found in the
- A) front drum.
- B) middle drum.
- C) mud drum.

2914. Water softeners
- A) remove impurities.
- B) make water caustic.
- C) make water acid.

2915. Which is the hardest scale?
- A) Calcium
- B) Magnesium
- C) Silica

2916. Engine bearings are composed of
- A) brass.
- B) bronze.
- C) babbitt.

2917. A third class stationary engine license covers
- A) 5,000 ft^2 heat surface and 20 hp.
- B) 7,500 ft^2 heat surface and 50 hp.
- C) 7,500 ft^2 heat surface and 100 hp.

2918. A pH of 6 is
- A) alkaline.
- B) acid.
- C) neutral.

2919. A pH of 8 is
    A) alkaline.
    B) acid.
    C) neutral.

2920. Fusible plugs melt at
    A) 450 °F.
    B) 350 °F.
    C) 600 °F.

2921. Temperature in a furnace is
    A) 1,600 °F.
    B) 2,500 °F.
    C) 3,600 °F.

2922. Valve length affects
    A) lead.
    B) cutoff.
    C) angle of advance.

2923. A compound engine always has two
    A) cylinders.
    B) or more cylinders.
    C) stages.

2924. When starting an engine, the piston is at
    A) head end, dead center.
    B) crank end, dead center.
    C) mid-position.

2925. The energy used in generating electricity is
    A) mechanical.
    B) chemical.
    C) friction.

2926. When warming up an oil-fired boiler, start the
    A) top burner.
    B) side burner.
    C) lower burner.

2927. Fill a fuel oil tank to
    A) near capacity.
    B) near capacity in summer, capacity in winter.
    C) capacity.

2928. Oil in an air compressor should be
    A) high flash point.
    B) low flash point.
    C) SAE 60.

2929. On an impulse-driven turbine, the fixed blades and steam jets are connected to the
A) shaft.
B) rotor.
C) casing.

2930. A good electrical conductor is
A) water.
B) leather.
C) rubber.

2931. One cubic foot of water from the Detroit River weighs
A) 62.4 lb.
B) 64.8 lb.
C) 57 lb.

2932. When a motor runs faster, friction hp is
A) the same.
B) more.
C) less.

2933. On dc, watts are equal to
A) volts x amps.
B) volts x ohms.
C) volts + amps.

2934. Turbine efficiency is
A) 20%.
B) 25%.
C) 30%.

2935. As pressure rises, the boiling point
A) decreases.
B) increases.
C) stays the same.

2936. Tubes are beaded in fire-tube boilers
A) to support tube sheet.
B) so tube ends will not burn off.
C) for better gas circulation.

2937. Tubes in a water-tube boiler are
A) mostly beaded.
B) mostly flared.
C) flared and beaded.

2938. On fire-tube boilers, tubes are
A) flared.
B) flared and beaded.
C) beaded.

2939. The greatest losses in a boiler are from
A) dry gases up stack.

B) leaks in setting.
C) unburned coal in ashes.

2940. A standard Sterling boiler has
A) 2 flue passages.
B) 3 flue passages.
C) 4 flue passages.

2941. After renewing water glass, open
A) steam valve first.
B) water valve first.
C) both water and steam valves at the same time.

2942. Coke breeze is burned by a(n)
A) underfeed stoker.
B) chain grate (grate bar type).
C) spreader stoker.

2943. Electric fuse plugs are marked in
A) volts.
B) amperes.
C) watts.

2944. What causes the greatest heat loss?
A) Blowing down boiler
B) Carbon dioxide
C) Carbon monoxide

2945. A lever on a safety valve is used to
A) raise valve off seat.
B) release boiler pressure.
C) do both A and B at the same time.

2946. Simplex and duplex pumps are
A) horizontal only.
B) vertical only (grate bar type)
C) horizontal or vertical depending on usage.

2947. What will cause a crack in a furnace wall?
A) Starting a fire too fast
B) Starting with a flash
C) Poor combustion and draft

2948. Steam vapor is
A) visible.
B) invisible.
C) liquid.

2949. Given the same conditions, which pump is most effective?
A) Reciprocating duplex
B) Rotary
C) Centrifugal

2950. Which pump uses the most steam?
A) Reciprocating
B) Simplex
C) Compound

2951. Litmus paper is used to test for
A) acid.
B) alkaline.
C) both A and B

2952. The pump used for pumping heavy oil is a
A) rotary.
B) centrifugal.
C) reciprocating.

2953. Compared to anthracite coal, fuel oil has
A) more Btu.
B) less Btu.
C) the same Btu.

2954. Which valve restricts flow the least?
A) Globe
B) Gate
C) Petcock

2955. Dry pipe in a boiler
A) is open at both ends.
B) has a small hole on top.
C) has a small hole in bottom.

2956. Soot blowers blow
A) between tubes.
B) on tubes.
C) toward stack.

2957. Current load is measured by a(n)
A) ammeter.
B) wattmeter.
C) voltmeter.

2958. When lighting off a boiler with an internal superheater,
A) open the drain.
B) bypass the superheater.
C) shut off the superheater.

2959. Which boiler has a blowdown tank?
A) Portable boiler
B) Low pressure boiler
C) Power boiler

2960. A return trap feeds
A) condensate only.
B) make-up water and condensate.
C) power boiler.

2961. Water can evaporate at
    A) 212 °F only.
    B) above 212 °F only.
    C) any temperature.

2062. Which valve has the least resistance?
    A) Gate
    B) Globe
    C) Right angle check

2963. Which valve is easiest to tell when open?
    A) Gate
    B) Globe
    C) CS & Y

2964. Purge a boiler of gas when
    A) starting.
    B) shutting down.
    C) operating.

2965. Which contains the most Btu per volume?
    A) Gas
    B) Oil
    C) Coal

2966. Acid-forming substances are removed from feed water by
    A) chemicals.
    B) heat.
    C) filtration.

2967. A manometer reads 2.8" of water. This corresponds to
    A) 14.6 psi.
    B) 14.8 psi.
    C) 14.7 psi.
    D) 11.9 psi.

2968. When washing out a boiler, it must be
    A) washed while the boiler is hot.
    B) allowed to set and dry out.
    C) washed as soon as possible after emptying.

2969. Which is the proper procedure when blowing-down a boiler?
    A) Intermittently
    B) Continuously
    C) Spasmodically

2970. A compound gauge
    A) measures pressure and vacuum.
    B) has two Bourdon tubes.
    C) shows absolute and gauge pressure.

2971. Fusible plugs are changed
A) every year.
B) every 5 years.
C) every 10 years.

2972. Superheater fins are made of
A) steel.
B) cast iron.
C) brass.
D) aluminum.

2973. An elliptical manhole plate runs
A) longitudinally.
B) radially.
C) with girth seam.

2974. To tighten up a screwed pipe, use a
A) crescent wrench.
B) stillson wrench.
C) monkey wrench.

2975. Pipe threads
A) have a taper of 70.
B) are always tapered to keep threads tight.
C) are not tapered (as with a machine bolt).

2976. The greatest variety of coal can be burned on a(n)
A) underfeed.
B) spreader.
C) chain grate.

2977. Fittings on a steam pipe are made of
A) steel.
B) cast iron.
C) carborundum.

2978. Flange joints are best suited because they
A) have no gaskets.
B) are easy to maintain.
C) have flexibility.

2979. What causes soft sludge to become scale?
A) High temperature
B) Feed water treatment is discontinued
C) Improper feed water treatment

2980. As a boiler deteriorates, the safety factor
A) increases.
B) decreases.
C) stays the same.

2981. Direct, gas-fired, makeup air heaters may be used as door heaters if the heaters are equipped with approved controls that prevent their operation unless the door is opened sufficiently to provide not less than 1 square foot of open door area per _______ cfm of heater fan delivery capacity.
   A) 200
   B) 300
   C) 500
   D) 700

2982. On a streamline, the strainer is
   A) before trap.
   B) behind trap.
   C) in trap.

2983. A non-return trap on a line that is to drain is placed
   A) above line.
   B) below line.
   C) equal with line.

2984. The steam gauge on a boiler operating at 100 psi should register pressures up to
   A) 100 psi.
   B) 150 psi.
   C) 200 psi.

2985. What is the vacuum in an injector?
   A) 14"
   B) 11-1/2"
   C) 9"

2986. A plunger pump is packed
   A) inside.
   B) outside.
   C) both inside and outside.

2987. What causes the expansion trap to work?
   A) Steam heat
   B) Water condensation
   C) Mechanical process

2988. When testing a pressure gauge,
   A) use a gauge of similar type.
   B) raise pressure until safety is about to pop.
   C) use a dead weight testing applicator.

2989. In cleaning a crank case or lubricator oil tank, it is necessary to use
   A) waste material.
   B) wiping cloth.
   C) steel wool.

2990. When a boiler uses coal, the type of soot blower to use is
A) steam hand lance.
B) regular steam blower.
C) compressed air blower.

2991. When possible, valves on a horizontal pipe
A) have the stem pointing upward.
B) have the stem pointing downward.
C) elbow.

2992. Highly contaminated steam
A) uses an open heater.
B) uses a closed heater.
C) dumps to sewer.

2993. Which contains the most water?
A) Saturated steam
B) Superheated steam
C) Wet steam

2994. Which oil burner is started first?
A) Center
B) Top
C) Bottom
D) Sides

2995. To air-clean electrical equipment, use
A) low moisture.
B) high moisture.
C) oxygen.

2996. Very large HRT boilers have
A) one manhole.
B) two manholes.
C) three manholes.

2997. A high velocity of air in oil firing causes
A) the flame to spark.
B) high combustion.
C) high rating.

2998. An HRT is a
A) one-pass boiler.
B) two-pass boiler.
C) three-pass boiler.

2999. When there is a temporary drop in load with a mechanical oil burner,
A) adjust fuel supply.
B) change burner tips.
C) adjust air.

3000. The greatest loss in an electrical wire is in
  A) amperes.
  B) ohms.
  C) volts.

3001. What would cause a motor to vibrate?
  A) Motor too small
  B) Bearings out of line
  C) Rotor out of balance

3002. Which water causes the most scale?
  A) Inside water treatment
  B) Zeolite
  C) Both A and B are the same

3003. When the upper valve on water glass is plugged up, the water
  A) rises slowly.
  B) rises quickly.
  C) stays the same.

3004. The blowdown tank outlet is
  A) the same size as blowdown line.
  B) 1-1/2 times as large as blowdown line.
  C) twice as large as blowdown line.

3005. Fluid in a system shall not enter a tee fitting through the
  A) front.
  B) back.
  C) top.
  D) side.

3006. What type of fuse has a changeable link?
  A) Plug
  B) Cartridge
  C) Fusetron

3007. Water corrodes most with what kind of pipe?
  A) Copper
  B) Steel
  C) Cast iron

3008. Vent the boiler when draining
  A) at atmospheric pressure.
  B) below atmospheric pressure.
  C) at either A or B.

3009. The container used to store mercury should be made of
  A) chamois.
  B) copper.
  C) glass or earthenware.

3010. To tighten the hex nut on the shoulder of a leaking valve, use a(n)
A) stillson wrench.
B) Allen wrench.
C) monkey wrench.

3011. The fuel bed in a traveling grate is
A) over 6 inches.
B) 2 to 6 inches.
C) 2 inches even.

3012. Water in an open heater is
A) at the boiling point.
B) above the boiling point.
C) lower than the boiling point.

3013. When blowing down a boiler and not in sight of water glass,
A) station another person to watch column.
B) drain boiler.
C) blow down until low water alarm blows.

3014. The easiest method of putting pipe together is
A) screwing it on.
B) bolting it on.
C) welding it together.

3015. The first thing an engineer does when taking over a shift is to check the
A) log book.
B) safety valve.
C) water column.

3016. If an engine is out of line,
A) there is excessive use of steam.
B) it will not handle overload.
C) connecting rods or guides will be wobbly.

3017. What types of stays are in Scotch marine boilers?
A) Head-to-head
B) Diagonal
C) Hollow

3018. Combustion gases passing through a boiler heat through
A) radiation.
B) conduction.
C) convection.

3019. An operating superheater is filled with
A) saturated steam.
B) dry steam.
C) water.

3020. To remove the lime coating inside an injector, use
A) muriatic acid.

B) sulfuric acid
C) a solution of muriatic acid.

3021. Which of these is vented to atmosphere?
A) Open heater
B) Closed heater
C) Both A and B

3022. Oil flash point compared to ignition is
A) 80 °F less.
B) 50 °F less.
C) 20 °F less.

3023. If water glass breaks,
A) shut off steam valve first.
B) shut off water valve first.
C) drain boiler.

3024. As feed water, which contains the most impurities?
A) Water being internally treated
B) Water being externally treated
C) Raw water

3025. A Scotch boiler is a(n)
A) fire-tube.
B) water-tube.
C) inclined tube.

3026. In electricity, resistance is measured in
A) resistors.
B) ohms.
C) volts.

3027. A Scotch marine boiler is
A) horizontal.
B) vertical.
C) locomotive.

3028. In drying a boiler after cleaning, use
A) quick lime.
B) soda ash.
C) solution of lime.

3029. The flash point of a lubrication oil used in a steam cylinder is
A) when it holds a steady flame.
B) above fire point.
C) below fire point.

3030. Which motor is the largest in size?
A) Single-phase
B) Two-phase
C) Three-phase

3031. In electricity, the current that flows in one continuous direction is
A) ac.
B) dc.
C) neither A nor B.

3032. Fire bed thickness with hand firing is
A) over 6 inches.
B) 2 to 6 inches.
C) 2 inches even.

3033. Open heater filter beds are made of
A) coke.
B) stones.
C) gravel.

3034. Silica will carry over with steam at
A) 250 psi.
B) 350 psi.
C) 450 psi.

3035. Steam has difficulty separating from water at
A) 1,000 psi.
B) 3,200 psi.
C) 4,500 psi.

3036. The best coal to use in a traveling grate stoker has
A) the lowest ash content.
B) 3% ash content.
C) 7% ash content or more.

3037. Water put into a steaming boiler through an economizer should not be less than
A) 70 °F.
B) 120 °F.
C) 212 °F.

3038. The best torque in starting a motor is
A) compound.
B) series.
C) shunt.

3039. A diffuser pump is
A) usually a centrifugal pump.
B) always a centrifugal pump.
C) a rotary pump.

3040. Water from a sodium zeolite system will
A) increase dissolved solids.
B) decrease dissolved solids.
C) have the same amount of dissolved solids, as it is only on ion exchanger.

3041. An interpole is used
    A) on a dc generator.
    B) on an ac generator.
    C) for phasing.

3042. Steam is at its highest temperature at
    A) 210 psi.
    B) 450 psi.
    C) 3206 psi.

3043. A motor with constant speed as load is applied is a
    A) shunt.
    B) series.
    C) compound.

3044. In a synchronous motor, dc strength affects the
    A) speed of motor.
    B) rotation of shaft.
    C) power factor.

3045. A condenser without injection water is a
    A) barometric.
    B) jet.
    C) surface.

3046. A centrifugal pump should
    A) take its water 3 feet or more above heater.
    B) take its water from the surface of the heater.
    C) have vapor free water.

3047. Boilers and pressure vessels must be fabricated to meet
    A) DOT standards.
    B) ASTM standards.
    C) ASME standards.
    D) none of the above.

3048. On a slide-valve engine with indirect valve and indirect valve gear, the eccentric
    A) leads crank 90° plus angle of advance.
    B) is behind crank 90° plus angle of advance.
    C) is behind crank 90° minus angle of advance.

3049. Convection is the principal mode of heat transfer in
    A) radiation.
    B) fluids.
    C) gases.

3050. Bypasses should be installed on all valves over
    A) 5".
    B) 8".
    C) 12".

3051. In waterwall circulation with headers below the mud drum, the take-off to the water-tube header is usually taken from
A) a water and steam drum.
B) the mud drum.
C) another header attached to the waterwall header.

3052. To get the best turbulence in a pulverized fuel furnace, use
A) tangential firing.
B) horizontal firing.
C) vertical firing.

3053. An 8" pipe is measured by
A) nominal inside diameter.
B) nominal outside diameter.
C) outside diameter minus the thickness.

3054. A major adjustment on a safety valve with two adjusting rings is made by
A) top ring only.
B) bottom ring only.
C) both rings.

3055. On the Ringelmann scale, maximum density is at
A) one.
B) five.
C) ten.
D) four.

3056. To test for CO_2 in boiler feed water, use a(n)
A) Orsat analyzer.
B) CO_2 recorder.
C) pH indicator.

3057. A superheater is kept cool by
A) high steam velocity.
B) steam in contact with tubes.
C) both A and B.

3058. The fine spray mist in an atomizing heater is created by high velocity
A) steam.
B) steam and water.
C) water coming into contact with heating steam.

3059. One gallon of No. 6 fuel oil weighs
A) 7 to 7-1/2 lb.
B) 7-1/2 to 8 lb.
C) 8 to 9 lb.

3060. Mean effective pressure of indicator card is measured from
A) atmospheric pressure line.
B) absolute pressure line.
C) exhaust pressure line.

3061. An altimeter used in heating systems is graduated in
    A) pounds.
    B) inches of mercury.
    C) feet.

3062. On an underfeed stoker, excess air is
    A) 25%.
    B) 35%.
    C) 45%.

3063. A boiler may support what percent of steam piping?
    A) Any amount
    B) Less than half
    C) None

3064. Water can contain the most dissolved solids at
    A) high temperature.
    B) low temperature.
    C) any temperature.

3065. Which must be resurfaced first?
    A) Valve
    B) Valve seat
    C) Either one

3066. In an air heater, as the velocity of flue gases increases, gas will
    A) increase in temperature.
    B) decrease in temperature.
    C) have little effect.

3067. A tandem compound engine has
    A) two pistons side by side.
    B) two pistons and one piston rod.
    C) neither A nor B.

3068. In a furnace that has waterwalls, there is
    A) no danger of incomplete combustion through cooling.
    B) danger of incomplete combustion through cooling.
    C) incomplete combustion if forced.

3069. Fire doors on a water-tube boiler should be
    A) self-locking.
    B) friction contact.
    C) spring closed.

3070. Most scale deposits will be found
    A) on the water inlet.
    B) in the rear bank of tubes.
    C) in the tubes over the fire.

3071. The thrust in a single suction centrifugal pump is to the
    A) suction.
    B) discharge.
    C) radial thrust.

3072. To change the direction of rotation in an induction motor (polyphase), change
   A) armature connections.
   B) the phases.
   C) any two ac power leads.

3073. As steam pressure goes up, the weight of water an injector will handle
   A) increases.
   B) decreases.
   C) remains the same.

3074. If a change is made to pulverized fuel when burning oil, the furnace will need
   A) a larger volume.
   B) less volume.
   C) nothing, as no change is needed.

3075. The advantage of a compound engine over a simple engine is
   A) fewer moving parts.
   B) less steam consumption.
   C) it is smaller.
   D) less condensation.

3076. The best time to blow down a boiler with waterwalls is
   A) when it is off line.
   B) when it is steaming lightly.
   C) during heavy loads.

3077. The centrifugal pump motor has the greatest load when the valve is
   A) throttled.
   B) open.
   C) closed.

3078. Turbines run most efficiently at
   A) high speed.
   B) low speed.
   C) variable speed with the load.

3079. With a 27 gallon expansion tank, how many Btu can be installed?
   A) 25,000
   B) 50,000
   C) 75,000
   D) 100,000

3080. Temperature of oil on a bearing is
   A) 120 °F.
   B) 150 °F.
   C) 180 °F.

3081. The dc motor with the most starting torque is the
    A) series.
    B) compound.
    C) shunt.

3082. A Jones stoker is
    A) side feed.
    B) underfeed.
    C) overfeed.

3083. The condenser that does not need a condenser pump is a
    A) surface.
    B) barometric.
    C) jet.

3084. The vertical hanger spacing for copper tubing that is less than 1-1/4" in diameter is
    A) 5 ft.
    B) 10 ft.
    C) 15 ft.
    D) 20 ft.

3085. At equal horsepower, a fire-tube, compared to a water-tube boiler, has
    A) more water space.
    B) less water space.
    C) same water space.

3086. Increased efficiency with air heaters is greatest with
    A) spreader stoker.
    B) pulverized coal.
    C) underfeed stoker.

3087. The use of primary air is necessary with
    A) pulverized coal.
    B) spreader stoker.
    C) Detroit Roto-Stoker.

3088. In thrust bearings, lubrication is due to
    A) oil bath.
    B) water pressure.
    C) wedge principle.

3089. What is the position of a piston valve engine with inside admission and direct valve gear at release?
    A) Valve and piston are traveling in same direction
    B) Valve and piston are traveling in opposite directions
    C) Inside admission has no relation to piston travel

3090. Cast iron flanges and nozzles on boilers of 100 psi
    A) may be used.
    B) may not be used.
    C) can be used if not exposed to fire.

3091. To reverse a dc motor, reverse
A) armature leads.
B) number of poles.
C) phases.
D) any two outside leads.

3092. A non-releasing Corliss engine has a
A) shaft governor.
B) flyball governor.
C) pendulum governor.

3093. A regenerative air preheater air duct is
A) the same size as gas duct.
B) smaller than gas duct.
C) larger than gas duct.

3094. The minimum size of a manhole plate is
A) 11" x 15".
B) 10" x 15".
C) 11" x 16".

3095. The maximum pressure to which cast iron can be used on a boiler is
A) 160 psi.
B) 250 psi.
C) 450 psi.

3096. The type of coal that is plastic is
A) coking.
B) caking.
C) volatile bituminous.

3097. Coal that produces the most smoke is
A) semi-bituminous.
B) bituminous.
C) subbituminous.

3098. Economizer operation is the most hazardous at a
A) light load.
B) heavy load.
C) banked load.

3099. On a Copes feed water regulator, the steam line from the boiler pitches
A) to the boiler.
B) to the regulator.
C) in no direction, as it is level.

3100. If a series generator became motorized, it would
A) reverse direction.
B) cause the windings to burn up.
C) use more current.

3101. On an underfeed stoker, the limiting factor of preheated air is
   A) rate of firing.
   B) type of metal castings.
   C) fusion temperature of coal.

3102. A dryer or bathroom exhaust shall vent to the
   A) attic.
   B) outdoors.
   C) basement.
   D) none of the above.

3103. Nuts on a safety valve gag should be tightened
   A) with a wrench.
   B) with a stillson wrench.
   C) fingertight.

3104. On a pulverizer, the fuel that needs the most primary air is
   A) anthracite.
   B) bituminous.
   C) lignite.

3105. To maintain level water and prevent priming, use a boiler with large
   A) water space to heating surface.
   B) steam to water space.
   C) drum internals.

3106. What causes corrosion in boiler piping?
   A) Dissolved gases
   B) Dissolved solids
   C) Ammonia

3107. A turbine is protected from excess pressure by a
   A) condenser.
   B) governor.
   C) relief valve.

3108. Grate that burns the most coal per hour per square foot is a
   A) traveling grate.
   B) single retort grate.
   C) hand-fired grate.

3109. Highest temperatures are obtained in heating fuel oil if the heater is located
   A) in tank.
   B) after pump.
   C) before pump.

3110. A manometer is used for draft because it
   A) has no moving parts.
   B) has no parts to foul up.
   C) gives more accurate readings.

3111. A venturi meter measures
  A) quantity of flow.
  B) quality of flow.
  C) steam density.

3112. Which seam is the strongest?
  A) Girth
  B) Longitudinal
  C) Head

3113. What is the easiest way to get oil into a boiler?
  A) Jet condenser
  B) Surface condenser
  C) Closed feed-water heater

3114. An air compressor uses gaskets of
  A) asbestos.
  B) rubber.
  C) air wick.

3115. A reaction turbine uses
  A) stationary blades only.
  B) rotating blades only.
  C) stationary and rotating blades.

3116. The time required for cut-off on a Corliss engine depends upon the
  A) speed of engine.
  B) releasing gear.
  C) positive pressure in dash pots.

3117. Stationary blades in a reaction turbine are mounted on the
  A) rotor.
  B) casing.
  C) shaft.

3118. Expansion of steam piping is figured on
  A) inches/hundred feet.
  B) thousandths of an inch/hundred feet.
  C) feet/hundred feet of pipe.

3119. The number to divide by to find grains per gallon is
  A) 17.1.
  B) 0.058.
  C) 0.58.

3120. Combustion of coal on underfeed stokers takes place in
  A) retort.
  B) top of fuel bed.
  C) extension grates.

3121. An underfeed stoker, fired properly, has
  A) fire in retort.
  B) coal in retort.
  C) ashes in retort.

3122. A steam connection to a water column should pitch
    A) to the column.
    B) to the boiler.
    C) nowhere, as it should be level.

3123. A sight glass on a motor bearing can have pipe from
    A) top of glass only.
    B) top and bottom of glass.
    C) bottom of glass only.

3124. Phosphorous in boiler metal makes the steel
    A) hot short.
    B) cold short.
    C) beneficial.

3125. In a dry bottom furnace, the water screen helps
    A) the slag to break into smaller pieces.
    B) protect refractory.
    C) reduce temperature of the molten slag.

3126. Air leakage through a boiler setting will lower
    A) furnace temperature.
    B) CO_2.
    C) combustion rate.

3127. If eccentric is advanced on a D-slide valve engine, cut-off will
    A) occur earlier.
    B) occur later.
    C) not be affected.

3128. On a turbine with carbon rings, the rings
    A) rotate with the shaft.
    B) do not rotate with the shaft.
    C) rotate slowly with the shaft (like oil rings).

3129. A pump in a fuel-oil system is protected by a
    A) fuel oil strainer.
    B) relief valve.
    C) safety head.

3130. Water circulation in a water-tube boiler, compared to a fire-tube, is
    A) less rapid.
    B) more rapid.
    C) the same.

3131. An induction motor
    A) is always a three-phase motor.
    B) can be a single-phase motor.
    C) cannot be reversed.

3132. The viscosity of a good grade of mineral oil for a centrifugal pump is
    A) 300 SSU at 70 °F.
    B) 500 SSU at 70 °F.
    C) 700 SSU at 70 °F.

3133. A new boiler should be started
- A) slowly (to dry out the brickwork).
- B) fast (to get heat through quickly).
- C) fast (to prevent gas temperatures below the dew-point).

3134. A marking not found on a safety valve is the
- A) ASME symbol.
- B) popping pressure.
- C) capacity.
- D) discharge pipe diameter.

3135. The use of induction motors in a plant will
- A) be an aid to the power factor.
- B) not be an aid to the power factor.
- C) not affect the power factor.

3136. A mercury column reading of 2" is equal to a pressure of approximately
- A) 0.5 psi.
- B) 1 psi.
- C) 2 psi.

3137. A cubic foot of water weighs 62.4 lb at 39 °F. At 212 °F, it weighs
- A) 59.8 lb.
- B) 61.0 lb.
- C) 62.4 lb.

3138. Waterwalls on a bent-tube boiler are fed from the
- A) steam and water drum.
- B) separate headers.
- C) mud drum.

3139. The reactionary force of steam escaping from the huddling chamber
- A) increases blow-back.
- B) decreases blow-back.
- C) increases popping pressure.

3140. A plain babbitt bearing should use an oil of
- A) 300 SSU at 70 °F.
- B) 500 SSU at 70 °F.
- C) 700 SSU at 70 °F.

3141. When the fire is killed on a boiler with low water, the first thing to do is
- A) close the steam stop valve.
- B) speed up the boiler feed pump.
- C) stop the boiler feed pump.
- D) let the pressure drop.

3142. Water-tube boiler drums are constructed
- A) single-course.
- B) two-course.
- C) three-course.

3143. If the combustion rate of fuel is doubled,
   A) combustion space must be doubled.
   B) the rate of heat must be doubled.
   C) the steam temperature is doubled.

3144. The expansion of 100-foot long steam pipe with 2 psig is about
   A) 1/2 inch.
   B) 1.0 inch.
   C) 1-1/2 inches.
   D) 2 inches

3145. The greatest temperature difference is between
   A) gas film and tube.
   B) tube and water and steam film.
   C) water film and tube film.

3146. To detect trap operation,
   A) listen with a stethoscope.
   B) feel the trap.
   C) use a contact thermocouple.

3147. An HRT boiler
   A) is set level.
   B) slopes from front to rear.
   C) slopes from rear to front.

3148. Thermosiphon arch in locomotive boilers
   A) decreases boiler capacity.
   B) decreases water circulation.
   C) increases water circulation.

3149. Lantern rings can seal
   A) inward.
   B) outward.
   C) both inward and outward.

3150. The head that a centrifugal pump can pump against is
   A) greater for water than for oil.
   B) greater for oil than for water.
   C) the same for oil and water.

3151. If the inside wedges on the crank end of a connecting rod are taken up, the piston will
   A) move forward.
   B) move toward crank end.
   C) not change position.

3152. Lantern rings are used for
   A) oil rings.
   B) packing.
   C) leak-off cavities.

3153. To change the voltage on an induction motor (squirrel cage),
- A) change the stator.
- B) change the rotor.
- C) both A and B.

3154. A resistance thermometer operates on what principle?
- A) Ohmmeter
- B) Resistance of coil of nickel and platinum compared to standard
- C) Resistance of two dissimilar wires due to voltage generated causes temperature rise

3155. What advantage has a compound engine over a simple engine?
- A) Fewer moving parts
- B) Less steam consumption
- C) Takes up less space

3156. What motor has the most torque?
- A) Series
- B) Compound
- C) Shunt

3157. How many Btu are necessary to generate one kW hour?
- A) 1,000
- B) 2,000
- C) 3,000

3158. What effect would increasing the phases have on frequency of motor?
- A) Increase
- B) Decrease
- C) No effect

3159. Steam escaping from a safety valve tends to
- A) Increase blow-back
- B) Decrease blow-back
- C) Not change blow-back

3160. Which has the most drop through an air preheater?
- A) Gas
- B) Air
- C) Both gas and air

3161. Which will give the lowest amount of heat by convection?
- A) Natural gas
- B) Oil
- C) Coal

3162. Which will give the lowest heat by radiation?
- A) Natural gas
- B) Oil
- C) Coal

3163. When taking a test with an Orsat analyzer,
- A) CO_2 is the gas analyzed first.

B) O_2 is the first gas found.
C) it does not matter which gas is analyzed first.

3164. When taking a test with an Orsat analyzer,
A) the water in the bottle must be pure distilled water.
B) plain tap water is sufficient.
C) any fluid could be used.

3165. When applied to coal, the terms *caking* and *coking*
A) mean the same thing.
B) refer to different qualities of coal.
C) refer to type of ash left after burning the coal.

3166. Why is smoke-free combustion so important?
A) Smoke indicates that combustion is incomplete and fuel has been wasted
B) Smoke contaminates air and makes neighborhood look bad
C) Smoke violates the smoke abatement code

3167. The Hargrove method is one of two distinct methods for
A) determining the grindability index of coal.
B) determining fusion temperature of coal.
C) analyzing flue gas.

3168. When pulverizing coal,
A) low-volatile coal should be pulverized finer than high-volatile coals.
B) high-volatile coal should be pulverized finer than low-volatile coals.
C) low- and high-volatile coals can be pulverized the same.

3169. In most cases, gaseous fuels are measured and purchased in terms of cubic feet. In some cases they are purchased by *therms*. One therm equals
A) 100,000 Btu.
B) 75,000 Btu.
C) 50,000 Btu.

3170. Manufactured gas (coal gas or coke over gas) has a heating value of
A) 500 to 600 Btu/ft^3.
B) 700 to 900 Btu/ft^3.
C) 900 to 1,100 Btu/ft^3.

3171. Gas burners should never be lighted
A) until furnace is purged at least 5 minutes at an air flow equal to 50% of full load.
B) until furnace is purged at least 2 minutes, being sure back damper is open.
C) at all.

3172. There is a wide range of heating values of various gases due to the
- A) percent of hydrocarbons present.
- B) amount of O_2 (oxygen) present.
- C) difference in the weight of various gases.

3173. When burning gas, the best type of flame is a
- A) short, non-luminous flame.
- B) pulsating flame.
- C) long, pale blue flame that is tipped with yellow.

3174. Modern, large-capacity units are built with
- A) one upper drum to lower cost.
- B) sometimes two drums to provide sufficient steam-relieving capacity.
- C) three drums in order to be able to install tubes at a 90° radius to drums.

3175. The most important element in determining dew point is
- A) absolute humidity of the air.
- B) relative humidity of the air.
- C) sulfur content of the coal.

3176. A mollier chart is used for determining
- A) enthalpy.
- B) entropy.
- C) steam conditions.

3177. When taking an indicator card, the first thing to do is the
- A) atmospheric line is drawn.
- B) card is taken.
- C) indicator should be warmed up.

3178. Engine clearance is the space between
- A) cross head and cylinder when at head end dead center.
- B) piston and cylinder with piston at head end dead center.
- C) piston and cylinder plus steam passages with piston on dead center.

3179. Steam is used extensively to
- A) increase the ratio of work done to heat used.
- B) decrease the amount of back pressure.
- C) decrease the size of cylinder needed for given work output.

3180. The mechanical equivalent of heat by the U.S. Bureau of Standards is
- A) 10 ft^2 heating surface.
- B) the number of Btu/minute or Btu/hour.
- C) the number of foot-lb in a Btu.

3181. How many Btu in one horsepower minute?
- A) 33,000
- B) 2,545
- C) 42.42

3182. Thermometers can indicate temperatures ranging from
  A) 150 to 1,000 °F.
  B) 20 to 3,000 °F.
  C) -10 to 212 °F.

3183. For measuring extremely high temperatures, use a
  A) hygrometer.
  B) pyrometer.
  C) neither A nor B.

3184. At what temperature does water, at maximum density, weigh more per unit of volume?
  A) 32 °F
  B) 63 °F
  C) 39 °F

3185. In boiler operation, furnace heat is transferred by
  A) conduction.
  B) radiation only.
  C) convection and radiation.

3186. Steam engines converted from noncondensing to condensing operation develop
  A) less horsepower.
  B) more horsepower.
  C) same horsepower.

3187. The average life of a low-speed steam engine is
  A) 17 years.
  B) 20 years.
  C) 28 years.

3188. A ring valve is used on
  A) injectors.
  B) simplex pumps.
  C) steam engines.

3189. Tempering of coal means the amount of
  A) moisture it contains as fired.
  B) moisture when found.
  C) carbon.

3190. Correct tempering of coal is
  A) 12 to 15 percent moisture.
  B) less than 3 percent moisture.
  C) neither A nor B.

3191. For equal horsepower, the HRT has
  A) more water space than water tube.
  B) less water space than water tube.
  C) same water space as water tube.

3192. Air is a(n)
 A) element.
 B) compound.
 C) mixture of elements.

3193. Which has the greatest effect on heating calculations with respect to latent heat?
 A) Bathrooms
 B) Bedrooms
 C) Infiltration
 D) Furniture

3194. The Dulong formula has to do with
 A) absolute temperature.
 B) heat values of fuel.
 C) heat balance of a plant.

3195. The highest efficiency that can be economically justified in a boiler is about
 A) 90%.
 B) 80%.
 C) 70%.

3196. For a given set of steam conditions and firing methods, it is cheaper to build
 A) one larger boiler.
 B) two smaller boilers.
 C) boilers of equal size.

3197. A micron is
 A) one millionth of a meter.
 B) one tenth of a meter.
 C) one thousandth of a meter.

3198. In burning fuel in suspension, the fuel is converted into a combustible by the
 A) firing equipment.
 B) heat of the furnace.
 C) pulverizer.

3199. Smoke and soot indicate incomplete combustion, which can be best controlled by
 A) increasing excess air.
 B) time, temperature and turbulence.
 C) reducing fire rate.

3200. The rotary pump is like the centrifugal pump in that they both
 A) are positive displacement pumps.
 B) give a non-pulsating flow.
 C) have and use poppet valves.

3201. The use of preheated feed water can overcome
    A) scale and corrosion.
    B) the excess of expansion and contraction of boiler metals.
    C) heat balance.

3202. Can a condensate line dump to a condensate tank?
    A) No
    B) Yes

3203. In ac motors, speed is determined by
    A) a rheostat.
    B) frequency of power and number of poles.
    C) the number of poles.

3204. Cavitation of pump impellers is caused by
    A) a positive suction head.
    B) a negative suction head at or near vaporizing pressure.
    C) too high a liquid temperature.

3205. In impulse-type turbines, the diaphragm
    A) is part of the rotor.
    B) holds the nozzles and is a stationary part of the turbine.
    C) holds rotating buckets.

3206. What percent of steam does useful work on engines?
    A) 10%
    B) 25%
    C) 16%

3207. Brake horsepower is
    A) more than indicated horsepower.
    B) less than indicated horsepower.
    C) the same as indicated horsepower.

3208. When mercury is used in a U tube draft gauge, meniscus is
    A) convex.
    B) concave.
    C) level.

3209. The bottom of meniscus is read when using
    A) oil.
    B) mercury.
    C) water.

3210. How many moving parts on a continuous flow steam trap?
    A) Three
    B) One
    C) Zero

3211. The top of meniscus is read when using
    A) oil.
    B) mercury.
    C) water.

3212. The safety valve on a closed feed water heater should be set
- A) 15 to 20 lb above boiler pressure.
- B) 15 to 20 lb below boiler pressure.
- C) the same at boiler pressure.

3213. The average amount of water lost to evaporation when using a spray pond is
- A) 3%.
- B) none.
- C) 7%.

3214. In a large industrial plant, an engine used just for power should be more efficient to operate
- A) condensing.
- B) non-condensing.
- C) heating.

3215. A properly staged volute pump is in
- A) radial balance.
- B) axial balance.
- C) radial and axial balance.

3216. For coal used in an average plant, the fusion point of ash is
- A) 2,100 °F.
- B) 3,100 °F.
- C) 4,100 °F.

3217. What is the ratio of heat absorption per square foot in an air preheater?
- A) Two
- B) Three
- C) Four

3218. A modern central-station boiler might contain
- A) 1 mile of tubes.
- B) 35 miles of tubes.
- C) 70 miles of tubes.

3218. Which pump has the most valves?
- A) Rotary
- B) Simplex
- C) Duplex

3219. The average weight of bituminous coal is
- A) 50 lb/ft^3.
- B) 40 lb/ft^3.
- C) 60 lb/ft^3.

3220. The actual capacity of a new air compressor is
- A) more than piston displacement.
- B) less than piston displacement.
- C) the same as piston displacement.

3221. Mineral wool for insulation is made from
    A) binet furnace slag.
    B) vegetable and animal fibers.
    C) cellular glass.

3222. The best insulating material for steam temperature condition is
    A) mineral wool.
    B) asbestos.
    C) plaster of Paris.

3223. Insulating materials made of animal and vegetable fibers should only be used for
    A) low temperatures.
    B) high temperatures.
    C) moderate temperatures.

3224. From an economic standpoint, the
    A) thicker the insulation, the greater the savings.
    B) thickness of insulation has no bearing on amount of heat saved.
    C) weight of the secondary surface would be most important

3225. The loss per lineal foot per hour of bare steam pipe is
    A) more with superheated steam.
    B) more with saturated steam.
    C) the same for superheated or saturated steam.

3226. Manganese in boiler metal
    A) makes metal cold short.
    B) makes metal hot short.
    C) is considered desirable.

3227. Steam that does not contain H_2O in any proportion is
    A) saturated steam.
    B) dry steam.
    C) wet steam.

3228. The furnace sheet on a firebox is supported by
    A) stay bolts.
    B) diagonal stays.
    C) corrugated sheet.

3229. Purification of steam and prevention of carry over are achieved by
    A) dry pipe.
    B) steam.
    C) drum internals.

3230. Heating water above the boiling point in an open heater with a continuous vent to atmosphere is considered inefficient because
    A) water flashes to steam.
    B) of loss of condensate to atmosphere.
    C) it causes excess pressure.

3231. A flyball governor on a Corliss engine controls
- A) angle of advance.
- B) cut-off.
- C) admission.

3232. Which gives the dirtiest steam?
- A) Submerged tubes
- B) Through tubes
- C) Neither A nor B

3233. The safety valve on a reheat cycle is needed
- A) because of the boiler and superheater.
- B) to relieve total amount of steam going through reheater.
- C) but only to be large enough to relieve 50% of saturated steam of boiler.

3234. Which of these companies does not make boilers?
- A) Fairbanks Morse
- B) Murray
- C) Keeler
- D) Wickes
- E) Cliff

3235. On high-pressure boilers of large capacity, the
- A) thickness of shell is the same throughout.
- B) shell is made thicker around tubes because tube ligament is weak.
- C) shell thickness does not make any difference.

3236. A bottle is
- A) used to feed water to a burner.
- B) a device used to slug feed chemicals to boiler.
- C) a dry drum on a boiler.

3237. If fresh coal and O_2 are added to a furnace, the most smoke will be
- A) near the flame.
- B) in retort.
- C) nearer the stack.

3238. The lowest water temperature that can be put into an economizer is
- A) 70 °F.
- B) 140 °F.
- C) 212 °F.

3239. Trisodium is
- A) sometimes used to adjust pH.
- B) never used.
- C) used to inhibit caustic embrittlement.

3240. Sodium sulfate and sodium carbonate ratio is maintained to
- A) inhibit caustic embrittlement.
- B) reduce scale.
- C) reduce O_2.

3241. In a closed feed-water heater, which has more pressure?
    A) Steam space
    B) Water space
    C) Water and steam spaces have equal pressures.

3242. When drum tops are at the same level on a Sterling boiler, circulation tubes are between
    A) all drums.
    B) 2 front drums.
    C) 2 back drums.

3243. Riser enters drum
    A) below water level.
    B) above water level.
    C) both above and below water level.

3244. If sulfur content of No. 6 oil increases, the Btu content
    A) decreases.
    B) increases.
    C) stays the same.

3245. After a turbine is shut down, the auxiliary most likely to be left running is the
    A) oil pump.
    B) water to gland cooling.
    C) air to injector.

3246. On a horizontal reciprocating machine, rotation direction is determined by the
    A) ac or dc motor.
    B) cut-off valve arrangement.
    C) power, whether it is from or to the shaft.

3247. Flue gas velocity is
    A) 4,000 to 6,000 fpm.
    B) 6,000 to 8,000 fpm.
    C) 8,000 to 10,000 fpm.

3248. Expansion of water gauge glass is
    A) 0.000643".
    B) 0.00034".
    C) 0.034".

3249. An auto-transformer has
    A) one winding.
    B) two windings.
    C) multi-windings.

3250. An oil burner uses ________ steam.
    A) dry
    B) wet
    C) superheated

3251. With power at zero, the
 A) voltage is zero.
 B) current is zero.
 C) both current and voltage are zero.

3252. Chlorides are tested
 A) before alkalinity.
 B) after alkalinity.
 C) before hardness.

3253. Automatic gauge glasses are
 A) round.
 B) flat.
 C) round and flat.

3254. An injection pump is used with a
 A) barometric condenser.
 B) surface condenser.
 C) low-level jet condenser.

3255. Diffuser pumps are
 A) always volute.
 B) usually volute.
 C) never volute.

3256. A regenerative air preheater gas duct is
 A) smaller than air duct.
 B) larger than air duct.
 C) neither A nor B.

3257. Fire brick refractory for high temperature is
 A) silicon carbide.
 B) plastic.
 C) tile brick.

3258. Tighten the adjusting nut on a Copes to
 A) raise the water in the boiler.
 B) lower the water in the boiler.
 C) neither A nor B.

3259. What should be considered in the location of a gas furnace?
 A) Noise, vibration, and type of floor
 B) Cost, size and type
 C) Gas pipe route, duct and vent

3260. A gasket suitable for condensate return is made of
 A) cork.
 B) neoprene.
 C) asbestos.
 D) rubber.

3261. Do not wipe oil tank or piping with
 A) sponges.

B) rags.
C) cotton waste.

3262. What percent of steam piping is supported by the boiler?
A) 10%
B) 0%
C) 20%

3263. Of the three listed here, which furnace has the greatest capacity?
A) Coal
B) Gas
C) Oil

3264. Viscosity of oil decreases
A) when heated.
B) to 250 °F and then levels off.
C) when cooled.

3265. It is best to heat oil with
A) wet steam.
B) dry steam.
C) superheated steam.

3266. Which of the following is not a type of steam trap?
A) Bucket trap
B) Float and thermostatic trap
C) P-trap
D) Impulse trap

3267. Which of the following is a type of steam heating system?
A) Direct expansion
B) Vacuum
C) Reverse return
D) Injection

3268. Which has water in the tubes being heated by the hot gases outside the tubes?
A) Water tube boiler
B) Fire tube boiler
C) Radiant boiler
D) Atmospheric boiler

3269. Which method of heat transfer is not employed by heat transfer units?
A) Radiation
B) Convection
C) Gravitation
D) Conduction

3270. Which of the following is considered a heating accessory?
 A) Boiler
 B) Piping
 C) Coils
 D) Pumps

3271. Which of the following is not a main part of a heating system?
 A) Boiler
 B) Piping
 C) Heat transfer surfaces
 D) Water

3272. What is the weight of one cubic foot of water at 50 feet?
 A) 40.52 lb
 B) 62.41 lb
 C) 60.33 lb
 D) 70.01 lb

3273. An expansion joint
 A) minimizes vibration.
 B) allows for normal movement of the pipe due to expansion and contraction.
 C) compensates for movement of pumps or compressors.
 D) makes removal of pipe easier.

3274. When a line voltage thermostat is used to control a hot water circulating pump, a relay is unnecessary unless the motor load exceeds the stat capacity.
 A) True
 B) False

3275. Electric heating systems are thermally-operated sequence controls or thermally-operated staging controls, which act as time-delays to energize the heating elements.
 A) True
 B) False

3276. A capacitor-start motor develops a lower starting torque than a permanent split capacitor motor.
 A) True
 B) False

3277. When using a double element time-delay fuse with a 20 amp motor, the maximum fuse size should be
 A) 60 amps.
 B) 20 amps.
 C) 30 amps.
 D) 25 amps.

3278. The holding circuit in a magnetic motor starter holds the main contacts
 A) open until the control circuit through the interlock is made.

B) closed until the control circuit through the interlock is broken.
C) open until the control circuit through the interlock is broken.
D) closed until the control circuit through the interlock is made.

3279. How many pounds of water can 72 Btu heat 6 °F?
A) 0.833
B) 12
C) 432
D) 72

3280. On a pneumatic control system, the main line pressure should be at
A) 25 psig.
B) 5 psig.
C) 15 psig.
D) 60 psig.

3281. A pneumatic control valve that requires air pressure on the bellows or diaphragm to close the valve is a
A) diverting valve.
B) direct-acting valve.
C) reverse-acting valve.
D) normally-closed valve.

3282. A reverse-acting pneumatic humidistat increases air pressure to a controlled device when the humidity
A) increases.
B) decreases.
C) remains static.
D) mixes.

3283. The bimetal strip in a thermostat is
A) the anticipator strip.
B) the differential strip.
C) the contactor strip.
D) none of the above.

3284. On a pneumatic control system, the thermostat is connected to the valve of a damper motor by a small diameter tubing that measures
A) 1/4" OD.
B) 3/8" OD.
C) 3/8" ID.
D) 1/4" ID.

3285. For a heating application using a normally-open valve, the controller is a(n)
A) submaster control.
B) indirect-acting run thermostat.
C) direct-acting run thermostat.
D) manual switch control.

3286. The term applied to a device used to accumulate the effect of two or more thermostats to operate a single device is
- A) cumulator.
- B) accumulator.
- C) compensator.
- D) coordinator.

3287. On a single-inlet blower fan, the driving side is on which side in relation to the inlet of the fan?
- A) Neither side
- B) Either side
- C) Opposite side
- D) Same side

3288. Uninsulated breaching for solid fuel burning furnaces may not be within _______ inches of combustible material.
- A) 18
- B) 10
- C) 24
- D) 6

3289. A standing seam used as a cross seam for a large duct is
- A) used when a flat surface is required.
- B) used to eliminate the need for angle reinforcement.
- C) used for light gauge duct.
- D) simple to make.

3290. Which measures CO_2 in flue gas of large coal units?
- A) Smoke comparator
- B) Thermometer
- C) Orsat analyzer
- D) Furnace peephole

3291. If 24 ft^3 of oxygen unite with the fuel, how many cubic feet of nitrogen will enter the combustion chamber?
- A) 24
- B) 48
- C) 75
- D) 96

3292. The temperature of flue gas in the stack is 550 °F. The flue gas contains 12% carbon dioxide (CO_2). What is the heat loss when burning semibituminous coal?
- A) 10.6%
- B) 11.8%
- C) 17.00%
- D) 18.00%

3293. The flue gas in the stack is 550 °F and the fuel gas contains 10% CO_2. By lowering the stack temperature to 350 °F and raising the CO_2 to 14%, how many pounds of coal per ton is saved (in this case, a ton is equal to 2240 pounds)?

A) 150
B) 226
C) 250
D) 300

3294. What percent of heat loss is caused by smoke?
A) 2 to 3
B) 5 to 8
C) 12 to 15
D) 25 to 50

3295. What causes clinkers to form in a coal-fired space heater?
A) Chimney soot
B) Shaking the grates
C) Too much draft
D) Too little draft

3296. Which of these stokers creates three zones of fire in the coal bed?
A) Spreader
B) Chain grate
C) Underfeed
D) Overfeed

3297. The air openings in the retort of an underfeed stoker are referred to as
A) venturis.
B) air slots.
C) air ports.
D) tuyeres.

3298. What is the capacity range of hoppers in pounds of coal?
A) 200 to 1,000
B) 300 to 1,500
C) 375 to 2,000
D) 400 to 2,500

3299. A boiler operates very efficiently at 135 pounds of coal per hour. What type stoker is used with this boiler?
A) Screw-feed
B) Underfeed spreader
C) Traveling-grate
D) Overfeed

3300. What largely determines the type of stoker to be used in a heating plant?
A) Fuel used
B) Number of deadplates
C) Number of tuyeres
D) Size of retort

Heating

Level 3

3301. When installing a stoker that will handle coal ranging in size from dust to one inch, which stoker would be selected?
   A) Underfeed
   B) Spreader
   C) Screw-feed
   D) Traveling grate

3302. Why are there many small holes in the grates used with a spreader stoker?
   A) Eliminate fine coal particles
   B) Maintain heavy fuel bed
   C) Allow even distribution of air
   D) Extract coal not burned in suspension

3303. When using a stoker that distributes the coal over the entire grate by means of a nozzle, what type stoker is being used?
   A) Mechanical spreader
   B) Traveling grate
   C) Overfeed
   D) Pneumatic spreader

3304. A red, smoky fire fed by an automatic stoker is an indication of
   A) insufficient fuel.
   B) a normal fire.
   C) too much air or fuel bed too deep.
   D) insufficient air or too much fuel.

3305. Proper combustion of coal in the firebox of a furnace or boiler depends upon the size, type, and condition of the
   A) windbox.
   B) hopper.
   C) chimney.
   D) grate.

3306. A boiler requires a chimney having a cross-sectional area of 48 inches at sea level. How many square inches is the cross-sectional requirement for this chimney at 5,000 feet above sea level?
   A) 48
   B) 57.6
   C) 96.4
   D) 100

3307. What is the minimum number of cubic feet of air required for complete combustion of 100 pounds of coal?
A) 1,000
B) 5,000
C) 10,000
D) 15,500

3308. What type of coal is referred to as stone coal?
A) Lignite
B) Subbituminous
C) Bituminous
D) Anthracite

3309. What type of coal is commonly referred to as soft coal?
A) Lignite
B) Anthracite
C) Bituminous
D) Subbituminous

3310. Chemically, subbituminous coal is distinguished from bituminous by its
A) sulfur content.
B) moisture content.
C) fixed carbon.
D) hydrogen ratio.

3311. If there are 362,500 Btu in 25 lb of dry bituminous coal, how many Btu are in 25 lb of dry subbituminous coal?
A) 36,250
B) 337,500
C) 350,000
D) 350,500

3312. To obtain samples for inspection from a car of coal, expose the coal to a minimum depth of how many feet?
A) Two
B) Three
C) Four
D) Six

3313. The time to take a sample of coal is
A) during loading or unloading.
B) when it is burning.
C) when it is in storage.
D) while it is in transit.

3314. A gross coal sample should be a minimum of how many pounds?
A) 400
B) 600
C) 800
D) 1,000

3315. Improper storage of coal can cause
   A) loss in weight.
   B) warping of furnace grates.
   C) spontaneous combustion.

3316. Stored bituminous coal should be limited to a maximum height of how many feet?
   A) 13
   B) 20
   C) 30
   D) 35

3317. What is the general size (in feet) of coal stockpiles?
   A) 300 x 56
   B) 300 x 75
   C) 400 x 56
   D) 400 x 75

3318. To continue stacking coal in the same pile, the stacker must be moved. What is the maximum number of feet the stacker can be moved?
   A) 6
   B) 8
   C) 10
   D) 12

3319. A storage pile of coal has developed a hotspot because of spontaneous combustion. What action is taken?
   A) Flood hotspot with water
   B) Smother with additional coal
   C) Remove hotspot with hand shovels
   D) Remove hotspot with power shovel

3320. What temperature in a coal stockpile indicates the presence of a hotspot?
   A) 110 °F
   B) 115 °F
   C) 120 °F
   D) 160 °F

3321. When unloading fuel from a truck, the truck and oil tank must be grounded to
   A) reduce fuel vapors.
   B) prevent spark caused by static electricity.
   C) keep the oil truck in place.
   D) speed the unloading.

3322. An oil storage tank should be large enough to hold a supply of oil sufficient for
   A) 3 days.
   B) 4 days.
   C) one week.
   D) one month.

3323. A fuel oil storage tank must be installed for a heating plant that burns 500 gallons per day. What is the minimum number of gallons that the storage tank should hold?
   A) 1,500
   B) 2,000
   C) 3,500
   D) 4,000

3324. A 200-foot gas pipe line should be how many inches lower at one end than the other?
   A) Two
   B) Four
   C) Six
   D) Seven

3325. The total flue area of an oil burner chimney is 100 in^2. What is the maximum number of square inches the flue area could be closed off with flue dampers?
   A) 80
   B) 85
   C) 90
   D) 95

3326. What shapes the oil spray pattern in an air turbine commercial oil burner?
   A) Fan housing
   B) Atomizer
   C) Air inlet
   D) Air butterfly valve

3327. Which of the following gas burners is equipped with an electric motor?
   A) Upshot
   B) Inshot
   C) Combination
   D) Premix

3328. Which of the following burners is used for firing both coal and gas?
   A) Premix
   B) Inshot
   C) Upshot
   D) Combination

3329. When one end of a stove poker is in a flame, the other end soon becomes too hot to hold. This transfer of heat is called
   A) insulation.
   B) conduction.
   C) convection.
   D) radiation.

3330. Coal that is damaged by heat or fire should be
   A) discarded.
   B) aerated and used at once.
   C) mixed with undamaged coal.
   D) stored in different area.

3331. What type of gas must be used as fast as it is produced?
   A) Liquefied petroleum
   B) Natural
   C) Manufactured
   D) Blast-furnace

3332. While operating a spreader-type stoker, fuel flow stops. How long can the fuel already on the grates keep the fire going?
   A) 1 hour
   B) 30 minutes
   C) 15 minutes
   D) 5 to 6 minutes

3333. The pneumatic spreader stoker can be used with boilers having up to how many square feet of heating space?
   A) 3,000
   B) 3,500
   C) 4,500
   D) 5,000

3334. A furnace has a connecting-duct area of 400 in^2. What size (in gallons per hour) burner nozzle should be used?
   A) One
   B) Two
   C) Four
   D) Five

3335. What is the operational gas pressure (in psi) used in high-pressure gas systems?
   A) 1 to 10
   B) 2 to 20
   C) 2 to 25
   D) 3 to 30

3336. The total cross-sectional area of all the warm-air ducts in a gravity warm-air heating system is 50 in^2; therefore, the total cross-sectional area of all the cold-air ducts must be at least _______ in^2.
   A) 25
   B) 50
   C) 70
   D) 80

3337. A furnace has an input rating of 1,000,000 Btuh. How many square inches of area in the free air opening must be provided to supply combustion air to the furnace?
   A) 1,000
   B) 10,000
   C) 100,000
   D) 1,000,000

3338. Water in a 15 horsepower boiler must be checked
   A) daily.
   B) twice a week.
   C) weekly.
   D) monthly.

3339. When two power boilers are connected to the same steam header, they must have a
   A) check/stop valve.
   B) non-return valve.
   C) non-return valve and a stop valve.

3340. Phenolphthalein is used to indicate
   A) acidity.
   B) sodium chloride.
   C) alkalinity.

3341. A leaking safety valve has
   A) increased blow back.
   B) decreased blow back.
   C) neither A nor B.

3342. Upon failure of a package boiler with electrically-interlocked draft damper, the damper will
   A) automatically close.
   B) automatically open.
   C) not be affected.

3343. The efficiency of an injector being used as a pump is
   A) 5%.
   B) 10%.
   C) 20%.

3344. When the counterbalance weight on a governor is increased, the speed of the engine
A) increases.
B) decreases.
C) remains the same.

3345. What would be found in the line ahead of the fuel oil meter?
A) Strainer
B) Relief valve
C) Pressure gauge

3346. If a safety valve is chattering,
A) increase blow back.
B) decrease blow back.
C) adjust spring tension.

3347. The temperature limit of an alcohol thermometer is
A) 50 °F.
B) 250 °F.
C) 450 °F.

3348. The temperature limit of a mercury thermometer is
A) 1,000 °F
B) 450 °F
C) 75 °F

3349. A manometer with one end open to atmospheric pressure reads
A) gauge pressure.
B) absolute pressure.
C) vacuum.

3350. In a standard Sterling boiler, the steam outlet is on the
A) first drum.
B) second drum.
C) rear drum near feed water inlet.

3351. The minimum size of water gauge shut-off is
A) 1/2".
B) 3/4".
C) 5/8".

3352. The minimum size of water column connection to boiler is
A) 1/2".
B) 3/4".
C) 1".

3353. On a standard Sterling boiler, there are water circulating tubes between
A) first and second drums.
B) second and rear drums.
C) all drums.

3354. An inverted bucket trap can work with
  A) steam pressure below rating.
  B) steam pressure above rating.
  C) any steam pressure.

3355. An Ogee ring is used on a(n)
  A) Ogee boiler.
  B) Scotch boiler.
  C) upright boiler.

3356. The travel of the valve on an engine is
  A) equal to the eccentricity.
  B) twice the eccentricity.
  C) half the eccentricity.

3357. In a wet-type vertical boiler, the fusible plug is located
  A) two-thirds up on outside tube.
  B) in upper tube sheet.
  C) nowhere, as there is no fusible plug.

3358. The closed feed water heater has
  A) a vent on water side.
  B) a vent on steam side.
  C) no vent.

3359. An open feed water heater
  A) is vented to atmosphere.
  B) has relief valve set at 15 lb.
  C) has a vacuum.

3360. The Wickes and Manning boilers are
  A) internally fired.
  B) externally fired.
  C) different in that one is internally fired and one is externally fired.

3361. The piston and valve are going in the same direction at
  A) cut-off.
  B) release.
  C) compression.
  D) admission.

3362. The amount of coal burned per square foot of grate (hand fire, natural draft) is approximately
  A) 20 lb.
  B) 35 lb.
  C) 50 lb.

3363. If a packing gland is running hot,
  A) tighten packing evenly.
  B) loosen packing evenly.
  C) pour water on it.

3364. The fan used to draw stack gases, etc., is
   A) a forced draft fan.
   B) an induced draft fan.
   C) neither A nor B.

3365. To increase engine speed, increase diameter of
   A) driving pulley to governor.
   B) driven pulley to governor.
   C) idler pulley.

3366. On a reciprocating pump, the diameter of the steam cylinder is
   A) larger than the water.
   B) smaller than the water.
   C) the same as the water.

3367. In jacking an engine over by hand, it is a good practice to have the steam valve
   A) open.
   B) closed.
   C) cracked.

3368. In an oil-burning system, the coarsest strainer is on
   A) suction side of pump.
   B) discharge side of pump.
   C) neither A nor B.

3369. For a constant speed motor that drives a pump, the pump speed may be raised by
   A) a hydraulic coupling.
   B) a Fordomatic.
   C) neither A nor B.

3370. A releasing Corliss engine governor is a
   A) shaft governor.
   B) loaded flyball governor.
   C) pendulum governor.

3371. The percentage of carbon in boiler steel is
   A) 1.25%.
   B) 0.25%.
   C) 2.25%.

3372. The main purpose of drum internals is to
   A) direct the flow of steam in drum.
   B) remove moisture from the steam.
   C) increase temperature of incoming feed water.

3373. A circular gauge glass will wear faster with a pH of
   A) 11.5.
   B) 8.5.
   C) 6.5.

3374. A circular gauge glass less than 18 inches long
   A) needs no provision for expansion.
   B) may have an automatic shut-off.
   C) requires no special provisions.

3375. Making an adjustment for wear on the connecting rod at crank end
   A) decreases clearance on head end.
   B) decreases clearance on crank end.
   C) shortens the connecting rod.

3376. Torching in a recuperative air preheater is from
   A) preheater plugging.
   B) hole in plate between flue gas and air.
   C) burning in preheater.

3377. Inside admission piston valve with indirect valve linkage will be set for proper lead 90°
   A) plus angle of advance behind the crank.
   B) plus angle of advance ahead of crank.
   C) minus angle of advance behind crank.

3378. All steam pipe is to be installed to drain to
   A) a steam trap.
   B) an expansion trap.
   C) the boiler.
   D) none of the above.

3379. A shaft governor with loose eccentric controls has a(n)
   A) valve travel.
   B) angle of advance.
   C) throw of eccentric.

3380. Phosphorous in boiler steel
   A) makes the steel hot short.
   B) makes the steel cold short.
   C) is a desirable quality.

3381. Sulfur in boiler steel
   A) will make the steel hot short.
   B) is desirable.
   C) will make the steel cold short.

3382. The largest thermal drop will be through the
   A) gas film.
   B) tube metal.
   C) scale.

3383. The mud drum in a Heinie is at the bottom
   A) of the front header.
   B) inside the drum.
   C) of the rear drum.

3384. A safety valve is a
  A) relief valve.
  B) pressure safety valve to release excess pressure of the boiler.
  C) neither A nor B.

3385. On an accumulation test, the pressure should not rise more than
  A) 3%.
  B) 6%.
  C) 10%.

3386. A duplex pump, when steam bound, is full of
  A) water.
  B) water and steam.
  C) steam.

3387. A vacuum chamber on a duplex pump causes the pump to
  A) race.
  B) pulsate.
  C) work smoothly at beginning and end of stroke.

3388. A preheater is located
  A) between boiler and economizer.
  B) between stack and economizer.
  C) on other side of stack.

3389. On a fire-tube boiler, the
  A) water is inside the tubes.
  B) fire and gases pass through the tubes.
  C) fire hits the outside of the tubes.

3390. Fusible plugs are used on boilers of
  A) 1,200 psi.
  B) 50 psi.
  C) 1,000 psi.

3391. Two safety valves are used on boilers that have
  A) more than 500 ft^2 of heating surface.
  B) less than 500 ft^2 of heating surface.
  C) less than 100 hp.

3392. Flexible stay bolts are used on
  A) locomotive-type boilers only.
  B) Heinie type or water-tube boilers.
  C) firebox water legs boilers.

3393. A dry pipe in a boiler
  A) catches impurities.
  B) catches the entrained condensate from the boiler.
  C) makes drier steam.

3394. A pressure gauge corresponds to
  A) the safety valve setting.
  B) 5 times SWP.
  C) 1-1/2 to 2 times safety valve setting.

3395. A gas shut-off valve at the furnace shall be
A) inside the unit.
B) adjacent to the unit.
C) 3 to 6 feet off the floor.

3396. The type of turbine used on a feed-water pump is
A) impulse.
B) reaction.
C) extraction.

3397. The general procedure when warming up a turbine is to
A) stop turbine.
B) turn over slowly.
C) turn over at maximum speed.

3398. Caking coal is
A) low in oxygen.
B) high in oxygen.
C) average.

3399. Lost motion in a duplex pump is
A) necessary.
B) unnecessary.
C) caused by wear.

3400. A vacuum chamber is used on the
A) discharge side.
B) suction side.
C) chest cover.

3401. An air preheater is used in conjunction with a(n)
A) induced draft fan.
B) forced draft fan.
C) economizer.

3402. A high-carbon coal is
A) bituminous.
B) anthracite.
C) lignite.

3403. A blow-off cock should have the plug held in place by
A) a gland and a nut.
B) no gland but with a nut.
C) a guard or gland marked in line on passage.

3404. Rotor speed on an impulse turbine should be
A) same speed or velocity of the steam.
B) 50% speed or velocity of the steam.
C) 87% speed or velocity of the steam.

3405. A duplex pump in good working order has a suction lift of
A) 22 ft.
B) 26 ft.
C) 34 ft.

3406. A pump in good working order should not have a slippage over
   A) 5%.
   B) 8%.
   C) 10%.

3407. A centrifugal governor
   A) increases or decreases the steam pressure in the steam chest.
   B) decreases the amount of steam flow.
   C) increases the amount of steam according to the load on engine.

3408. When the water in a boiler is dangerously low,
   A) speed up the pumps.
   B) remove all combustible matter.
   C) leave the feed pump and fuel supply alone.

3409. When the water supply in a boiler is low,
   A) shut off the fuel supply.
   B) restore the water level.
   C) leave the pump alone.

3410. Low level cut-off valves are used to shut off
   A) fuel when water level is low.
   B) water when water level is low.
   C) air when water level is low.

3411. In case of tube failure in a boiler,
   A) increase the forced draft.
   B) open the doors.
   C) increase induced draft.

3412. The water and steam in a Copes feed-water regulator
   A) move rapidly.
   B) move slowly.
   C) remain motionless.

3413. The safety valve should be set when the
   A) water level is high.
   B) water level is low.
   C) drum is drained.

3414. Lead is found on an engine when the
   A) piston is on dead center.
   B) valve is in mid-travel.
   C) piston is in mid-travel.

3415. If a pump is vapor-bound, it
   A) stops.
   B) short strokes and is noisy.
   C) races.

3416. A Wickes vertical boiler is supported by
A) lugs on lower drum.
B) steam drum.
C) mud ring.

3417. An oil separator is used on the exhaust line of a(n)
A) engine.
B) oil heater.
C) turbine.

3418. Caking coals are burned on a(n)
A) underfeed stoker.
B) traveling grate.
C) Detroit Roto-Stoker.

3419. Correct preheated air in stoker firing
A) slows firing rate.
B) aids combustion.
C) prevents slag formation.

3420. Vertical water-tube boilers are supported by
A) lugs on lower drum and allowed to expand upwards.
B) lugs on lower drum and slings connected to top drum.
C) hangers.

3421. A dash pot on a governor
A) helps close valve on Corliss engine.
B) prevents hunting.
C) helps change speed.

3422. The drum of a straight tube box-header boiler is
A) parallel with tubes.
B) parallel with header.
C) perpendicular to tubes.

3423. Boiler hp, according to most city codes, is
A) 34.5 lb/hr at 212 °F.
B) 10 ft^2 heating surface.
C) 33,000 Btu.

3424. Water with an excess of hydroxyl ions in solution will be
A) acid.
B) neutral.
C) alkaline.

3425. Continuous blow down
A) removes oil.
B) removes impurities on surface water.
C) controls accumulation.

3426. Removing irregular-size coal causes
A) more uniform combustion.
B) better firing.
C) an even fuel bed.

3427. A commutator is found on a(n)
    A) alternating current motor.
    B) dc motor.
    C) transformer.

3428. Which type of coal has the highest O_2 content?
    A) Anthracite
    B) Bituminous
    C) Semi-anthracite

3429. The flash point, in comparison to ignition, is
    A) higher.
    B) even.
    C) lower.

3430. Piston rod is connected to connecting rod by the
    A) crosshead.
    B) guide.
    C) suitable pin.

3431. A column of water 1 in^2 x 1 ft equals 0.433 psi. A column of water 5 in^2 x 1 ft has a pressure of
    A) 0.433 psi.
    B) 2.31 psi.
    C) 105 psi.

3432. Under the same conditions, are a duplex pump and centrifugal pump equally effective?
    A) Yes, they are just the same
    B) No, the duplex is more effective
    C) No, the centrifugal is more effective.

3433. Water space in a vertical fire-tube boiler is less
    A) for more steam space.
    B) to superheat system.
    C) for less carry over.

3434. Shaft governors employ
    A) centrifugal force.
    B) centrifugal and inertia force.
    C) inertia force.

3435. The temperature at which a D slide valve is limited to is dependent upon
    A) the warping of the valve.
    B) lubrication of the valve.
    C) action of the valve.

3436. The piping connections allowed between boiler and water column are
- A) damper regulator and steam gauge only.
- B) none at all.
- C) anything that does not allow an appreciable amount of steam flow.

3437. Lead is given an engine to
- A) compensate for angularity.
- B) help the piston get started.
- C) cause compression.

3438. The lowest part of the water glass should be at least how many inches above the lowest safe level?
- A) 1"
- B) 2"
- C) 3"

3439. If the high-low water alarm is whistling,
- A) blow down boiler.
- B) determine the true water level.
- C) bank boiler.

3440. Valves between column and boiler should be
- A) non-rising stem gate valve.
- B) outside screw and yoke type.
- C) globe.

3441. Tighten packing on piston rod when the engine is
- A) stopped.
- B) running.
- C) leaking excessively.

3442. When trammeling an engine,
- A) find mid-position.
- B) find head and crank end dead center.
- C) square the valve.

3443. The best way to prevent spontaneous combustion is to
- A) increase O_2.
- B) decrease O_2.
- C) leave everything the same.

3444. An injector has
- A) high thermal efficiency.
- B) an economical way to feed water to boiler.
- C) higher suction left than pump.

3445. A 350 hp boiler used for heating is down for 4 weeks. The best thing to do is
- A) leave banked fire with slight pressure in it.
- B) take it out of service, using all service precautions.
- C) leave it alone.

3446. What will cause flash-back when starting an oil burner?
   A) Mixture of oil vapor, air and ignition
   B) Too much oil
   C) Too much air

3447. In a reciprocating pump, water fills the air chamber until it is
   A) 3/4 full.
   B) 1/2 full.
   C) 1/4 full.

3448. A vertical fire-tube boiler has
   A) submerged flues and Dutch oven.
   B) enclosed firebox and submerged flues.
   C) exposed tubes and Dutch oven.

3449. When changing from forced to induced draft, use
   A) a larger fan.
   B) a smaller fan.
   C) same fan.

3450. Fire cracks on an HRT are at
   A) girth seam.
   B) front head seam.
   C) rear head seam.

3451. How much larger must the fan be for an induced fan rather than a forced fan?
   A) 50%
   B) 25%
   C) Same size

3452. Chain grates are
   A) grate units attached on bars.
   B) continuous chain conveyors.
   C) grate bars forming chain.

3453. In balanced draft,
   A) a forced draft fan puts in more air than induced fan.
   B) both fans are equal.
   C) an induced fan pulls more products of combustion.

3454. The grate bar key stoker is
   A) underfeed.
   B) chain grate.
   C) traveling grate.

3455. A slide valve engine is considered to be
   A) high speed.
   B) low speed.
   C) medium speed.

3456. A steam engine with two steam cylinders has to be
- A) compound.
- B) not necessarily compound.
- C) conventional.

3457. A safety valve releases saturated steam from boilers at
- A) 3% above the maximum working pressure.
- B) 6% above the maximum working pressure.
- C) 10% above the maximum working pressure.

3458. Which has a coking arch?
- A) Chain grate
- B) Traveling grate
- C) Underfeed

3459. Dual stacks cannot be located in a
- A) kitchen.
- B) hallway.
- C) bathroom.
- D) bedroom.

3460. A flame rod in oil firing
- A) lights off a burner.
- B) is a safety device if flame goes out.
- C) is for use as a torch.

3461. What affects natural water circulation in a boiler?
- A) Outgoing steam velocity
- B) Feed water coming in
- C) Difference of density of steam and water

3462. For steam passing through stationary blades of an impulse turbine, the velocity
- A) increases.
- B) decreases.
- C) stays the same.

3463. A duplex pump valve
- A) has a lap.
- B) has no lap.
- C) uses steam extensively

3464. Pressure used on a D slide valve is
- A) 100 to 200 psi.
- B) 200 to 300 psi.
- C) 300 to 400 psi.

3465. Flange gaskets on high-pressure steam applications are
- A) thin.
- B) medium.
- C) thick.

3466. In balanced draft, pressure in the furnace is __________ when compared to atmosphere.
   A) higher
   B) lower
   C) the same

3467. In a traveling grate, burn
   A) low-volatile coal.
   B) high-volatile coal.
   C) high-volatile coking coal.

3468. The minimum blow back on a safety valve is
   A) 2 to 8 lb.
   B) 6 lb.
   C) 10 lb.

3469. An Adamson ring is used on
   A) Scotch marine boilers.
   B) water-tube boilers.
   C) a Manning boiler.

3470. An Ogee ring is used on
   A) Scotch marine boilers.
   B) water-tube boilers.
   C) Manning boilers.

3471. A safety valve on a long pipe would
   A) not seat.
   B) open easier.
   C) open later.
   D) chatter.

3472. When an engine gets a sudden load, it
   A) speeds up.
   B) slows down.
   C) runs the same.

3473. It is possible to get more superheated steam by
   A) adding more moisture to steam.
   B) adding more pressure to steam.
   C) throttling.

3474. In an oil-fired boiler when boiler is lighting off, there is
   A) no smoke.
   B) a light brown haze.
   C) gray smoke.

3475. If a chimney is smoking heavily,
   A) there is a lack of O_2.
   B) the fuel bed is too heavy.
   C) the wrong fuel is being used.

3476. An air chamber on a duplex pump can be found on
- A) the suction side.
- B) the discharge side.
- C) both the suction and discharge sides.

3477. A horizontal D slide valve engine is set for equal
- A) lead.
- B) cut-off.
- C) lead and cut-off.

3478. If a safety valve is leaking,
- A) tighten down and spring until leaking stops.
- B) close the stop valve.
- C) remove from boiler and repair.

3479. How much air is needed for combustion using No. 6 oil?
- A) 12 lb
- B) 15 lb
- C) 18 lb

3480. A minimum blow-off line is
- A) 1".
- B) 2-1/2".
- C) 3/4".

3481. On a hydrostatic lubricator, the sight glass is full of
- A) water.
- B) half water and half oil.
- C) oil.

3482. In case of serious overpressure on an oil- or gas-fired boiler,
- A) lift safety valve by hand.
- B) shut off fuel and lift safety valve by hand.
- C) shut off fuel.

3483. When the centrifugal pump is running, what stops water from going back to the suction side?
- A) Casing
- B) Impeller
- C) Wearing ring

3484. If superheated, the volume of a given weight of steam
- A) increases.
- B) decreases.
- C) stays the same.

3485. With traveling grate stokers or chain grates, ignition should start and be stable after entering furnace
- A) as soon as possible after entering boiler.
- B) less than 1/4 distance after entering.
- C) about 1/4 distance after entering.

3486. Lead affects the following events in what order?
A) Admission, cut-off, release and compression
B) Admission, release, cut-off and compression
C) Admission, compression, exhaust and cut-off

3487. Fire-tube boilers are usually not used above
A) 150 lb.
B) 300 lb.
C) 450 lb.

3488. In case of low water in a Scotch marine boiler, what is damaged first?
A) Crown sheet
B) Top-row tubes
C) Flues

3489. Entrance to the blowdown tank should be
A) not less than blowdown pipe from boiler.
B) 1-1/2 times larger than blowdown pipe from boiler.
C) 2 times larger than blowdown pipe from boiler.

3490. When using an extension light inside a boiler, it
A) should be grounded.
B) should not be grounded.
C) doesn't make any difference whether or not it is grounded.

3491. In a centrifugal pump, which way does the impeller curve?
A) Backwards to the rotation of the shaft
B) Backwards against the shaft
C) They do not curve

3492. When water leaves a centrifugal pump, it will have high
A) velocity and low pressure.
B) pressure and low velocity.
C) pressure and high velocity.

3493. In burning breeze fuel, it would burn best on
A) chain grate.
B) spreader.
C) underfeed.

3494. On a boiler, heat loss is due to
A) too much O_2.
B) too much H.
C) heat going up the stack.

3495. A simplex pump is
A) single-acting.
B) double-acting.
C) both single- and double-acting.

3496. With a magnetic valve, there is circulation
- A) when valve closes.
- B) when valve opens.
- C) at all times.

3497. On an induction motor, which part is stationary?
- A) Rotor
- B) Stator
- C) Commutator

3498. At what part of the stroke does steam cut off on a pump?
- A) 1/3
- B) 1/2
- C) 3/4

3499. On a steam engine that has no lap or lead, cut-off occurs at
- A) or near end of stroke.
- B) 3/4 of stroke.
- C) 5/8 of stroke.

3500. On a fuel-oil system, the pump should be protected by a strainer
- A) before pump.
- B) after pump.
- C) before and after pump.

3501. Excess oil in an oil-fired system returns to the
- A) tank.
- B) suction side of pump.
- C) sewer.

3502. On an engine, a hydrostatic lubricator lubricates the
- A) crank.
- B) crosshead.
- C) steam chest.

3503. A small vertical steam engine crank is lubricated by
- A) a splash system.
- B) a forced oil pump.
- C) gravity.

3504. On a gas-fired system, the magnetic valve is held open by
- A) gas pressure in line.
- B) electricity.
- C) gas pressure, which keeps it open, and electricity, which closes it.

3505. If a boiler is foaming badly, first
- A) blow down the boiler.
- B) determine true water level.
- C) open safety valve by hand.

3506. On a throttling governor, steam cut off from engine
- A) reduces steam pressure in steam chest.

B) alters cut-off.
C) shuts off steam from engine.

3507. What type of stoker handles any size coal best?
A) Spreader
B) Underfeed
C) Chain grate

3508. On a simplex pump, the steam valve is
A) mechanically operated.
B) operated by pilot valve.
C) steam thrown.

3509. A safety device on a line to prevent overload is a(n)
A) open switch.
B) closed switch.
C) circuit breaker.

3510. On a duplex pump, the steam valve is
A) steam operated.
B) mechanically operated.
C) operated by pilot.

3511. In feed water treatment, ppm means
A) pounds per minute.
B) parts per minute.
C) parts per million.

3512. Water temperature to the suction of boiler feed pump should be
A) 180 °F.
B) 200 °F.
C) cool enough so as not to cause the pump to become vapor-bound.

3513. A new turbine entering commercial service should be tested within
A) 2 weeks.
B) 2 months.
C) 1 year.

3514. On a non-return trap, what opens the discharge valve?
A) Steam heat
B) Cooling effect of water
C) It is mechanically operated

3515. An automatic injector will restart itself when there is an interruption in
A) water.
B) steam.
C) steam or water.

3516. Filtration is a
A) sedimentation process.
B) mechanical process.
C) chemical process.

3517. Filter material is most often found in a(n)
  A) open heater.
  B) discharge line from pump.
  C) suction line from pump.

3518. All conditions being equal, it is important to pay closer attention during blow down during
  A) internal treatment.
  B) external treatment.
  C) both A and B, depending on the system used.

3519. On a Copes feed-water regulator, the
  A) water level cannot be changed readily.
  B) water level can be changed readily.
  C) regulator has a bypass.

3520. A Copes feed-water regulator
  A) should be installed at a 45° angle.
  B) is operated by difference in temperature of water and steam.
  C) is operated by a difference in pressure.

3521. Combustion on a stoker grate may be inhibited by the temperature
  A) of flue gases.
  B) of preheated air.
  C) at which fuel ignites.

3522. A safety valve seat sticks less often with a seat of
  A) 45 °F.
  B) 90 °F.
  C) flat.

3523. Wet steam on an oil burner atomizer
  A) causes intermittent firing.
  B) carbons the tip.
  C) aids combustion.

3524. On a locomotive boiler. the hot gases pass
  A) under the drum and through tubes to stack.
  B) through the furnace then through tubes to stack.
  C) through the furnace then under tubes to stack.

3525. If the lost motion on a duplex pump is decreased, the stroke is
  A) increased.
  B) decreased.
  C) not affected.

3526. A valve stamped SWP 125 is good for
  A) any fluid or gas.
  B) water but not steam.
  C) steam but not water.

3527. Underfeed stokers carry a fuel bed of
  A) 6 to 12 inches.

B) 2 to 6 inches.
C) 2 inches or less.

3528. Which can burn the most coal per square foot in grate surface per hour?
A) Chain grate
B) Single retort stoker
C) Hand firing

3529. When there is a temporary drop in load with a mechanical oil burner,
A) adjust fuel supply.
B) change burner tips.
C) adjust air.

3530. Which engine has more parts that can wear?
A) Tandem engine
B) Single-cylinder engine
C) Cross-compound engine

3531. What prevents the steam pressure in a steam turbine chest from getting too high?
A) Relief valve
B) Governor
C) Throttle

3532. What type of trap is used on high-pressure systems for greatest effect?
A) Thermostatic
B) Ball float
C) Inverted bucket

3533. When starting a rotary pump, leave the discharge
A) open.
B) closed.
C) partly open.

3534. A non-return valve can be
A) double-acting.
B) single-acting.
C) both double- and single-acting.

3535. When the bottom valve on a water glass is plugged up, the water
A) rises slowly.
B) rises fast.
C) stays the same.

3536. The percent of overspeed trip-off turbine is
A) 1%.
B) 5%.
C) 10%.

3537. On an engine with a flywheel, the flywheel
A) balances the engine.
B) drives a pulley.
C) absorbs energy from the engine.

3538. When an engine develops more power, friction horsepower
A) increases.
B) decreases.
C) stays the same.

3539. Using feed water to wash outgoing steam helps to
A) reduce moisture in steam.
B) reduce the amount of solid concentration.
C) heat feed water.

3540. Butting or attaching another safety valve that is leaking is
A) dangerous.
B) permissible.
C) all right if pressure does not rise above maximum.

3541. Increasing the area of the valve disc underside on a safety valve will cause it to
A) increase blow back.
B) suddenly pop.
C) slowly open.

3542. Holes in a strainer, in relation to pipe size, should be
A) less.
B) slightly more.
C) equal.

3543. Excess steam pressure is removed from turbine casing by a
A) relief valve.
B) governor.
C) condenser.

3544. Tube ends of waterwalls usually terminate in
A) mud drums.
B) cross boxes.
C) headers.

3545. Which of these is not considered carry over?
A) Slugs of water
B) 80% steam
C) Saturated steam

3546. The connection between piston rod and connecting rod is a
A) crank pin.
B) cross head.
C) connecting rod.

3547. A crane uses a
A) vertical boiler.
B) horizontal boiler.
C) water-tube boiler.

3548. A common slide valve has
A) 2 lap edges.
B) 3 lap edges.
C) 4 lap edges.

3549. To increase horsepower of a Corliss engine, never
A) adjust governor.
B) adjust valve travel.
C) decrease back pressure.

3550. Tension relief on a single-element feed-water regulator is to
A) keep tension on bronze pivots.
B) relieve pressure produced by adjusting nut.
C) avoid excessive pressure on valve when boiler is cold.

3551. The recommended way to start a centrifugal pump with no head is suction
A) and discharge valves closed.
B) open and discharge closed.
C) closed and discharge open.

3552. Across the moving blades of a reaction turbine there is
A) a perceptible drop in pressure.
B) no perceptible drop in pressure.
C) an increase in absolute velocity.

3553. Phasing is
A) dependent upon the number of poles.
B) not dependent on the number of poles
C) dependent upon the speed of the generator.

3554. The speed of piston travel on head end to mid-point in relation to crank end is
A) faster.
B) slower.
C) the same.

3555. Water circulation in a boiler tends to increase with the
A) increase in pressure.
B) decrease in pressure.
C) increase in load.

3556. A pressure relief valve shall be installed on the ________ side of a hydronic system.
A) high
B) downstream
C) low

3557. More primary air would be used in a pulverizer with
A) semi-bituminous coal.
B) bituminous coal.
C) subbituminous coal.

3558. Correct a pulverizer fire by
A) using water.
B) increasing the fuel-to-air ratio.
C) using CO_2.

3559. A pulverizer fire can be recognized by the
A) outlet fuel-air temperature.
B) smoke at air tempering damper.
C) neither A nor B.

3560. The difference between a traveling grate and a chain grate is,
A) the chain grate is a traveling chain with bars.
B) the traveling grate is a traveling chain with bars, while a chain grate has chain coming from above.
C) the chain grate is made of tiny chain bars while both are endless chains.

3561. Cushion valves, as used on a duplex pump, are on the
A) steam side.
B) water side.
C) both steam and water sides.

3562. A shaft governor will assume its new position within
A) 16 revolutions.
B) 32 revolutions.
C) 64 or more revolutions.

3563. Air temperature in a stoker furnace depends upon the
A) type of fuel being burned.
B) combustion rates.
C) steel castings.

3564. To burn one pound of typical fuel oil, the theoretical amount of air needed is
A) 12 to 14 lb.
B) 14 to 16 lb.
C) 16 to 18 lb.

3565. The amount of dissolved solids cannot be reduced by
A) removing carbonates.
B) removing sulfates.
C) running water through a pressure filter.

3566. Increased chlorides in a boiler occur when
A) increasing the pressure.
B) decreasing the steaming rate.
C) running water to the boiler from a zeolite tank without flushing.

3567. When steam is throttled in a reducing station from 200 to 15 psi,
   A) a higher temperature and superheat are reached.
   B) temperature of steam is increased and the steam is superheated.
   C) steam temperature is increased and some of the steam is condensed.
   D) steam will have a higher temperature than that corresponding to pressure, and the steam is superheated.

3568. In using a feed water test for chlorides, take
   A) a fresh sample and neutralize it.
   B) a fresh sample and do not neutralize it.
   C) the test after the P and M test.

3569. If using fuel with a low fusing point, the result will be
   A) water screen.
   B) dry bottom.
   C) wet bottom.

3570. A device that has nothing to do with thrust is a
   A) Kingsbury bearing.
   B) shrouding.
   C) dummy piston.

3571. A constant head chamber is used on a
   A) pressure regulator to control water level.
   B) device to determine true water level in a boiler.
   C) blowdown tank to ensure full head of water.

3572. When laying-up a boiler, the chemical to use in pans is
   A) slaked lime.
   B) unslaked lime.
   C) sodium sulfite.

3573. The ASME symbol on an HRT is
   A) on the side of the boiler near the front head.
   B) on the front head above the center line.
   C) any place on the side in a conspicuous place.

3574. An open feed water heater can be used as a de-aerator because
   A) an open heater can heat water to 210 °F.
   B) gases won't stay in solution when heated near boiling point.
   C) an open heater is open to the atmosphere.

3575. Boiler horsepower can be calculated based on the
   A) pressure and temperature of the boiler.
   B) heating area only.
   C) heat absorbing capacity of the boiler.

3576. In feed water, a heat recovery of 20 is equal to an efficiency of
 A) 1%.
 B) 2%.
 C) 3%.

3577. What stoker has the largest combustion space for pounds of fuel burned?
 A) Spreader
 B) Underfeed
 C) Traveling grate

3578. At 150 psi, the temperature is
 A) 358 °F.
 B) 330 °F.
 C) 256 °F.

3579. Stacks should be protected from lightning by
 A) a proper ground.
 B) a lightning rod.
 C) stack design.

3580. A well-lubricated reciprocating engine should be limited to a friction of
 A) 10%.
 B) 20%.
 C) 30%.

3581. Tightening up the adjusting nut of a single-element feed-water regulator results in a
 A) raised water level.
 B) lowered water level.
 C) faster-operating regulator.

3582. The safe working pressure of a bent-tube boiler is calculated based on the
 A) ligament of the tubes.
 B) longitudinal seam.
 C) girth seam.

3583. The common cause of tube bagging is
 A) not enough water circulation.
 B) overheating of the tubes.
 C) scale on the inside of the tubes.

3584. A staybolt drilled 1/2" past the inside plate means the staybolt length is
 A) over 8".
 B) under 8".
 C) used on fire-tube boilers only.

3585. For each 1 ft^3 of natural gas, how much air is needed to support combustion?
 A) 5 ft^3

B) 10 ft^3
C) 15 ft^3

3586. Vertical water-tube boilers are supported by
    A) legs on lower drum and allowed to expand upward.
    B) hangers on upper drum and allowed to expand upward.
    C) hangers on upper drum and allowed to expand downward.

3587. A shaft bearing, when renewed, should be fitted by
    A) scraping.
    B) a machine fit.
    C) filing.

3588. Flow velocity is increased by increasing the speed of a centrifugal pump
    A) at the square of the pump speed.
    B) at the cube of the pump speed.
    C) directly proportional to the pump speed.

3589. For most efficient operation, pressure in open heaters is regulated by
    A) regulating the vent valve.
    B) throttling steam from the engine.
    C) throttling steam from auxiliaries.

3590. With adjusting ring attached by threads to the disc of a safety valve, ensure disc does not rotate when adjusting blow back by adjusting when
    A) valve is detached from boiler.
    B) the boiler is near popping.
    C) there is no pressure on boiler.

3591. When attached to valve seat by threads, turning the adjusting ring to the right
    A) increases blow back.
    B) decreases blow back.
    C) decreases popping pressure.

3592. With an increase in load, a convection superheater's temperature will
    A) increase.
    B) decrease.
    C) not change.

3593. With a decrease in load, a radiant superheater's temperature will
    A) increase.
    B) decrease.
    C) remain the same.

3594. Adding weight to the balls of a flyball governor
    A) increases speed of governor.
    B) decreases speed of governor.
    C) causes no change in speed of governor.

3595. Priming can be noticed by
   A) water hammer.
   B) vibration of machinery.
   C) condensation in steam lines.

3596. In a stoker-fired boiler, pressure in combustion chamber is
   A) below atmospheric.
   B) above atmospheric.
   C) balanced draft.

3597. If a given weight of steam is superheated, the volume
   A) increases.
   B) decreases.
   C) remains the same.

3598. The primary air in a pulverized-firing furnace is used for
   A) mill drying.
   B) carrying the fuel.
   C) increasing combustion rate.

3599. Which flanges must be tightened the hardest?
   A) Smooth surface
   B) Rough surface
   C) Serrated surface

3600. In a properly fitted bearing, the bearing touches
   A) 0% of journal.
   B) 80% of journal.
   C) 100% of journal.

3601. Compared to a simple engine, a compound engine requires a flywheel size that is
   A) larger.
   B) smaller.
   C) the same size.

3602. In a generator with 4 poles and 1,800 rpm, the frequency is
   A) 90 cycles.
   B) 120 cycles.
   C) 60 cycles.

3603. On a cross compound engine, later cut-off on a high-pressure cylinder
   A) increases power on high-pressure cylinder.
   B) decreases power on high-pressure cylinder.
   C) decreases power on low-pressure cylinder.

3604. In an impulse turbine, pressure through moving blades
   A) increases.
   B) decreases.
   C) is the same.

3605. When pressure is put on a steam gauge, the gauge forms a(n)
   A) oval.
   B) circle.
   C) square.

3606. What does a flat gauge use for sealing purposes?
   A) Mica
   B) Rubber gasket
   C) Shellac

3607. In an impulse turbine, velocity through stationary blades
   A) increases.
   B) decreases.
   C) remains the same.

3608. The weakest part of a welded, bent-tube, multi-drum boiler is
   A) tube ligaments.
   B) tubes.
   C) seams.

3609. When cutting in an open feed-water heater to service, open
   A) steam inlet.
   B) steam inlet and then drain.
   C) drain and then steam inlet.

3610. If flue gas temperature increases, CO_2
   A) increases.
   B) decreases.
   C) remains the same.

3611. The weight on a Copes feed-water regulator aids in
   A) opening the valve.
   B) closing the valve.
   C) stabilizing the regulator.

3612. If the speed of an engine suddenly increases, the water level in the boiler
   A) increases.
   B) decreases.
   C) remains the same.

3613. When starting multiple-retort, underfeed, stoker-fired boilers equipped with superheaters,
   A) start a small fire in middle of grates.
   B) start a small fire across full width of grates.
   C) warm up furnace with oil or gas burners.

3614. The absolute steam velocity across the moving blades of reaction
   A) increases.
   B) decreases.
   C) does not change.

3615. If a fire occurs in an operating pulverizer,
   A) stop the fuel feed to the pulverizer.
   B) increase the hot air flow to the pulverizer.
   C) increase the raw fuel feed to the pulverizer.

3616. If water goes out of sight on a steaming boiler, first pull the fire, then
   A) increase pump speed.
   B) wait for steam pressure to drop.
   C) close steam stop valve.

3617. The gravel in a zeolite softener is
   A) above the bed.
   B) below the bed.
   C) above and below the bed.

3618. Torching in recuperative air preheaters is caused from
   A) fouling of the flue gas surfaces.
   B) burning of the metal in the preheater.
   C) a hole in the plate between flue gas and air passages.

3619. A generator with a power factor of 80 is most efficient at
   A) 80% power factor.
   B) 100% power factor.
   C) 60% power factor.

3620. It is necessary to pulverize fuel to pass through 200 mesh
   A) to make sure ignition is rapid and sure.
   B) because larger size particles plug up the burner.
   C) to decrease size of furnace.

3621. Most pulverizer specifications require an output of not more than 2% plus 50 mesh fuel because
   A) more than that causes excessive slagging in furnace.
   B) more than that causes excessive wear of burner.
   C) burner would plug up if larger sizes were used.

3622. All things being equal, it is possible to check oil viscosity more quickly by using
   A) Staybolt Universal Viscosimeter.
   B) Staybolt Furel Viscosimeter.
   C) no special instrument, as there is not much difference in time required.

3623. High-pressure gas burners used in large furnace installations have pressures from
   A) 20 to 30 psi.
   B) 40 to 50 psi.
   C) 60 to 100 psi.

3624. The common term used for so-called tangential firing is
   A) corner firing.
   B) angle firing.
   C) mixture firing.

3625. In a non-steaming economizer, water temperature should be raised to
    A) 34 to 45 °F of saturation temperature.
    B) 45 to 55 °F of saturation temperature.
    C) 55 to 65 °F of saturation temperature.

3626. Theoretical efficiency of an engine is the ratio of heat
    A) rejected to heat received.
    B) received to heat rejected.
    C) extracted to heat received.

3627. Define the term adiabatic.
    A) Steam expanded to greater volume than normal for given pressure
    B) Steam compressed to greater pressure than input pressure
    C) Expanded or compressed steam that does not lose or gain any heat by heat transfer

3628. A condensing engine
    A) operates with back pressure less than atmosphere.
    B) operates with back pressure equal to or greater than atmospheric pressure.
    C) condenses steam before exhausting.

3629. At what speed does a low-speed engine operate?
    A) 200 rpm or less
    B) 110 to 200 rpm
    C) 125 rpm or less
    D) 40 rpm or less

3630. At what throttle pressure does a low-pressure engine operate?
    A) 150 psi or more
    B) 80 to 150 psi
    C) 80 psi or less
    D) 40 psi or less

3631. How does a shaft governor control speed?
    A) Throttles input steam
    B) Controls valve action
    C) Opposes speed by opposing direction of rotation of flywheel

3632. How does a flyball governor control speed?
    A) Throttles input steam
    B) Controls valve action
    C) Changes pressure of steam

3633. What is an indicator diagram used for?
    A) Analysis of what is taking place in engine cylinder
    B) Indicates how much water is entering with steam
    C) Indicates the speed of the engine

3634. A double-acting engine
- A) has two pistons.
- B) takes steam on both strokes.
- C) takes steam on one stroke with flywheel momentum carrying engine through return stroke.

3635. When an engine is running over, it means that
- A) oil is running over.
- B) the engine is running over the speed designed for.
- C) the top of the flywheel is turning away from the cylinder.

3636. An intercooler is used with a(n)
- A) open heater.
- B) closed heater.
- C) air compressor.

3637. Which operates first on an element Copes feed-water regulator?
- A) Thermostatic tube
- B) Diaphragm
- C) Orifice

3638. At what temperature does silica carry over with steam?
- A) 650 °F
- B) 250 °F
- C) 450 °F

3639. For a given pressure, the temperature and density of a saturated vapor are
- A) fixed.
- B) varied, with the vapor being a few degrees higher.
- C) insignificant.

3640. A soot blower connection, attached to the outlet from the superheater or reheater and used for safety valve connection,
- A) is not allowed by code.
- B) is allowed by code.

3641. Expansion and contraction of soft-rolled steel per lineal inch 1 F.A.H. increases or decreases
- A) 0.000005.
- B) 0.00005.
- C) 0.005.

3642. Practically all of the earth's heat is derived, either directly or indirectly, from the
- A) sun.
- B) volatile matter under the earth's surface.
- C) ocean trade winds.

3643. Large condensing steam engines use how much less steam than non-condensing engines?
- A) 10 to 20 percent
- B) 20 to 40 percent
- C) 40 to 50 percent

3644. Slide valves are employed in engines where
   A) simplicity and low price are most important.
   B) steam consumption needs to be kept down.
   C) economy of the engine is important.

3645. Proper load for any slide valve should be
   A) determined by an indicator.
   B) 3-1/4" for each foot of stroke.
   C) determined by quiet operation.

3646. Economy resulting from a feed water heater using exhaust steam is
   A) 5 to 10 percent.
   B) 11 to 14 percent.
   C) 14 to 20 percent.

3647. A primary or vacuum heater is located
   A) between engine and condenser.
   B) between feed pump and boiler.
   C) in various places.

3648. Heavy fuel oil must be continually circulated past a burner
   A) in order to return oil not burned to tank.
   B) because temperature at burner would change with firing rate, thus a variation of oil viscosity would affect controls.
   C) to keep carbon from forming in furnace.

3649. A full fluid film of oil between bearing surfaces
   A) moves only with rotating bearings.
   B) does not move at all.
   C) moves with temperature changes.

3650. In thrust bearings, lubrication due to floating pads is accomplished by the
   A) wedge principle.
   B) oil bath.
   C) oil pressure.

3651. The critical temperature and pressure of steam is
   A) 3,206 psi.
   B) 3,105 psi.
   C) 4,001 psi.

3652. An indicator spring may be used indefinitely and must be
   A) tested periodically.
   B) replaced every 6 months.
   C) replaced every year.

3653. A modern central-station boiler might contain
   A) 0.75 lb.
   B) 2.5 lb.
   C) 3.0 lb.

3654. The best place to install a throttling calormotor is
   A) a straight length of pipe in which steam flow is descending.
   B) a straight length of pipe in which steam flow is rising.
   C) some pipe bond in horizontal length of pipe.

3655. Economizers used in waste heat boilers, in relation to degree of temperature drop,
   A) are more efficient than direct-fired boilers.
   B) are less efficient than direct-fired boilers.
   C) have the same effect.

3656. The conductivity motor is used to determine
   A) the amount of ionizing boiler water salts.
   B) measure of suspended matter such as organic material and nonionized solids.
   C) the amount of hardness in ppm.

3657. For the same fuel burning rate, if feed-water temperature is lowered, the
   A) degree of superheat steam increases.
   B) temperature of superheat is not affected by the temperature of feed water.
   C) degree of superheat steam decreases.

3658. Gas weights and temperatures entering superheater of equal capacity will
   A) vary approximately to the heat absorbed per lb of steam.
   B) not change.
   C) change when feed-water temperature is lowered 50 °F.

3659. The reheater or resuperheater must be installed to receive
   A) radiant heat only.
   B) convection heat only.
   C) both radiant and convection heat.

3660. The reheater usually reheats superheated steam near
   A) the original temperature.
   B) the original pressure.
   C) both temperature and pressure to original

3661. To acquire the ultimate maximum temperature required for the combustion of fuel oil, the fuel oil is heated
   A) on suction side of pump.
   B) on discharge side of pump.
   C) under hood of tank.

3662. The constant-speed dc motor having the most torque is
   A) series.
   B) shunt wound.
   C) compound wound.

3663. A solid connecting rod with the inside wedges taking up wear would
 A) decrease clearance volume.
 B) lengthen rod.
 C) shorten rod.

3664. In a piston valve engine with inside admission and indirect valve, when the valve cuts off steam,
 A) valve and piston are traveling in same direction.
 B) valve and piston are traveling in opposite direction.
 C) neither A nor B.

3665. A compound engine with crank set 90° apart would
 A) need a receiver.
 B) exhaust directly to low-pressure cylinder.
 C) not need a receiver.

3666. What is required to absorb 1 lb of CO_2?
 A) 10 lb of sodium sulfate
 B) 8 lb of sodium sulfite
 C) 5 lb of calcium sulfate

3667. What is the ppm of hardness is most river water?
 A) 0.10
 B) 100
 C) 397

3668. A regenerator air preheater cold duct, as compared to a gas duct, would be
 A) same size.
 B) smaller than gas duct.
 C) larger than gas duct.

3669. An Adamson ring aids in
 A) rigidity of boiler.
 B) supporting the flues.
 C) cleaning dirt from between boilers.

3670. Retarders are in
 A) fire-tube boiler tubes.
 B) water-tube boiler tubes.
 C) blow-down line to slow water to tank.

3671. Thermal conductance is
 A) Moht.
 B) Thom.
 C) Mode.

3672. Bifurcated tube construction is used on
 A) economizers.
 B) waterwalls.
 C) neither A nor B.

3673. The frequency of a synchronous motor having 4 poles and turning at 1,800 rpm is
  A) 25 cycles.
  B) 60 cycles.
  C) 90 cycles.

3674. In laying up a boiler, what chemical is used to control O_2?
  A) Sodium sulfite
  B) Sodium sulfate
  C) Sulfate

3675. What does a 600 hp boiler require by code?
  A) One electric pump
  B) One steam pump and one electric pump
  C) Two electric pumps

3676. A steam blanket can be found in
  A) internal downcomers.
  B) steam on top of water.
  C) steam in risers.

3677. A constant head chamber is used for
  A) liquid level indicator.
  B) height of water in blow-down tank.
  C) height of water in water column.

3678. When does a piston travel fastest?
  A) First half of stroke
  B) Second half of stroke
  C) The piston always travels at the same speed

3679. What condenser doesn't require supply pump once flow has started?
  A) Barometer condenser
  B) Surface condenser
  C) Jet condenser

3680. All amounts being equal, which item removes more hardness?
  A) Soda ash
  B) Caustic soda
  C) Sanosite

3681. If a series generator became motorized, it would
  A) reverse the rotation of the field.
  B) leave the field the same.
  C) burn up windings.

3682. Which causes most blow down?
  A) Sulfate
  B) Carbonate
  C) Silica

3683. If one generator is driven by a turbine engine and one is driven by a reciprocating engine, which would need more poles?
    A) Reciprocating engine
    B) Turbine engine
    C) They would need the same amount

3684. On a Corliss releasing engine, it is not considered good practice (when engine is running a maximum load) to increase power by
    A) varying cut-off.
    B) adjusting linkage to governor.
    C) lowering exhaust pressure.

3685. All steel pipe shall be
    A) 3/4 bore.
    B) 1/2 bore.
    C) full bore.

3686. On a Copes water regulator, tension relief is
    A) an aid in closing valve.
    B) to reduce strain on linkage when boiler is down.
    C) an aid in opening valve.

3687. On an underfeed stoker, the factor that determines the temperature to which preheated air can be used is
    A) rate of firing.
    B) type of metal in stoker.
    C) fusion temperature of coal.

3688. Stays are used on which type of boiler?
    A) Wickes vertical
    B) Keeler CP boiler
    C) Wickes A type

3689. A P reading is usually referred to as the amount of sodium
    A) sulfate in boiler water.
    B) carbonate in boiler water.
    C) sulfite in boiler water.

3690. On a centrifugal pump, if the discharge valve is throttled while the pump is running,
    A) overload the driver.
    B) unload the driver.
    C) there is no effect on the driver.

3691. What is the speed of a regenerative-type air preheater?
    A) 7 to 10 rpm
    B) 2 to 3 rpm
    C) 10 to 25 rpm

3692. When the release takes place on an engine with a piston-type valve, the valve is moving in
    A) same direction as piston.
    B) opposite direction as piston.

3693. What is the oil temperature coming from the bearing of a turbine?
A) 120 °F
B) 140 °F
C) 180 °F

3694. In a plate-type air heater, the curved vanes are on the
A) air side.
B) gas side.
C) air and gas sides.

3695. A steam flow meter with a reservoir
A) prevents pulsation.
B) ensures a full head of water.
C) ensures a full head of oil.

3696. On a forced-circulation boiler, what ensures that there is water in all the tubes?
A) Sizing of tubes
B) Tube orifice
C) Pump size

3697. Bubble caps are used as
A) evaporators.
B) remote sight glass.
C) sediment at bottom of water chamber.

3698. Riser tubes from waterwalls discharge steam and water
A) below water line.
B) above water line.
C) at water line.

3699. The size and shape of recirculators are to
A) prevent carry over.
B) superheat steam.
C) equalize water level.

3700. In a 4 drum Sterling boiler, water is
A) equal in all drums.
B) high in rear drums.
C) lower in rear drums.

3701. Water velocity in risers, as compared to downcomers, is
A) greater in risers.
B) greater in downcomers.
C) the same in both risers and downcomers.

3702. A centrifugal pump with single suction is in hydraulic
A) axial balance.
B) radial balance.
C) neither A nor B.

3703. A centrifugal pump with double suction is in hydraulic
  A) axial balance.
  B) radial balance.
  C) neither A nor B.

3704. An evaporator should have a
  A) safety valve.
  B) drain.
  C) neither A nor B.

3705. If bicarbonates are added to a boiler containing carbonates, there will be
  A) biborato CO_2.
  B) increased carbonates.
  C) neither A nor B.

3706. Surface blowoff should not exceed
  A) 1".
  B) 1-1/2".
  C) 2-1/2".

3707. A 20" x 40" plate supporting 10 tons has how much pressure resting on it?
  A) 250 psi
  B) 25 psi
  C) 15 psi

3708. When safety valves are mounted,
  A) the spindle must be vertical and in an upright position.
  B) they can be located below water level provided steam line is taken from highest steam space
  C) neither A nor B.

3709. Carbonate hardness is
  A) 7.
  B) made up of sulfates.
  C) made up of bicarbonates of calcium and magnesium.

3710. An impact tube measures
  A) pressure.
  B) velocity pressure.
  C) total pressure.

3711. Waterwell expansion is
  A) downward.
  B) upward.
  C) either upward or downward.

3712. When draining an economizer, temperature should be
  A) above 200 °F.
  B) below 400 °F.
  C) below steam temperature.

3713. When flue gas temperature gets too hot at the preheater,
A) bypass air around heater.
B) bypass gas around heater.
C) either A or B.

3714. When two safety valves of different sizes are placed on the same line, the smaller must discharge
A) 50% of the larger.
B) 75% of the larger.
C) 25% of the larger.

3715. To remove condensate from a condenser, use a
A) vacuum pump.
B) siphon.
C) steam trap.

3716. The basic principle in adjusting the blow back on a safety valve is
A) spring pressure.
B) valve disc area.
C) kinetic energy of steam.

3717. On a compound engine, steam goes from high-pressure cylinder to
A) low-pressure cylinder and then to receiver.
B) receiver and then to low-pressure cylinder.
C) neither A nor B.

3718. The primary purpose of not having alkalinity too high is to prevent
A) caustic embrittlement.
B) carry over.
C) neither A nor B.

3719. When a return trap dumps, it requires
A) same volume of steam to water.
B) more volume of steam to water.
C) less volume of steam to water.

3720. A furnace, when ash is abrasive or clinker, uses
A) chrome refractory fire brick.
B) glazed brick.
C) neither A nor B.

3721. Is there a piping connection between the boiler and a water column?
A) A piping damper regulator, on the steam gauge only
B) No piping whatsoever
C) Piping that does not allow an appreciable amount of steam flow

3722. Reaction turbines are
A) velocity compounded.
B) pressure compounded.
C) neither A nor B.

3723. Can cast iron flanges be connected to a boiler?
A) Yes

B) Yes, but only on boilers under 100 psig
C) No, they can never be used

3724. Double blow-down valves in one casting
A) are not permitted.
B) may be used if the operation of one valve does not affect the other valve.
C) neither A nor B.

3725. Water screens used to cool the ash are
A) tubes connected to mud drum.
B) water pumped separately.
C) in regular boiler circulation.

3726. A Copes tension relief
A) prevents valve from hunting.
B) relieves valve from starting in cold start.
C) aids in thermostatic expansion.

3727. The perpendicular lines on an indicator card are taken from
A) atmospheric line.
B) exhaust line.
C) neither A nor B.

3728. Zeolite softeners remove
A) temporary hardness.
B) permanent hardness.
C) some temporary and some permanent hardness.

3729. A proportioning pump is used on
A) zeolite softener.
B) hot soda lime.
C) neither A nor B.

3730. Weight of shaft on horizontal engines is
A) 80%.
B) 90%.
C) 100%.

3731. Underfeed stoker use of preheated air is determined by
A) air temperature.
B) furnace temperature.
C) metal of grates.

3732. The impeller vanes or blades on a centrifugal pump are
A) curved forward.
B) curved backwards.
C) straight.

3733. A tandem compound engine has
A) a large flywheel.
B) the same rod for both pistons.
C) a receiver.

3734. Control CO_2 by using
- A) sodium sulfite.
- B) sodium sulfate.
- C) neither A nor B.

3735. The vertical hanger spacing for steel pipe is
- A) 5 feet.
- B) 10 feet.
- C) 15 feet.
- D) 20 feet.

3736. In boiler water, the ratio of CO_2 to sulfite is
- A) the same.
- B) 8 lb sulfite to 1 lb of CO_2.
- C) neither A nor B.

3737. A 600 hp boiler needs a(n)
- A) steam and electric pump (coal-fired).
- B) steam pump only.
- C) electric pump only (gas, oil, pulverized).

3738. A water column used for pressures of 350 psi is made of
- A) steel.
- B) cast iron.
- C) neither A nor B.

3739. On a modern high-pressure boiler, the
- A) solid refractory walls have the thickest insulation.
- B) air-cooled walls don't need insulation.
- C) air-cooled walls have the thickest insulation.

3740. To raise alkalinity, it is best to use
- A) monosodium phosphate.
- B) disodium phosphate.
- C) trisodium phosphate.

3741. Plate-type air preheated baffles are used on the
- A) gas side.
- B) air side.
- C) neither A nor B.

3742. Steam at 100 psi that is reduced to 15 psi is
- A) wet.
- B) saturated.
- C) superheated.

3743. There is less discharge on a centrifugal pump with
- A) oil.
- B) water.
- C) either oil or water.

3744. A lower discharge ring on a safety valve
- A) decreases blow back.

B) increases blow back.
C) has the same blow back.

3745. A compound engine
A) needs a receiver.
B) does not need a receiver.

3746. Tubes are measured in
A) ID.
B) OD.
C) monical.

3747. What is the minimum height of a steam return pipe?
A) 12 inches
B) 18 inches
C) 24 inches
D) None of the above

3748. A boiler-fired underfeed stoker that is changed to oil would
A) decrease capacity.
B) increase capacity.
C) not affect capacity.

3749. The type of governor used in a non-releasing Corliss valve gear is a
A) loaded flyball.
B) shaft.
C) simple pendulum.

3750. A shaft governor works on
A) centrifugal force.
B) inertia.
C) both centrifugal force and inertia.

3751. To warm up a water tank on an open feed water heater,
A) run steam through vent.
B) throttle steam to warm up slowly.
C) open quickly.

3752. To increase horsepower of a Corliss engine, never
A) adjust governor.
B) adjust valve travel.
C) decrease back pressure.

3753. Tension relief on a single-element regulator is to
A) keep tension on a bronze pivot.
B) relieve pressure produced by adjusting nut.
C) avoid excessive pressure on valve when boiler is cold.

3754. When using a fuel pump in oil heating, the number of psi that pump pressure must drop to close nozzle valve is known as
A) delivery.
B) valve differential.
C) head of oil.
D) lift.

3755. If a fuel system is one in which the tank is buried below the burner level and the oil passes through a line filter, the _______ must show a reading. If it does not, an air leak is present.
A) pressure gauge
B) fuel gauge
C) vacuum gauge
D) sight glass

3756. Which is a reason for a steam boiler getting too much water?
A) Leaking hot water coil in boiler
B) Faulty swing check in return header
C) Trap plugged up
D) Return pump not operating

3757. When the pitch of the thread on a pipe is 1/4", how many turns are required to thread 2-1/2" of the pipe?
A) 8
B) 10
C) 12
D) 13

3758. What procedure should be followed when a beam interferes with the installation of a pipe line and the piping must be looped over the beam?
A) Provide for an anchor to prevent vibration
B) Provide for venting air from high point of pipe
C) Use oversized fittings and pipe to prevent restriction
D) Make sure that an expansion joint is used

3759. Which type of valve is used in a horizontal line requiring complete drainage?
A) Globe
B) Check
C) Gate
D) Relief

3760. Which valve should not be used for throttling purposes?
A) Plug
B) Needle
C) Gate
D) Globe

3761. All piping systems shall be capable of withstanding a hydrostatic test pressure of how many times the designed pressure?
A) 3
B) 1-1/2
C) 2
D) 2-1/2

3762. Which is not generally used to take care of thermal expansion in pipe lines?
   A) Loop in the line
   B) Packed slip-joints
   C) Bellows-type joints
   D) Change in size of line

3763. What is required between the steam gauge and the boiler?
   A) A trap or siphon with a valve of the T or lever handle type
   B) A vacuum breaker check valve
   C) Cross fittings to facilitate cleaning

3764. When air is passed through a water spray, some moisture in the air is given up if the water temperature is
   A) below the dew point of the air.
   B) below the temperature of the air.
   C) above the dew point of the air.
   D) above the temperature of the air.

3765. What factors determine the size of an expansion tank?
   A) Amount of space the system water requires in its expanded state
   B) Amount of space the system air requires in its expanded state
   C) Amount of air in the system
   D) Operating pressure in the system

3766. When condensate backs up in the return lines due to a lack of proper head between the dry return and boiler water level, the water line in a steam boiler will
   A) rise several inches.
   B) rise.
   C) not vary.
   D) drop.

3767. The condenser that doesn't need a condensate pump is the
   A) surface.
   B) barometric.
   C) jet.

3768. A secondary feed water heater is of the
   A) open type.
   B) closed type.
   C) either open or closed type.

3769. The pressure drop between steam inlet and air take off in a well designed contact condenser is
   A) zero.
   B) small.
   C) large.

3770. Of the metals listed, the one most popularly used for condenser tubes is
A) admiralty metal.
B) cast iron.
C) monel metal.

3771. The two primary functions of a steam surface condenser are to reduce back pressure on a turbine and to
A) reheat the feed water.
B) recover full temperature condensate.
C) remove gases from the steam and water cycle.

3772. A two-stage air ejector with inter- and after-condensers uses
A) air for cooling.
B) condensate for cooling.
C) steam for cooling.

3773. To remove condensate from an after-condenser, use a
A) vacuum trap.
B) siphon.
C) steam trap.

3774. How many moving parts are there in a continuous flow steam trap?
A) Three
B) One
C) Zero

3775. An atomizing heater is
A) open.
B) closed.
C) barometric.

3776. In a condenser, submergence means the
A) distance above pump suction.
B) depth in hot well.
C) pump center line to level of water in hot well.

3777. The condenser that can't use an air ejector or air pump is a
A) jet.
B) barometric (counter current).
C) barometric (parallel current).

3778. A horizontal evaporator has its normal water level
A) at the centerline.
B) above the centerline.
C) below the centerline.

3779. Condenser tube spacing is closest near the
A) steam inlet.
B) air cooler section.
C) hot well.

3780. When installing condenser tubes, a ferrule is used
  A) for expansion purposes.
  B) to protect the tube sheets.
  C) to prevent galvanic action.

3781. Air is removed from the exhaust steam by a(n)
  A) air ejector.
  B) steam ejector.
  C) vacuum breaker.

3782. A closed water system is treated for algae with
  A) chlorine.
  B) soda ash.
  C) fluorine.

3783. Two-stage air ejectors are used because they
  A) are easier to operate.
  B) will obtain a lower vacuum.
  C) are more economical.

3784. Tube support plates for a condenser are constructed of
  A) cast steel or admiralty metal.
  B) cast iron or copper bearing steel.
  C) arsenical copper or copper nickel alloys.

3785. The maximum velocity of condensate flow from the hot well outlet to the hot well pump is
  A) 2 feet/second.
  B) 4 feet/second.
  C) 6 feet/second.

3786. The cross-sectional area taken up by condenser tubes in a condenser is
  A) 20 to 25%.
  B) 30 to 40%.
  C) 40 to 45%.

3787. The heaviest tube concentration in a surface condenser is
  A) at the steam inlet.
  B) at the air outlet.
  C) just above the water in the hot well.

3788. The trap used on a live steam separator in a steam line is a(n)
  A) thermostatic.
  B) inverted bucket.
  C) float.

3789. Maximum condenser tube size is
  A) 1/2".
  B) 1-1/4".
  C) 1-1/2".

3790. Minimum condenser tube size is
- A) 1/2".
- B) 5/8".
- C) 1-1/2".

3791. The melting temperature of a fusible plug is
- A) 212 °F.
- B) 450 °F.
- C) 600 °F.

3792. Surface condensers ensure steam flow over the full length of tubes by
- A) use of an air ejector.
- B) holes where there are no tubes in the tube support sheet.
- C) an annular chamber above the hot well.

3793. Atmospheric relief on a condenser should be
- A) set at two differential settings.
- B) notched so it can be adjusted.
- C) screw adjusted (can be lifted manually).

3794. The velocity of steam entering a condenser tube bank is
- A) 100 feet/second.
- B) 150 feet/second.
- C) 200 feet/second.

3795. The velocity of cooling water entering the condenser tubes is
- A) 8 feet/second.
- B) 12 feet/second.
- C) 20 feet/second.

3796. Double-effect evaporators
- A) are connected in series.
- B) are connected in parallel.
- C) use the vapor from the first effect as the heating medium for the second effect.

3797. An open heater, if used for pressures under 15 psig, is usually made of
- A) cast iron.
- B) steel.
- C) malleable iron.

3798. When starting up an open heater fill it with
- A) evaporated water.
- B) city water.
- C) condensate.

3799. The loop seal on a two-stage air ejector is located on the
- A) after condenser.
- B) before condenser.
- C) inner condenser.

3800. Surface condensers are
   A) single flow.
   B) single and double flow.
   C) single, double and multi-flow.

3801. Tube spacing in a condenser
   A) increases as the volume of steam is reduced.
   B) decreases as the volume of steam is reduced.
   C) is proportional to the volume of steam.

3802. After the heating medium passes through a two-stage air ejector, it is discharged to the
   A) sewer.
   B) storage tank.
   C) condenser hot well.

3803. Single-stage condensate pumps operate economically within a compression ratio of
   A) 8 to 1.
   B) 10 to 1.
   C) 5 to 1.

3804. Water velocity is limited in the tubes to
   A) 4.5 to 6.5 feet/second.
   B) 6.5 to 8 feet/second.
   C) 9 to 10.5 feet/second.

3805. The steam consumption of a two-stage air ejector at 29" Hg is
   A) 11.5 lb/hr of steam/lb of dry air, with mixture temperature of 7.5 °F lower than the saturated temperature of the steam entering main condenser.
   B) 12.6 lb/hr, with conditions as above.
   C) 13.6 lb/hr with conditions as above.

3806. Normal tube construction in a horizontal evaporator
   A) has the tubes slightly bowed and connected to ridged tube sheets.
   B) is straight at normal operation.
   C) is vertical.

3807. An evaporator should have a
   A) drain.
   B) safety valve.
   C) pressure gauge.

3808. In a closed heater used at over 600 °F, the tube sheets are made of
   A) admiralty metal.
   B) muntz.
   C) copper nickel.

3809. To operate most efficiently, evaporators must have a
- A) relief valve.
- B) blow down line.
- C) sight glass.

3810. Evaporators are usually cleaned
- A) with acid.
- B) by thermal shock.
- C) mechanically.

3811. An evaporator is usually used in a
- A) heating plant
- B) power plant running condensing
- C) power plant running non-condensing

3812. On an evaporator with floating heads, the tubes are
- A) bowed 90% to the heat.
- B) slightly bowed in the center and 90% to the tube sheet on the ends.
- C) not supported.

3813. Water leaving a de-aerating heater has
- A) higher pH.
- B) lower pH.
- C) no change.

3814. On a closed heater, the
- A) tubes are stronger than the shell
- B) shell is stronger than the tubes
- C) tubes and shell are of equal strength.

3815. If the stack gas temperature is reduced 100%, what is the boiler efficiency gain?
- A) 2 to 4%
- B) 10 to 15%
- C) 15 to 20%

3816. In a modern boiler, the overall heat absorbed per ft^2 of waterwall surface per hour is
- A) 20,000 to 30,000 lb/hr.
- B) 30,000 to 50,000 lb/hr.
- C) 50,000 to 80,000 lb/hr.

3817. In a modern boiler, the overall heat absorbed per ft^2 of superheater surface per hour is
- A) 4,000 to 8,000 lb/hr.
- B) 8,000 to 12,000 lb/hr.
- C) 12,000 to 16,000 lb/hr.

3818. In a modern boiler, the overall heat absorbed per ft^2 of economizer surface per hour is
- A) 3,000 to 4,500 lb/hr.

B) 6,000 to 8,000 lb/hr.
C) 1,000 to 2,000 lb/hr.

3819. In a modern boiler, the overall heat absorbed per ft^2 of boiler surface per hour is
A) 2,000 to 5,000 lb/hr.
B) 6,000 to 8,000 lb/hr.
C) 10,000 to 12,000 lb/hr.

3820. In a modern boiler the overall heat absorbed per ft^2 of preheater surface per hour is
A) 500 to 700 lb/hr.
B) 1,000 to 1,500 lb/hr.
C) 2,000 to 2,500 lb/hr.

3821. In a modern, high-pressure, high-temperature boiler equipped with a preheater under normal load conditions, the air pressure drop through the preheater is
A) 5" of water.
B) 2" of mercury.
C) 2.5 psi.

3822. What is the flue gas rise in temperature between soot blowing?
A) 1%
B) 5%
C) 10%

3823. In an average plant, the savings in fuel by the installation of a feed water heater is
A) 14%.
B) 5%.
C) 25%.

3824. Pressure in an open heater is regulated by
A) varying the amount of steam from pumps and other auxiliaries.
B) throttling steam from engine.
C) vent valve.

3825. Under light loads when flue gas will be cooled too much going through an air heater, bypass
A) the gas around air heater.
B) the air around air heater.
C) both gas and air around air heater.

3826. If feed water temperature is raised 20 °F by waste heat, the efficiency gain is
A) 2%.
B) 1%.
C) 3%.

3827. If furnace temperature rises above normal over a period of time, expect
- A) low excess air.
- B) soot build up on tubes.
- C) high excess air.

3828. Sudden temperature increase in an air preheater on light loads, can be caused by
- A) fire.
- B) normal operation.
- C) bypass dampers being closed.

3829. Combustible material is likely to leave the furnace and deposit on heating surfaces and in fly-ash hoppers during
- A) light loads.
- B) heavy loads.
- C) medium loads.

3830. Leaking at flange or rolled joint in an economizer is often caused by
- A) high sulfur content.
- B) high temperature at start up.
- C) temperature shock at start up.

3831. The relieving capacity of reheater safety valves is
- A) included as part of the relieving capacity of the superheater.
- B) included as part of the relieving capacity of the boiler and superheater.
- C) not included as part of the relieving capacity of the boiler and superheater.

3832. The total relieving capacity of all safety valves on the reheater outlet shall be
- A) not more than 15% of the required total.
- B) not less than 15% of the required total.
- C) not less than 25% of the required total.

3833. The relieving capacity of safety valves installed on a reheater must be equal to
- A) 25% of the total steam flow through the reheater.
- B) 75% of the total steam flow through the reheater.
- C) 100% of total steam flow through the reheater.

3834. A reheater or superheater safety valve connection
- A) can have nothing requiring a steam flow attached to it.
- B) has a dry pipe connected near the outlet.
- C) can have a soot blower connection attached to it.

3835. Live steam reheaters are normally used on
- A) compound turbines.
- B) compound engines.
- C) simple engines.

3836. Cooling towers are
  A) forced draft only.
  B) induced draft only.
  C) forced, induced and natural draft.

3837. When steam is wire drawn, it decreases in
  A) pressure and temperature.
  B) temperature and increases in pressure.
  C) pressure and increases in temperature.

3838. Reheat in conjunction with turbines results in high steam
  A) pressure without high steam temperature.
  B) temperature without high steam pressure.
  C) temperature and high steam pressure.

3839. A reheater in a turbine-cycle will result in a gain of
  A) 2 to 4% in the heat rate.
  B) 4 to 6% in the heat rate.
  C) 6 to 8% in the heat rate.

3840. The greater the allowable leaving loss,
  A) the smaller the turbine capacity.
  B) the greater the turbine capacity.
  C) capacity is not affected.

3841. The heat exchanger showing the greatest gain in economy is the
  A) live steam separator.
  B) air preheater.
  C) economizer.

3842. Steam for evaporators may be obtained from
  A) one or more heater bleed lines.
  B) boiler feed water at full or reduced pressure.
  C) exhaust from turbine.
  D) all of the above.
  E) none of the above.

3843. Drain coolers are
  A) found on continuous blow lines of high-pressure boilers.
  B) found on regenerative feed water system.
  C) used to cool condensate from high pressure drip lines before going to storage tank.

3844. Circulating pumps supply about
  A) 50 lb of water for each lb of steam condensed.
  B) 75 lb of water for each lb of steam condensed.
  C) 100 lb of water for each lb of steam condensed.

3845. Live steam reheaters are
  A) usually of the shell and tube type.
  B) bare tube.
  C) tubes with fins shrunk on them .

3846. Water circulation in a boiler increases as
- A) load increases.
- B) load decreases.
- C) feed water temperature increases.

3847. To open a valve on a de-aerator,
- A) do not open.
- B) open steam valve rapidly.
- C) open steam valve slowly.

3848. The specific heat of dry steam is approximately
- A) 0.51.
- B) 1.0.
- C) 1.5.

3849. Pipe straps should not be used for pipe over
- A) 1 inch.
- B) 2 inches.
- C) 2-1/2 inches.
- D) 3 inches.

3850. A chimney having natural draft uses
- A) conduction.
- B) convection.
- C) radiation.

3851. The factor of safety is
- A) working pressure times 5.
- B) ratio between bursting pressure and safe work pressure.
- C) ratio between working pressure and bursting pressure.

3852. The reactionary force of steam escaping from the huddling chamber of a safety valve is to
- A) increase blow back.
- B) decrease blow back.
- C) neither A nor B.

3853. When saturated steam is throttled in a reducing station to 15 psi,
- A) a higher temperature and superheat is reached.
- B) steam will have a higher temperature.
- C) a higher temperature than that corresponding to pressure and superheat is reached.

3854. When suspended, the mud drum to a conventional Sterling type boiler
- A) expands downward and backward.
- B) expands upward.
- C) hangs freely.

3855. A centrifugal pump varies
- A) indirectly at the rotation speed.
- B) directly at the rotation speed.
- C) to the cube of the rotation speed.

3856. One pound of dry saturated steam has the most latent heat at
 A) atmospheric pressure and corresponding temperature.
 B) 450 psi.
 C) critical pressure and corresponding temperature.

3857. Carbon ring shafts used on certain types of turbines
 A) rotate with the shaft.
 B) hold stationary.
 C) rotate slowly like oil rings.

3858. How much steam piping weight will a boiler support?
 A) Any amount
 B) Less than half of weight
 C) None

3859. The specific heat of industrial fuel oil in Btu is
 A) 0.511.
 B) 1.0.
 C) 1.5.

3860. The boiler that best handles a fluctuating load is the
 A) large steam storage capacity.
 B) small water storage capacity.
 C) large waterwall capacity.

3861. Reciprocating pumps are particularly adapted for
 A) high capacities and low pressures.
 B) low capacities and high pressures.
 C) high capacities and high pressures.

3862. An open feed water heater can be successfully used as a de-aerator because
 A) the preheated water treatment solution is injected in the heater.
 B) the oxygen in the water boils off at that temperature.
 C) gases cannot stay in the solution when water is heated to the boiling point.

3863. When steam moves across the moving blades of an impulse turbine, there is
 A) a perceptible pressure drop.
 B) no perceptible pressure drop.
 C) an increase in absolute velocity.

3864. The most effective method of producing intense turbulence in a furnace with pulverized fuel is with
 A) vertical firing.
 B) horizontal firing.
 C) tangential firing.

3865. Flange bolts and mild carbon steel bolts are used with applications below
   A) 900 °F and 55,000 psi.
   B) 700 °F and 30,000 psi.
   C) 500 °F and 10,000 psi.

3866. Pressure regulators used with a 6" line would have a
   A) needle valve.
   B) balanced valve.
   C) gate valve.
   D) globe valve.

3867. The horizontal hanger spacing for copper tubing that is less than 1-1/4" in diameter is
   A) 6 ft.
   B) 8 ft.
   C) 12 ft.
   D) 15 ft.

3868. The satisfactory operation of the combined area of holes in a dry pipe for efficient operation must be
   A) no greater than the area of the steam pipe.
   B) equal to the area of the steam pipe.
   C) no less than twice the area of the steam pipe.

3869. A diffuser type pump
   A) is usually a centrifugal pump.
   B) is always a centrifugal pump.
   C) can be a centrifugal pump.

3870. The compound engine is located
   A) one behind the other in line operating on same piston.
   B) in line one behind the other with both engines acting on different pistons.
   C) side by side and they both act on the same piston.

3871. Most of the scale deposit on a water tube boiler is found
   A) in the bank of tubes next to the fire.
   B) near the water line.
   C) in the rear bank of tubes.

3872. In making blow back adjustments on a safety valve, it is considered good practice to never to turn the adjustment more than
   A) 5 notches without testing.
   B) 10 notches without testing.
   C) 15 notches without testing.

3873. Water delivered by a centrifugal pump varies
   A) directly as the rotation speed.
   B) adversely as the rotation speed.
   C) adversely as the square of the rotation speed.

3874. Softening of boiler feed water without evaporators is accomplished by
  A) coagulation.
  B) mechanical means.
  C) chemical means.

3875. In proper operation, the governor will assume its new position within
  A) 1 to 2 revolutions.
  B) 10 revolutions.
  C) 32 revolutions.

3876. In a direct-fired pulverized coal system, primary air is not used for
  A) drying the coal.
  B) conveying the coal.
  C) increasing the combustion rate of the coal.

3877. It is a recommended practice to bypass most stop valves after they have approached the size of
  A) 2 inches.
  B) 5 inches.
  C) 8 inches or over.

3878. The principal reason for omitting the center row of boiler tubes in an HRT type boiler is to
  A) facilitate water circulation.
  B) facilitate replacement of worn out tubes.
  C) permit clearing of the external surface of the lower tubes.

3879. The feed water discharged into an HRT boiler is
  A) above the normal water line.
  B) below the normal water line.
  C) near the bottom of the shell in the rear of the boiler.

3880. On an open feed water heater in a centrifugal solution system, the pump suction should
  A) receive only vapor-free liquid.
  B) be at least 3 feet above the normal heater water level.
  C) draw from the water surface inside the heater.

3881. The vertical hanger spacing for copper tubing larger than 1-1/2" in diameter is
  A) 5 ft.
  B) 10 ft.
  C) 15 ft.
  D) 20 ft.

3882. As the steam flows through the moving blades of a reactionary turbine there is
  A) a perceptible pressure drop.
  B) no perceptible pressure drop.
  C) an increase in absolute velocity.

3883. Complete combustion requires that all carbon be converted to
A) CO and CO_2.
B) ash and CO_2.
C) CO_2.

3884. The fire door required on a combustion chamber of a water-tube boiler is
A) friction contact.
B) an outward opening door with a self-locking latch.
C) spring closed.

3885. A venturi meter can be used for determining
A) quantity of the fluid.
B) viscosity of the fluid.
C) specific gravity of the fluid.

3886. Of the tube seam used in the construction of the boiler drum shell, the
A) longitudinal seam is the strongest.
B) girth seam is the strongest.
C) ligament of the tubes is the strongest.

3887. If the adjusting ring of safety valve is secured by threads to the valve seat, lowering the adjusting ring will
A) increase blow back.
B) decrease blow back.
C) not affect blow back.

3888. The heat developed by a centrifugal pump varies
A) directly as the square of the rotation speed.
B) inversely of the rotation speed.
C) directly as the rotation speed.

3889. The dissolved salts of calcium magnesium give water the quality called
A) pH.
B) ionization.
C) hardness.

3890. Increasing the weight of the counterbalance with a flyball governor will
A) increase the operating speed of the engine.
B) not change the operating speed of the engine.
C) decrease the operating speed of the engine.

3891. For an underfeed stoker of the retort type, fuel should start burning
A) before the coal passes from the feeding cylinder.
B) in the retort.
C) after the coal passes the retort.

3892. The test used on a safety valve to check for sufficient capacity is the
A) hydrostatic test.

B) accumulation test.
C) air pressure test before installing.

3893. The principal reason for using a manometer to measure draft pressures instead of a Bourdon tube gauge is it
A) has no moving parts.
B) is not affected by ambient temperature.
C) is not affected by corrosion.

3894. Radial stays are used on
A) HRT boilers.
B) locomotive boilers.
C) economical boilers.

3895. If the adjusting ring of the safety valve is secured by threads and it is necessary to make blow back adjustments, the best procedure to follow to prevent the safety valve disk from rotating is to make adjustments when
A) the valve is detached from the boiler.
B) the boiler is near popping pressure.
C) there is no pressure on the boiler.

3896. As the temperature of water decreases, its capacity to absorb dissolved gases
A) increases.
B) decreases.
C) remains the same.

3897. A shaft governor with a loose eccentric controls the speed of the engine by changing the
A) angle of advance.
B) throw of the eccentric.
C) steam pressure in the chest.

3898. For proper operation of an overfeed stoker, the ends of the grate bars adjacent to the dump grates should be
A) thin.
B) covered with a hot fuel bed.
C) covered with a thin fuel bed and ash.

3899. Which gasket material is most suitable for flange joints used in steam piping at 300 psi?
A) A bolt of thin sheets of fiber
B) Rubber inserted gaskets
C) Jacketed asbestos

3900. Chimney tops should be protected against lightning by
A) including factor of safety in the design.
B) installing a fuse of the proper size.
C) being properly grounded.

3901. What is usually the weakest part of the bent tube boiler?
A) Longitudinal seam of the drum
B) Circumference seam of the drum
C) Tube ligament of the drum
D) Girth seam of the drum

3902. Superheater tubes are prevented from overheating by
A) a high velocity steam flow.
B) intermittent contact between steam heating surfaces.
C) the moisture in the steam.
D) high velocity steam flow in intermittent contact between steam and heating surface.

3903. A vent condenser is found on a(n)
A) air-ejector from a condenser.
B) open feed water heater.
C) vacuum heater line.

3904. Test for carbon dioxide in a boiler water cycle using a(n)
A) Orsat analyzer.
B) de-aerator.
C) pH indicator.

3905. In a well-constructed engine that is properly adjusted and lubricated, the total loss due to friction should not be more than
A) 10% of the total loss.
B) 20% of the total loss.
C) 30% of the total loss.

3906. Which type of fuel burning system has a water screen?
A) Oil unit
B) Pulverized fuel unit
C) Neither A nor B

3907. Expansion pipes are usually expressed and calculated on the basis of
A) inches per hundred feet.
B) thousandths of an inch per 100 feet.
C) feet per hundred feet of pipe.

3908. The steam line connected to a water column should
A) pitch to the column.
B) pitch to the boiler.
C) be level.

3909. When the adjusting ring on a safety is secured by threads to the valve, raising the adjusting ring will
A) increase blow back.
B) decrease blow back.
C) not affect blow back.

3910. The indicated card diagram for the steam cylinder of a duplex pump is usually
A) trapezoid.

B) rectangular.
C) similar to a D-Slide engine.

3911. The boiling temperature of water at a pressure of 150 psi is approximately
A) 358 °F.
B) 330 °F.
C) 256 °F.

3912. Turbine oil tanks, pipes, etc., should never be cleaned with
A) a sponge.
B) a wiping cloth.
C) cotton waste.

3913. In a furnace having a combination of refractory and water-cooled walls,
A) there is no danger of incomplete combustion because of excessive heat.
B) there is danger of incomplete combustion.
C) incomplete combustion is forced.

3914. A joint mixture of 3 listed in use with rubber gaskets is
A) shellac.
B) graphite with oil.
C) graphite with water.

3915. It is considered good practice with bent tube, water-tube boilers to have
A) all of the tube ends flared.
B) all of the tube ends rolled and beaded.
C) some of the tube ends rolled and beaded.

3916. An automatic gauge glass shut off (of the approved type) is permitted for use
A) only with a flat gauge glass.
B) only with a circular gauge glass.
C) with either a flat or circular gauge glass.

3917. The purpose of a pump governor on a direct-acting steam pump for boiler feed service is to
A) control the speed of the pump.
B) maintain a constant pressure on the feed line.
C) prevent the pump from running away.

3918. If a few drops of phenolphthalein indicator is added to a sample of boiler water in a column that remains unchanged, the water is considered
A) alkaline.
B) acid.
C) acid or neutral.

3919. Unnecessary steam leakage past the various rows of moving blades in a steam turbine is prevented by
   A) gaskets.
   B) diaphragms.
   C) shaft seals.

3920. If a stoker-fired boiler using coal as fuel uses No. 6 fuel oil as an alternate, boiler capacity would
   A) be reduced.
   B) remain unchanged.
   C) possibly be increased.

3921. Which refractory is most suitable for use in stoker-fired boilers in areas of excessive abrasion or clinker formation?
   A) Silica carbide refractory
   B) Plastic fire boiler
   C) Glazed tile fire boiler

3922. Auto transformers have
   A) a single coil.
   B) a double coil.
   C) two or more coils.

3923. The usual method of supplying headers situated below the mud drum is by having the downcomers circulate water
   A) with intermittent line directly from a top steam and water drum.
   B) from the mud drum.
   C) directly from the feed water tank.

3924. The double-blowdown valve is one casing, but when used as the only valve on a boiler blow-off line is
   A) not permitted.
   B) permissible provided that one valve does not affect the operation of the other.
   C) permissible provided both valves are of the quick opening and quick closing type.

3925. In a tray-type, open, de-aerator, feed water heater, the de-aerator trays are located
   A) below the heating trays.
   B) above the heating trays.
   C) in the vent condenser.

3926. The primary purpose of softening boiler water is to
   A) increase boiler water alkalinity.
   B) remove its hardness and constituents by chemical means.
   C) change the temporary hardness characteristics.

3927. Waste heat boilers, as compared to an underfeed boiler,
   A) reduce the flue gas temperature to that of the surrounding air.
   B) have more clean out openings.
   C) have more fuel.

3928. When operating, steam pressure causes the cross section of the Bourdon tube to be
    A) oval.
    B) square.
    C) circle.

3929. The horizontal hanger spacing for copper tubing greater than 1-1/2 inches in diameter is
    A) 6 ft.
    B) 8 ft.
    C) 10 ft.
    D) 15 ft.

3930. One lb of saturated vapor contains the most total heat at
    A) the critical pressure with the corresponding temperature.
    B) 0 psia with the corresponding water temperature.
    C) 140 psia with the corresponding water temperature.
    D) 450 psia.

3931. The cut-off point of a detached Corliss engine depends upon the
    A) speed of the engine.
    B) position of the knock-off cam.
    C) positive pressure in the dash pots.

3932. Increased efficiency due to superheaters is greater in
    A) engines and turbines.
    B) boilers.
    C) auxiliaries.

3933. End thrust in centrifugal pumps is greater in
    A) the plunger type pump.
    B) single suction pumps.
    C) double suction pumps.

3934. Before a boiler is put in service after being cleaned with an acid solution, it should be
    A) deactivated.
    B) cleaned with a neutralizer.
    C) cleaned with Apexior.

3935. Safety necessitates that soot blowers not be operated unless the boiler is steaming at a capacity of at least
    A) 80%.
    B) 50%.
    C) 30%.

3936. To magnify the readings with a liquid manometer, the liquid can be replaced by another with
    A) a greater specific gravity.
    B) a lesser specific gravity.
    C) magniflux.

3937. If steam of 98% quality is being generated by a boiler, the carry over is
A) less than 2 percent.
B) 2 percent.
C) over 4 percent.

3938. A shaft governor with a slot eccentric controls the speed of an engine by changing
A) the angle of advance.
B) the eccentric.
C) steam pressure in the steam chest.

3939. It is not good practice to heat No. 6 fuel oil above 230 °F because it can cause
A) vaporization of the oil.
B) the sulfur to leave the oil.
C) carbonization in the heater tubes and on the burner tips.

3940. The fuel oil temperatures can be obtained practically
A) only if the oil preheater is on the pump side.
B) only if the oil preheater is on the pump discharge.
C) when the pump suction is taken from beneath the hood of the fuel oil tank.

3941. Boiler reading of horsepower developed is dependent upon
A) normal operating temperature and pressure.
B) only the extent of the heating surface.
C) the rate of heat transfer to the boiler.

3942. What information is not required on safety valve identification?
A) ASME symbol
B) Popping pressure
C) Capacity
D) Discharge pipe diameter

3943. A gauge glass cylinder bearing oil reservoir may have
A) pipe from top of glass only.
B) pipe from top and bottom.
C) only a bottom connection from the governor.

3944. The principal chemical factors that produce corrosion in piping and apparatus ahead and beyond the boiler are
A) dissolved gas and free mineral acids.
B) soda ash.
C) ammonia.

3945. The bladings most commonly used in a reactionary type turbine are made of
A) stationary blades that utilize stationary work.
B) moving and stationary blades.
C) moving blades.

3946. When starting up a stoker-fired boiler installation that has a superheater,
   A) build a small fire in middle of grate.
   B) build a small fire for the full width of the grate.
   C) warm up furnace with oil or gas burners.

3947. Large steam boilers have _______ try cocks.
   A) three
   B) two
   C) one
   D) no

3948. The rotation direction of a 3-phase induction motor may be changed by
   A) interchanging any two phases.
   B) installing interpoles.
   C) doubling the poles.

3949. If transferred from the fire to the boiler water, the largest temperature drop is
   A) water-tube wall.
   B) the gas film next to the tube.
   C) water mixed in with film next to the tube.

3950. A round gauge glass will thin more rapidly when the pH reading is
   A) 11.6.
   B) 8.5.
   C) 6.5.

3951. Lantern rings are used to
   A) space packing rings.
   B) provide shaft lubrication.
   C) aid in sealing.

3952. The horizontal hanger spacing for PVC pipe is
   A) 4 ft.
   B) 10 ft.
   C) 15 ft.
   D) 20 ft.

3953. When a solid-type connecting rod with inside adjusting wedges has its bearings adjusted for wear,
   A) head-end clearance is decreased.
   B) crank-end clearance is decreased.
   C) the connecting rod is shortened.

3954. In a steam atomizing oil burner, the most efficient steam to use is
   A) wet steam.
   B) dry steam.
   C) superheated steam.

3955. Solid, non-flexible stay bolts shall be drilled to a distance of 1/2" beyond the inside plate provided they
A) are more than 8" in length.
B) are less than 8" in length.
C) could be used on fire tube boilers only.

3956. When laying up a boiler using the wet method, the boiler should have a residual sulfite content of
A) 100 ppm.
B) 50 ppm.
C) 300 ppm.

3957. Which factor is used to convert grains per gallon to parts per million?
A) 17.1
B) 0.058
C) 0.58

3958. Which of the following has nothing to do with thrust in a turbine?
A) Kingsbury bearing
B) Shroud
C) Dummy piston

3959. No. 6 fuel oil has an API gravity of
A) 14 to 16.
B) 24 to 32.
C) 26 to 39.

3960. When the fusion temperature of the ash is high,
A) higher combustion rates may be maintained.
B) lower combustion rate should be maintained.
C) combustion rates cannot be increased.

3961. The oil temperature leaving a bearing of an operating turbine should be approximately
A) 120 °F.
B) 150 °F.
C) 180 °F.

3962. A boiler plate contains approximately how much carbon?
A) 0.01%
B) 0.025%
C) 0.03%

3963. The temperature of a radiant superheater will
A) drop with an increase in load.
B) increase with an increase in load.
C) decrease with a decrease in load.

3964. In comparing the duct area of flue gas and air passages of a regenerative heater,
A) flue gas passages are larger.
B) air passages are larger.
C) flue gas and air passages are of the same area.

3965. Silica will distill with boiler steam when pressures range from
   A) 100 to 200 psig.
   B) 200 to 300 psig.
   C) 400 psig and up.

3966. The cross-compound engine having crank at 90% will
   A) not have a receiver.
   B) exhaust directly to low-pressure cylinder.
   C) have a receiver.

3967. The proper volume for a spreader-type stoker is
   A) more than that of an underfeed type.
   B) less than that of an underfeed type.
   C) approximately the same as that of an underfeed type.

3968. When applying a safety valve gag for an hydrostatic pressure test, the gag should be set
   A) fingertight.
   B) with a wrench.
   C) with a Stillson wrench.

3969. When laying up a boiler using the dry method, the chemical placed in the tray is
   A) slaked lime.
   B) unslaked lime.
   C) sodium sulfite.

3970. A regenerative preheater rotates at a speed of about
   A) 7 rpm.
   B) 3 rpm.
   C) 10 rpm.

3971. A chloride test should be taken
   A) before or after the P or M test on the same sample.
   B) after neutralizing the sample.
   C) on a sample that has been taken from the boiler without being neutralized.

3972. The condenser that in normal operation does not require a water injection pump is a
   A) surface condenser.
   B) barometric condenser.
   C) low-level jack condenser.

3973. A general service wear of a centrifugal pump is required to handle either very hot or cold water. The best practice requires that a
   A) long fiber graphite cotton packing be used.
   B) flaxed packing be used.
   C) graphite asbestos packing be used.

3974. An interpole is found in use with
 A) a dc motor.
 B) an ac motor.
 C) phasing.

3975. A cross box forming a mud drum is found in use with a
 A) sinuous header.
 B) box header.
 C) boiler feed with very muddy water.

3976. Separation of water from steam would be very difficult if operating pressure were
 A) 600 psi.
 B) 3000 psig.
 C) 45 psig.

3977. The steam lead to an automatic thermostatic type feed water regulator should
 A) slope to the element.
 B) slope to the boiler.
 C) be level.

3978. On a heat pump, the defrost thermostat usually ___________ the defrost cycle.
 A) starts
 B) starts and/or ends
 C) ends

3979. In a pulverized fuel plant, a high degree of fineness is desired when burning
 A) anthracite.
 B) bituminous.
 C) lignite.

3980. After a turbine is shut down, the piece of equipment most likely to be kept running is the
 A) auxiliary oil pump.
 B) water to gland sealing.
 C) air ejector.

3981. The direction of rotation of a dc shunt wound motor may be reversed by
 A) changing armature connection.
 B) changing the phases.
 C) reversing the armature leads.

3982. A Wickes vertical water-tube boiler has
 A) bent tubes.
 B) two gas baffles.
 C) straight tubes.

3983. A nozzle-type safety valve operates on the
   A) huddling chamber principle.
   B) reaction principle.
   C) pilot valve.

3984. Multi-staged centrifugal pumps are usually
   A) involute.
   B) volute.
   C) diffuse.

3985. Complete removal of 1 pound of oxygen by chemical means require approximately
   A) 8 lb of sodium sulfate.
   B) 8 lb of sodium sulfite.
   C) 16 lb of sodium sulfate.

3986. The minimum site of an elliptical manhole permitted by ASME code is
   A) 11" x 15".
   B) 10" x 15".
   C) 11" x 16".

3987. By raising the blow back ring on a safety valve, the
   A) velocity of the safety valve is increased.
   B) capacity of the safety valve is decreased.
   C) popping off point is not affected.

3988. Tightening the adjusting nuts on a Copes feed-water regulator
   A) raises the drum water level.
   B) lowers the drum water level.
   C) increases the power of the tension release.

3989. After water is treated with a sodium zeolite softener, the dissolved solids are
   A) increased.
   B) the same.
   C) reduced.

3990. Release on an engine that is controlled with an inside admission piston valve
   A) occurs when the valve and piston are moving in opposite directions.
   B) occurs when the valve and piston are moving in the same directions.
   C) is not controlled by the valve.

3991. In pulverized fuel burning, fusion ash coals indicate the use of
   A) dry bottom furnace.
   B) tangential firing.
   C) slag tapped furnaces.

3992. In pulverized fuel burning, a fire during operation is usually first detected by
   A) smoke at tempering air inlet.
   B) an increase in outlet fuel air temperature.
   C) smoky flame at fire.

3993. The most stable speed regulation is attained with a
   A) shunt wound motor.
   B) series wound motor.
   C) compound motor.

3994. Water columns may be constructed of cast iron for use up to
   A) 160 psi.
   B) 250 psi.
   C) 450 psi.

3995. Throttling the discharge of a centrifugal pump will
   A) overload the driver.
   B) decrease load of driver.
   C) have no effect on the driver.

3996. Sodium sulfate is used in boiler feed water to
   A) prevent scale in boilers.
   B) eliminate oxygen.
   C) inhibit caustic embrittlement.

3997. Horsepower and type of engine being the same, when comparing a simple engine to a cross-compound engine, the
   A) cross-compound engine requires a larger flywheel.
   B) simple engine requires a larger flywheel.
   C) flywheel will be same size in both cases.

3998. The chain grate and traveling grate stokers differ in that
   A) one is stationary and the other one moves.
   B) the bigger one is supported by an overhead chain.
   C) one is constructed of small grate bars even though both constitute an endless chain.

3999. A sudden drop in superheater temperature is generally due to
   A) a leaky baffle.
   B) water carry over.
   C) dirty convection temperature.

4000. The highest starting torque is obtained with a
   A) series wound motor.
   B) shunt wound motor.
   C) compound motor.

Answers

Refrigeration and Air Conditioning, Level 1

1 B	25 B	49 C	73 A	97 A
2 C	26 B	50 C	74 B	98 B
3 A	27 A	51 C	75 B	99 C
4 B	28 A	52 C	76 A	100 C
5 C	29 A	53 A	77 B	101 C
6 A	30 A	54 B	78 C	102 C
7 A	31 C	55 C	79 C	103 A
8 C	32 A	56 A	80 A	104 A
9 B	33 C	57 B	81 A	105 B
10 A	34 B	58 A	82 C	106 C
11 B	35 C	59 B	83 A	107 C
12 C	36 C	60 B	84 B	108 A
13 B	37 C	61 C	85 C	109 B
14 A	38 B	62 B	86 A	110 A
15 C	39 B	63 B	87 C	111 C
16 C	40 C	64 B	88 C	112 A
17 B	41 B	65 C	89 A	113 C
18 B	42 B	66 C	90 B	114 A
19 C	43 C	67 C	91 B	115 A
20 C	44 C	68 B	92 C	116 C
21 A	45 A	69 B	93 A	117 B
22 C	46 B	70 A	94 B	118 B
23 A	47 C	71 C	95 B	119 A
24 A	48 A	72 C	96 B	120 B

121 C
122 C
123 B
124 B
125 C
126 C
127 A
128 C
129 C
130 C
131 C
132 C
133 B
134 B
135 B
136 B
137 C
138 B
139 B
140 A
141 A
142 C
143 B
144 B
145 C
146 A
147 C
148 B
149 C
150 A
151 C
152 C
153 B
154 C
155 C
156 B
157 A
158 C
159 C
160 C
161 B
162 D
163 C
164 D
165 A
166 B
167 A
168 B
169 A
170 B
171 D
172 B
173 A
174 A
175 D
176 A
177 D
178 C
179 B
180 B
181 D
182 A
183 A
184 A
185 D
186 A
187 B
188 A
189 D
190 B
191 C
192 B
193 D
194 C
195 A
196 A
197 D
198 D
199 D
200 A
201 B
202 B
203 D
204 C
205 C
206 B
207 B
208 A
209 A
210 B
211 A
212 A
213 A
214 A
215 B
216 B
217 B
218 A
219 B
220 A
221 B
222 A
223 B
224 A
225 C
226 A
227 B
228 B
229 A
230 D
231 B
232 B
233 C
234 B
235 C
236 D
237 B
238 D
239 C
240 B
241 A
242 C
243 A
244 B
245 C
246 C
247 B
248 A
249 B
250 A
251 A
252 C
253 B
254 B
255 C
256 D
257 D
258 B
259 B
260 D
261 A
262 B
263 D
264 C
265 A

266 D	295 D	324 A	353 B	382 A
267 B	296 C	325 C	354 C	383 C
268 C	297 A	326 C	355 D	384 D
269 D	298 A	327 A	356 D	385 B
270 C	299 C	328 C	357 B	386 A
271 C	300 B	329 D	358 C	387 A
272 A	301 C	330 A	359 B	388 C
273 D	302 A	331 D	360 B	389 C
274 A	303 A	332 C	361 A	390 D
275 B	304 B	333 A	362 C	391 D
276 D	305 A	334 C	363 D	392 B
277 B	306 A	335 B	364 C	393 A
278 B	307 A	336 D	365 B	394 A
279 D	308 B	337 B	366 D	395 A
280 D	309 A	338 A	367 C	396 A
281 A	310 C	339 B	368 D	397 A
282 D	311 C	340 C	369 D	398 A
283 A	312 A	341 B	370 A	399 B
284 D	313 B	342 D	371 B	400 A
285 A	314 C	343 D	372 B	401 A
286 C	315 C	344 B	373 B	402 B
287 B	316 B	345 C	374 B	403 B
288 D	317 D	346 A	375 B	404 B
289 D	318 A	347 D	376 B	405 C
290 C	319 D	348 D	377 C	406 C
291 B	320 A	349 C	378 B	407 B
292 B	321 D	350 A	379 A	408 C
293 C	322 C	351 B	380 C	409 A
294 B	323 C	352 A	381 B	410 B

411 C
412 B
413 C
414 C
415 C
416 A
417 B
418 A
419 B
420 B
421 C
422 A
423 A
424 B
425 B
426 C
427 D
428 A
429 C
430 A
431 D
432 B
433 A
434 B
435 A
436 C
437 B
438 D
439 A
440 D
441 D
442 B
443 D
444 C
445 B
446 A
447 C
448 D
449 B
450 A
451 A
452 A
453 D
454 B
455 D
456 A
457 A
458 B
459 A
460 B
461 C
462 A
463 C
464 C
465 B
466 D
467 A
468 B
469 B
470 B
471 C
472 C
473 A
474 B
475 C
476 B
477 B
478 A
479 B
480 A
481 D
482 C
483 A
484 A
485 D
486 C
487 A
488 B
489 B
490 C
491 C
492 B
493 A
494 B
495 A
496 A
497 B
498 C
499 C
500 B
501 A
502 B
503 D
504 A
505 B
506 D
507 B
508 D
509 B
510 D
511 A
512 A
513 D
514 B
515 A
516 B
517 B
518 B
519 B
520 B
521 A
522 C
523 C
524 B
525 D
526 C
527 B
528 A
529 C
530 C
531 B
532 D
533 A
534 B
535 C
536 D
537 D
538 C
539 D
540 B
541 B
542 B
543 B
544 B
545 C
546 C
547 C
548 B
549 D
550 B
551 C
552 C
553 C
554 B
555 C

556 D
557 B
558 D
559 D
560 C
561 D
562 D
563 B
564 D
565 B
566 C
567 B
568 C
569 A
570 E
571 E
572 B
573 A
574 C
575 B
576 C
577 D
578 D
579 C
580 C
581 D
582 B
583 C
584 B
585 C
586 C
587 B
588 A
589 C
590 C
591 B
592 B
593 C
594 C
595 C
596 C
597 B
598 C
599 B
600 C
601 C
602 B
603 B
604 A
605 B
606 A
607 B
608 C
609 C
610 B
611 A
612 B
613 B
614 B
615 A
616 C
617 B
618 B
619 C
620 C
621 B
622 C
623 C
624 B
625 B
626 C
627 B
628 A
629 C
630 C
631 B
632 B
633 B
634 B
635 C
636 B
637 C
638 B
639 B
640 C
641 B
642 C
643 C
644 A
645 B
646 B
647 B
648 A
649 B
650 B
651 C
652 B
653 C
654 A
655 B
656 A
657 A
658 B
659 A
660 B
661 A
662 B
663 C
664 C
665 B
666 C
667 A
668 C
669 B
670 C
671 B
672 A
673 A
674 C
675 B
676 B
677 B
678 A
679 B
680 B
681 A
682 B
683 B
684 B
685 C
686 C
687 B
688 B
689 A
690 B
691 B
692 A
693 A
694 B
695 B
696 A
697 B
698 A
699 A
700 A

Refrigeration and Air Conditioning, Level 2

701 A	728 B	755 C	782 C	809 A
702 A	729 A	756 C	783 B	810 B
703 A	730 A	757 C	784 A	811 A
704 A	731 A	758 A	785 A	812 C
705 B	732 B	759 C	786 A	813 B
706 C	733 A	760 A	787 A	814 B
707 B	734 B	761 C	788 B	815 B
708 B	735 B	762 A	789 A	816 C
709 A	736 A	763 B	790 A	817 C
710 C	737 C	764 A	791 B	818 C
711 B	738 A	765 C	792 B	819 A
712 B	739 A	766 A	793 C	820 B
713 C	740 A	767 B	794 B	821 C
714 B	741 B	768 B	795 A	822 B
715 B	742 A	769 C	796 B	823 C
716 A	743 B	770 B	797 B	824 B
717 B	744 C	771 B	798 A	825 B
718 C	745 B	772 A	799 B	826 C
719 A	746 B	773 B	800 B	827 C
720 B	747 B	774 B	801 A	828 B
721 A	748 A	775 C	802 A	829 C
722 A	749 C	776 A	803 A	830 B
723 B	750 C	777 C	804 C	831 C
724 A	751 A	778 C	805 A	832 C
725 C	752 C	779 C	806 A	833 A
726 B	753 B	780 A	807 B	834 A
727 A	754 B	781 A	808 C	835 C

836 A
837 C
838 C
839 C
840 A
841 B
842 A
843 B
844 C
845 A
846 A
847 A
848 C
849 B
850 B
851 A
852 C
853 B
854 A
855 A
856 A
857 B
858 B
859 A
860 C
861 A
862 C
863 C
864 B
865 B
866 C
867 A
868 B
869 B
870 C
871 C
872 A
873 C
874 A
875 C
876 A
877 A
878 B
879 B
880 C
881 C
882 C
883 B
884 B
885 B
886 A
887 A
888 B
889 A
890 A
891 A
892 D
893 C
894 B
895 A
896 D
897 C
898 A
899 A
900 A
901 A
902 B
903 A
904 D
905 A
906 C
907 C
908 C
909 B
910 A
911 C
912 C
913 C
914 A
915 A
916 B
917 A
918 A
919 B
920 A
921 C
922 C
923 C
924 A
925 C
926 B
927 B
928 B
929 B
930 C
931 B
932 A
933 B
934 A
935 B
936 C
937 B
938 B
939 A
940 B
941 A
942 A
943 B
944 B
945 A
946 A
947 A
948 C
949 A
950 A
951 C
952 B
953 B
954 C
955 B
956 C
957 C
958 A
959 C
960 B
961 A
962 C
963 B
964 B
965 A
966 B
967 D
968 B
969 C
970 B
971 A
972 B
973 B
974 B
975 C
976 C
977 B
978 D
979 A
980 C

981 C
982 D
983 C
984 B
985 D
986 D
987 A
988 B
989 D
990 B
991 C
992 B
993 D
994 B
995 C
996 A
997 A
998 D
999 B
1000 B
1001 D
1002 B
1003 A
1004 C
1005 A
1006 B
1007 A
1008 A
1009 D

1010 A
1011 D
1012 D
1013 B
1014 A
1015 C
1016 C
1017 A
1018 D
1019 D
1020 C
1021 C
1022 C
1023 D
1024 A
1025 A
1026 A
1027 A
1028 C
1029 C
1030 B
1031 A
1032 D
1033 B
1034 B
1035 B
1036 B
1037 A
1038 B

1039 A
1040 B
1041 A
1042 B
1043 B
1044 B
1045 C
1046 C
1047 A
1048 C
1049 A
1050 A
1051 A
1052 A
1053 A
1054 C
1055 A
1056 C
1057 C
1058 B
1059 A
1060 A
1061 C
1062 D
1063 A
1064 A
1065 B
1066 B
1067 A

1068 D
1069 B
1070 A
1071 B
1072 B
1073 B
1074 D
1075 B
1076 A
1077 A
1078 B
1079 A
1080 A
1081 B
1082 C
1083 A
1084 B
1085 D
1086 A
1087 B
1088 B
1089 B
1090 A
1091 B
1092 D
1093 D
1094 C
1095 D
1096 D

1097 B
1098 A
1099 B
1100 C
1101 C
1102 C
1103 C
1104 D
1105 D
1106 C
1107 B
1108 D
1109 C
1110 B
1111 B
1112 A
1113 B
1114 A
1115 C
1116 A
1117 B
1118 B
1119 B
1120 A
1121 A
1122 B
1123 A
1124 A
1125 B

1126 A
1127 A
1128 B
1129 B
1130 B
1131 A
1132 B
1133 A
1134 B
1135 B
1136 A
1137 B
1138 B
1139 A
1140 B
1141 C
1142 A
1143 C
1144 A
1145 B
1146 C
1147 C
1148 C
1149 D
1150 C
1151 A
1152 A
1153 B
1154 B
1155 B
1156 B
1157 C
1158 C
1159 A
1160 B
1161 C
1162 C
1163 A
1164 A
1165 A
1166 A
1167 B
1168 C
1169 C
1170 B
1171 B
1172 D
1173 B
1174 A
1175 B
1176 C
1177 C
1178 C
1179 C
1180 B
1181 D
1182 C
1183 B
1184 B
1185 B
1186 B
1187 D
1188 A
1189 C
1190 A
1191 A
1192 B
1193 B
1194 D
1195 B
1196 A
1197 A
1198 D
1199 B
1200 D
1201 A
1202 B
1203 B
1204 B
1205 A
1206 B
1207 B
1208 C
1209 B
1210 A
1211 B
1212 D
1213 A
1214 A
1215 C
1216 E
1217 B
1218 E
1219 D
1220 D
1221 B
1222 C
1223 D
1224 A
1225 D
1226 C
1227 D
1228 A
1229 A
1230 C
1231 C
1232 A
1233 C
1234 A
1235 C
1236 C
1237 B
1238 B
1239 C
1240 D
1241 A
1242 A
1243 A
1244 B
1245 C
1246 E
1247 E
1248 E
1249 C
1250 A
1251 A
1252 C
1253 C
1254 C
1255 C
1256 C
1257 D
1258 B
1259 C
1260 D
1261 D
1262 C
1263 D
1264 C
1265 B
1266 A
1267 B
1268 C
1269 B
1270 B

1271 B	1297 B	1323 A	1349 A	1375 A
1272 B	1298 C	1324 A	1350 C	1376 A
1273 B	1299 B	1325 A	1351 B	1377 C
1274 C	1300 B	1326 A	1352 A	1378 C
1275 C	1301 A	1327 A	1353 A	1379 B
1276 C	1302 A	1328 D	1354 A	1380 A
1277 B	1303 B	1329 A	1355 A	1381 B
1278 B	1304 C	1330 C	1356 A	1382 B
1279 D	1305 D	1331 D	1357 A	1383 A
1280 D	1306 C	1332 C	1358 A	1384 A
1281 D	1307 A	1333 C	1359 B	1385 A
1282 C	1308 D	1334 A	1360 B	1386 C
1283 C	1309 C	1335 C	1361 B	1387 C
1284 B	1310 B	1336 C	1362 C	1388 A
1285 C	1311 A	1337 A	1363 A	1389 B
1286 C	1312 A	1338 D	1364 C	1390 B
1287 D	1313 A	1339 B	1365 A	1391 A
1288 C	1314 B	1340 D	1366 A	1392 C
1289 C	1315 A	1341 A	1367 A	1393 A
1290 C	1316 C	1342 A	1368 B	1394 B
1291 B	1317 C	1343 D	1369 A	1395 B
1292 B	1318 A	1344 A	1370 A	1396 A
1293 B	1319 B	1345 A	1371 A	1397 B
1294 C	1320 C	1346 B	1372 C	1398 C
1295 C	1321 D	1347 B	1373 B	1399 B
1296 C	1322 B	1348 B	1374 D	1400 B

Refrigeration and Air Conditioning, Level 3

1401 C
1402 B
1403 C
1404 B
1405 C
1406 A
1407 A
1408 B
1409 C
1410 C
1411 B
1412 A
1413 C
1414 C
1415 A
1416 B
1417 B
1418 A
1419 C
1420 C
1421 C
1422 B
1423 C
1424 C
1425 B
1426 C
1427 B
1428 B
1429 C
1430 C
1431 B
1432 C
1433 A
1434 C
1435 B
1436 C
1437 A
1438 A
1439 B
1440 A
1441 A
1442 B
1443 A
1444 A
1445 C
1446 A
1447 C
1448 B
1449 C
1450 C
1451 B
1452 C
1453 C
1454 C
1455 B
1456 A
1457 A
1458 B
1459 B
1460 B
1461 A
1462 B
1463 C
1464 B
1465 C
1466 B
1467 B
1468 B
1469 B
1470 A
1471 B
1472 A
1473 A
1474 A
1475 A
1476 B
1477 A
1478 B
1479 A
1480 B
1481 B
1482 C
1483 A
1484 A
1485 B
1486 A
1487 B
1488 B
1489 B
1490 A
1491 A
1492 A
1493 B
1494 A
1495 C
1496 B
1497 A
1498 C
1499 C
1500 A
1501 A
1502 B
1503 A
1504 C
1505 B
1506 C
1507 C
1508 B
1509 B
1510 A
1511 C
1512 A
1513 D
1514 A
1515 B
1516 C
1517 C
1518 D
1519 B
1520 A
1521 C
1522 C
1523 B
1524 A
1525 C
1526 A
1527 B
1528 B
1529 B
1530 C
1531 B
1532 A
1533 B
1534 A
1535 B

1536 C
1537 C
1538 C
1539 B
1540 C
1541 C
1542 B
1543 A
1544 A
1545 A
1546 C
1547 C
1548 A
1549 B
1550 B
1551 A
1552 C
1553 C
1554 A
1555 A
1556 C
1557 B
1558 A
1559 B
1560 B
1561 A
1562 B
1563 C
1564 C
1565 C
1566 B
1567 B
1568 C
1569 A
1570 B
1571 B
1572 A
1573 A
1574 B
1575 C
1576 C
1577 A
1578 B
1579 A
1580 B
1581 C
1582 B
1583 B
1584 A
1585 B
1586 B
1587 B
1588 B
1589 A
1590 A
1591 D
1592 C
1593 C
1594 A
1595 B
1596 B
1597 B
1598 D
1599 B
1600 A
1601 B
1602 D
1603 D
1604 B
1605 A
1606 A
1607 A
1608 A
1609 A
1610 A
1611 B
1612 B
1613 A
1614 A
1615 A
1616 C
1617 B
1618 B
1619 C
1620 A
1621 C
1622 A
1623 B
1624 C
1625 A
1626 A
1627 A
1628 B
1629 B
1630 C
1631 A
1632 C
1633 B
1634 C
1635 C
1636 B
1637 A
1638 A
1639 A
1640 B
1641 A
1642 C
1643 A
1644 A
1645 C
1646 B
1647 D
1648 C
1649 D
1650 C
1651 D
1652 A
1653 A
1654 B
1655 A
1656 B
1657 A
1658 A
1659 C
1660 B
1661 D
1662 C
1663 B
1664 D
1665 D
1666 C
1667 C
1668 C
1669 D
1670 C
1671 B
1672 C
1673 D
1674 A
1675 D
1676 B
1677 A
1678 C
1679 B
1680 C

1681 A
1682 B
1683 C
1684 A
1685 A
1686 A
1687 B
1688 A
1689 A
1690 A
1691 B
1692 A
1693 A
1694 A
1695 A
1696 B
1697 A
1698 A
1699 A
1700 B
1701 B
1702 B
1703 B
1704 B
1705 B
1706 A
1707 A
1708 B
1709 B
1710 A
1711 B
1712 A
1713 A
1714 B
1715 A
1716 A
1717 B
1718 B
1719 B
1720 A
1721 A
1722 B
1723 B
1724 B
1725 A
1726 A
1727 A
1728 A
1729 A
1730 B
1731 B
1732 B
1733 B
1734 B
1735 B
1736 A
1737 B
1738 B
1739 A
1740 A
1741 A
1742 B
1743 A
1744 B
1745 A
1746 A
1747 B
1748 A
1749 B
1750 A
1751 A
1752 A
1753 C
1754 B
1755 C
1756 A
1757 A
1758 A
1759 B
1760 C
1761 B
1762 A
1763 B
1764 C
1765 C
1766 C
1767 B
1768 A
1769 B
1770 B
1771 B
1772 C
1773 C
1774 B
1775 A
1776 C
1777 B
1778 B
1779 C
1780 A
1781 A
1782 C
1783 B
1784 D
1785 B
1786 B
1787 A
1788 B
1789 A
1790 B
1791 C
1792 A
1793 C
1794 D
1795 C
1796 C
1797 D
1798 B
1799 A
1800 B
1801 C
1802 A
1803 D
1804 B
1805 C
1806 A
1807 C
1808 D
1809 D
1810 C
1811 A
1812 A
1813 A
1814 B
1815 B
1816 B
1817 B
1818 B
1819 B
1820 B
1821 A
1822 B
1823 A
1824 A
1825 A

1826 C
1827 C
1828 B
1829 A
1830 C
1831 B
1832 B
1833 A
1834 B
1835 D
1836 A
1837 A
1838 D
1839 D
1840 A
1841 B
1842 D
1843 A
1844 C
1845 C
1846 C
1847 B
1848 B
1849 A
1850 B
1851 D
1852 B
1853 D
1854 C
1855 A
1856 A
1857 A
1858 C
1859 A
1860 C
1861 C
1862 B
1863 B
1864 B
1865 D
1866 B
1867 C
1868 C
1869 A
1870 D
1871 B
1872 C
1873 C
1874 E
1875 E
1876 D
1877 D
1878 B
1879 C
1880 A
1881 A
1882 D
1883 E
1884 B
1885 A
1886 D
1887 A
1888 A
1889 B
1890 A
1891 B
1892 D
1893 B
1894 E
1895 E
1896 A
1897 C
1898 C
1899 D
1900 A
1901 C
1902 B
1903 C
1904 A
1905 A
1906 B
1907 A
1908 B
1909 D
1910 D
1911 B
1912 A
1913 D
1914 C
1915 A
1916 A
1917 A
1918 C
1919 B
1920 C
1921 A
1922 D
1923 B
1924 A
1925 B
1926 B
1927 B
1928 C
1929 A
1930 A
1931 A
1932 A
1933 D
1934 A
1935 A
1936 B
1937 B
1938 D
1939 C
1940 D
1941 C
1942 C
1943 D
1944 B
1945 D
1946 D
1947 C
1948 A
1949 B
1950 C
1951 C
1952 D
1953 A
1954 B
1955 D
1956 C
1957 C
1958 D
1959 C
1960 D
1961 A
1962 D
1963 D
1964 C
1965 D
1966 B
1967 B
1968 A
1969 C
1970 D

1971 A
1972 B
1973 D
1974 B
1975 B
1976 A
1977 B
1978 B
1979 D
1980 B
1981 A
1982 C
1983 D
1984 A
1985 A
1986 B
1987 D
1988 D
1989 B
1990 B
1991 D
1992 A
1993 A
1994 B
1995 D
1996 D
1997 C
1998 C
1999 D
2000 C

Heating, Level 1

2001 B
2002 C
2003 C
2004 B
2005 D
2006 A
2007 D
2008 A
2009 B
2010 C
2011 D
2012 C
2013 D
2014 C
2015 C
2016 A
2017 B
2018 C
2019 C
2020 C
2021 D
2022 C
2023 B
2024 C
2025 B
2026 B
2027 D
2028 B
2029 C
2030 C
2031 A
2032 D
2033 D
2034 A
2035 B
2036 C
2037 C
2038 C
2039 B
2040 A
2041 A
2042 D
2043 B
2044 C
2045 C
2046 B
2047 A
2048 B
2049 B
2050 D
2051 B
2052 D
2053 A
2054 D
2055 E
2056 C
2057 B
2058 B
2059 C
2060 C
2061 C
2062 D
2063 C
2064 A
2065 E
2066 A
2067 C
2068 E
2069 A
2070 E
2071 B
2072 E
2073 A
2074 A
2075 C
2076 C
2077 C
2078 B
2079 B
2080 B
2081 A
2082 D
2083 A
2084 C
2085 D
2086 C
2087 B
2088 D
2089 B
2090 D
2091 E
2092 D
2093 E
2094 E
2095 D
2096 C
2097 B
2098 C
2099 C
2100 C

2101 C
2102 E
2103 B
2104 A
2105 B
2106 D
2107 C
2108 D
2109 A
2110 B
2111 C
2112 C
2113 C
2114 D
2115 C
2116 A
2117 C
2118 A
2119 B
2120 D
2121 D
2122 C
2123 A
2124 B
2125 C
2126 E
2127 B
2128 C
2129 D
2130 B
2131 C
2132 C
2133 A
2134 C
2135 B
2136 A
2137 C
2138 B
2139 C
2140 B
2141 B
2142 C
2143 C
2144 A
2145 C
2146 C
2147 A
2148 D
2149 C
2150 A
2151 C
2152 B
2153 C
2154 D
2155 B
2156 C
2157 A
2158 B
2159 B
2160 A
2161 A
2162 B
2163 D
2164 A
2165 D
2166 B
2167 C
2168 C
2169 A
2170 B
2171 C
2172 A
2173 D
2174 B
2175 C
2176 D
2177 B
2178 A
2179 A
2180 C
2181 A
2182 D
2183 A
2184 C
2185 B
2186 A
2187 D
2188 D
2189 C
2190 C
2191 C
2192 B
2193 A
2194 D
2195 C
2196 A
2197 B
2198 D
2199 C
2200 B
2201 D
2202 B
2203 C
2204 A
2205 B
2206 C
2207 A
2208 A
2209 B
2210 C
2211 B
2212 A
2213 B
2214 B
2215 A
2216 A
2217 D
2218 C
2219 B
2220 D
2221 C
2222 B
2223 A
2224 C
2225 B
2226 B
2227 B
2228 C
2229 D
2230 B
2231 B
2232 C
2233 B
2234 A
2235 B
2236 A
2237 B
2238 C
2239 B
2240 B
2241 A
2242 B
2243 D
2244 C
2245 C

2246 B
2247 B
2248 B
2249 A
2250 A
2251 C
2252 A
2253 B
2254 B
2255 B
2256 B
2257 C
2258 A
2259 B
2260 B
2261 B
2262 C
2263 A
2264 B
2265 B
2266 A
2267 B
2268 B
2269 C
2270 C
2271 C
2272 C
2273 B
2274 A
2275 C
2276 A
2277 C
2278 B
2279 C
2280 C
2281 A
2282 B
2283 A
2284 C
2285 C
2286 B
2287 A
2288 A
2289 A
2290 B
2291 B
2292 A
2293 B
2294 A
2295 A
2296 A
2297 C
2298 C
2299 C
2300 B
2301 B
2302 D
2303 A
2304 C
2305 D
2306 B
2307 C
2308 A
2309 A
2310 A
2311 A
2312 B
2313 B
2314 C
2315 C
2316 A
2317 C
2318 A
2319 D
2320 D
2321 A
2322 B
2323 D
2324 A
2325 A
2326 D
2327 A
2328 B
2329 A
2330 C
2331 C
2332 C
2333 B
2334 C
2335 A
2336 A
2337 A
2338 B
2339 D
2340 C
2341 B
2342 A
2343 C
2344 B
2345 D
2346 B
2347 A
2348 D
2349 D
2350 D
2351 B
2352 C
2353 D
2354 B
2355 C
2356 B
2357 A
2358 B
2359 D
2360 B
2361 B
2362 B
2363 A
2364 C
2365 C
2366 C
2367 C
2368 B
2369 B
2370 A
2371 C
2372 B
2373 B
2374 D
2375 B
2376 B
2377 B
2378 D
2379 D
2380 D
2381 B
2382 D
2383 B
2384 B
2385 C
2386 B
2387 A
2388 D
2389 C
2390 B

2391 C	2420 B	2449 D	2478 C	2507 B
2392 C	2421 C	2450 D	2479 D	2508 C
2393 A	2422 C	2451 A	2480 A	2509 D
2394 C	2423 C	2452 B	2481 B	2510 A
2395 C	2424 C	2453 D	2482 C	2511 D
2396 D	2425 D	2454 D	2483 B	2512 A
2397 C	2426 C	2455 D	2484 B	2513 B
2398 D	2427 C	2456 D	2485 A	2514 C
2399 A	2428 C	2457 B	2486 A	2515 A
2400 A	2429 A	2458 A	2487 A	2516 C
2401 B	2430 B	2459 B	2488 A	2517 C
2402 A	2431 C	2460 D	2489 D	2518 A
2403 B	2432 A	2461 D	2490 D	2519 A
2404 D	2433 B	2462 A	2491 B	2520 A
2405 C	2434 C	2463 A	2492 D	2521 B
2406 A	2435 D	2464 A	2493 C	2522 B
2407 B	2436 D	2465 D	2494 D	2523 D
2408 A	2437 B	2466 D	2495 C	2524 C
2409 B	2438 C	2467 C	2496 D	2525 B
2410 C	2439 A	2468 D	2497 D	2526 C
2411 C	2440 A	2469 C	2498 D	2527 C
2412 A	2441 A	2470 C	2499 D	2528 D
2413 B	2442 A	2471 B	2500 C	2529 C
2414 C	2443 A	2472 C	2501 D	2530 A
2415 D	2444 A	2473 C	2502 C	2531 B
2416 B	2445 A	2474 A	2503 C	2532 C
2417 B	2446 D	2475 B	2504 C	2533 D
2418 D	2447 D	2476 C	2505 A	2534 A
2419 B	2448 D	2477 A	2506 B	2535 B

2536 C
2537 C
2538 A
2539 C
2540 A
2541 C
2542 A
2543 A
2544 C
2545 B
2546 D
2547 D
2548 C
2549 C
2550 B
2551 C
2552 C
2553 B
2554 B
2555 D
2556 C
2557 C
2558 A
2559 B
2560 B
2561 C
2562 A
2563 B
2564 B
2565 C
2566 B
2567 A
2568 C
2569 D
2570 C
2571 C
2572 B
2573 C
2574 C
2575 A
2576 B
2577 D
2578 D
2579 A
2580 D
2581 A
2582 B
2583 C
2584 C
2585 D
2586 B
2587 B
2588 C
2589 B
2590 B
2591 A
2592 D
2593 C
2594 C
2595 C
2596 B
2597 D
2598 A
2599 D
2600 D

Heating, Level 2

2601 B
2602 B
2603 C
2604 D
2605 B
2606 C
2607 B
2608 D
2609 C
2610 B
2611 C
2612 A
2613 B
2614 C
2615 A
2616 B
2617 D
2618 A
2619 A
2620 A
2621 A
2622 B
2623 A
2624 A
2625 C
2626 B
2627 C
2628 D
2629 B
2630 A
2631 B
2632 D
2633 D
2634 B
2635 A
2636 A
2637 A
2638 A
2639 A
2640 C
2641 D
2642 D
2643 A
2644 B
2645 D
2646 D
2647 B
2648 A
2649 B
2650 A
2651 A
2652 B
2653 B
2654 D
2655 A
2656 D
2657 D
2658 B
2659 C
2660 D
2661 A
2662 D
2663 D
2664 A
2665 C

2666 B
2667 D
2668 B
2669 B
2670 D
2671 D
2672 A
2673 C
2674 D
2675 C
2676 A
2677 B
2678 B
2679 B
2680 D
2681 A
2682 D
2683 C
2684 B
2685 C
2686 C
2687 A
2688 C
2689 D
2690 D
2691 A
2692 B
2693 C
2694 D

2695 B
2696 B
2697 B
2698 A
2699 A
2700 A
2701 C
2702 D
2703 D
2704 C
2705 B
2706 A
2707 D
2708 D
2709 C
2710 D
2711 D
2712 B
2713 C
2714 A
2715 C
2716 B
2717 B
2718 A
2719 C
2720 A
2721 D
2722 A
2723 A

2724 C
2725 B
2726 B
2727 B
2728 A
2729 B
2730 C
2731 D
2732 B
2733 A
2734 B
2735 C
2736 C
2737 B
2738 D
2739 D
2740 A
2741 A
2742 A
2743 D
2744 D
2745 C
2746 A
2747 B
2748 C
2749 B
2750 C
2751 C
2752 B

2753 A
2754 C
2755 D
2756 C
2757 D
2758 A
2759 D
2760 A
2761 C
2762 B
2763 D
2764 A
2765 B
2766 C
2767 A
2768 D
2769 A
2770 C
2771 A
2772 A
2773 B
2774 C
2775 D
2776 D
2777 C
2778 C
2779 C
2780 C
2781 C

2782 D
2783 A
2784 A
2785 A
2786 C
2787 A
2788 A
2789 C
2790 D
2791 B
2792 B
2793 A
2794 B
2795 C
2796 D
2797 C
2798 A
2799 D
2800 D
2801 A
2802 B
2803 A
2804 B
2805 C
2806 B
2807 C
2808 B
2809 A
2810 A

2811 B	2840 A	2869 B	2898 A	2927 B
2812 B	2841 A	2870 A	2899 B	2928 A
2813 C	2842 B	2871 B	2900 A	2929 C
2814 B	2843 A	2872 B	2901 B	2930 A
2815 C	2844 B	2873 A	2902 A	2931 A
2816 A	2845 A	2874 B	2903 A	2932 C
2817 C	2846 A	2875 C	2904 A	2933 A
2818 C	2847 B	2876 B	2905 A	2934 C
2819 B	2848 A	2877 B	2906 B	2935 B
2820 A	2849 B	2878 B	2907 A	2936 B
2821 A	2850 A	2879 A	2908 C	2937 B
2822 C	2851 A	2880 A	2909 C	2938 B
2823 C	2852 A	2881 C	2910 C	2939 A
2824 B	2853 A	2882 B	2911 C	2940 B
2825 B	2854 A	2883 B	2912 C	2941 A
2826 C	2855 B	2884 A	2913 A	2942 B
2827 A	2856 B	2885 A	2914 A	2943 B
2828 C	2857 B	2886 C	2915 C	2944 C
2829 A	2858 A	2887 B	2916 C	2945 A
2830 B	2859 B	2888 B	2917 C	2946 C
2831 B	2860 A	2889 B	2918 B	2947 A
2832 A	2861 C	2890 B	2919 A	2948 B
2833 A	2862 A	2891 A	2920 A	2949 B
2834 C	2863 C	2892 A	2921 B	2950 C
2835 A	2864 C	2893 C	2922 B	2951 A
2836 C	2865 B	2894 A	2923 B	2952 C
2837 A	2866 B	2895 C	2924 C	2953 A
2838 C	2867 C	2896 B	2925 A	2954 B
2839 B	2868 B	2897 A	2926 C	2955 B

2956 A	2985 A	3014 B	3043 A	3072 C
2957 A	2986 C	3015 C	3044 C	3073 B
2958 A	2987 A	3016 C	3045 C	3074 A
2959 C	2988 C	3017 A	3046 C	3075 D
2960 A	2989 B	3018 C	3047 C	3076 A
2961 C	2990 C	3019 B	3048 A	3077 B
2962 A	2991 A	3020 C	3049 C	3078 A
2963 C	2992 C	3021 A	3050 B	3079 D
2964 A	2993 C	3022 C	3051 C	3080 B
2965 B	2994 A	3023 B	3052 A	3081 A
2966 A	2995 A	3024 C	3053 A	3082 B
2967 A	2996 B	3025 A	3054 A	3083 B
2968 C	2997 B	3026 B	3055 B	3084 B
2969 A	2998 B	3027 A	3056 C	3085 A
2970 A	2999 B	3028 A	3057 C	3086 B
2971 A	3000 B	3029 C	3058 A	3087 A
2972 B	3001 C	3030 C	3059 B	3088 C
2973 A	3002 A	3031 B	3060 C	3089 B
2974 B	3003 A	3032 B	3061 C	3090 B
2975 B	3004 A	3033 A	3062 B	3091 A
2976 A	3005 D	3034 C	3063 C	3092 A
2977 A	3006 B	3035 B	3064 B	3093 B
2978 B	3007 B	3036 C	3065 A	3094 A
2979 C	3008 A	3037 C	3066 B	3095 B
2980 B	3009 C	3038 B	3067 B	3096 B
2981 C	3010 C	3039 B	3068 A	3097 B
2982 A	3011 B	3040 A	3069 A	3098 C
2983 B	3012 C	3041 A	3070 C	3099 A
2984 B	3013 A	3042 C	3071 A	3100 A

3101 B
3102 B
3103 C
3104 B
3105 B
3106 A
3107 C
3108 B
3109 B
3110 C
3111 A
3112 B
3113 A
3114 A
3115 C
3116 C
3117 B
3118 A
3119 A
3120 B
3121 B
3122 A
3123 C
3124 B
3125 C
3126 B
3127 A
3128 B
3129 B
3130 B
3131 B
3132 A
3133 A
3134 D
3135 B
3136 B
3137 A
3138 B
3139 A
3140 A
3141 A
3142 A
3143 B
3144 C
3145 A
3146 B
3147 B
3148 C
3149 C
3150 A
3151 A
3152 C
3153 A
3154 C
3155 B
3156 A
3157 C
3158 C
3159 A
3160 A
3161 B
3162 A
3163 B
3164 A
3165 A
3166 A
3167 A
3168 A
3169 A
3170 A
3171 B
3172 A
3173 A
3174 B
3175 B
3176 A
3177 A
3178 C
3179 A
3180 C
3181 C
3182 C
3183 B
3184 C
3185 C
3186 B
3187 C
3188 A
3189 A
3190 C
3191 A
3192 C
3193 C
3194 B
3195 A
3196 A
3197 A
3198 C
3199 A
3200 B
3201 B
3202 B
3203 B
3204 C
3205 B
3206 C
3207 B
3208 A
3209 C
3210 B
3211 B
3212 A
3213 A
3214 A
3215 B
3216 B
3217 C
3218 C
3219 A
3220 B
3221 A
3222 B
3223 A
3224 B
3225 A
3226 C
3227 B
3228 A
3229 C
3230 B
3231 B
3232 C
3233 A
3234 C
3235 B
3236 C
3237 A
3238 B
3239 A
3240 A
3241 B
3242 B
3243 A
3244 A
3245 A

3246 C	3257 C	3268 A	3279 B	3290 C
3247 B	3258 B	3269 B	3280 C	3291 D
3248 B	3259 C	3270 A	3281 B	3292 D
3249 A	3260 C	3271 B	3282 A	3293 B
3250 A	3261 C	3272 B	3283 D	3294 A
3251 C	3262 B	3273 B	3284 A	3295 C
3252 B	3263 B	3274 A	3285 C	3296 C
3253 A	3264 B	3275 A	3286 A	3297 D
3254 A	3265 B	3276 B	3287 C	3298 C
3255 C	3266 C	3277 D	3288 A	3299 A
3256 A	3267 B	3278 B	3289 B	3300 A

Heating, Level 3

3301 B	3316 A	3331 D	3346 A	3361 D
3302 C	3317 A	3332 D	3347 B	3362 A
3303 D	3318 A	3333 D	3348 A	3363 B
3304 D	3319 D	3334 A	3349 A	3364 B
3305 C	3320 D	3335 C	3350 C	3365 B
3306 B	3321 B	3336 B	3351 A	3366 A
3307 D	3322 C	3337 A	3352 C	3367 B
3308 D	3323 C	3338 B	3353 A	3368 A
3309 C	3324 B	3339 C	3354 A	3369 A
3310 B	3325 A	3340 C	3355 C	3370 B
3311 B	3326 B	3341 A	3356 B	3371 B
3312 A	3327 D	3342 A	3357 B	3372 B
3313 A	3328 D	3343 A	3358 C	3373 A
3314 D	3329 B	3344 A	3359 A	3374 B
3315 C	3330 B	3345 A	3360 C	3375 B

3376 C
3377 B
3378 C
3379 B
3380 B
3381 A
3382 A
3383 B
3384 B
3385 B
3386 B
3387 C
3388 B
3389 B
3390 B
3391 A
3392 A
3393 C
3394 C
3395 B
3396 A
3397 B
3398 A
3399 A
3400 B
3401 A
3402 B
3403 C
3404 A
3405 A
3406 A
3407 C
3408 B
3409 B
3410 A
3411 C
3412 B
3413 B
3414 A
3415 C
3416 A
3417 A
3418 A
3419 B
3420 A
3421 B
3422 A
3423 B
3424 C
3425 C
3426 C
3427 B
3428 B
3429 C
3430 A
3431 A
3432 B
3433 A
3434 B
3435 A
3436 C
3437 C
3438 B
3439 B
3440 B
3441 B
3442 B
3443 B
3444 A
3445 B
3446 A
3447 C
3448 B
3449 A
3450 A
3451 A
3452 C
3453 C
3454 C
3455 C
3456 A
3457 B
3458 C
3459 C
3460 B
3461 C
3462 A
3463 B
3464 A
3465 A
3466 B
3467 B
3468 A
3469 A
3470 C
3471 D
3472 B
3473 C
3474 B
3475 A
3476 B
3477 C
3478 C
3479 A
3480 A
3481 A
3482 C
3483 C
3484 A
3485 C
3486 A
3487 B
3488 B
3489 A
3490 A
3491 A
3492 B
3493 A
3494 C
3495 C
3496 B
3497 B
3498 C
3499 A
3500 A
3501 A
3502 C
3503 A
3504 B
3505 B
3506 A
3507 B
3508 A
3509 C
3510 B
3511 C
3512 C
3513 B
3514 B
3515 C
3516 B
3517 A
3518 A
3519 C
3520 B

3521 B
3522 A
3523 A
3524 B
3525 B
3526 A
3527 A
3528 A
3529 B
3530 C
3531 C
3532 C
3533 A
3534 C
3535 A
3536 C
3537 A
3538 A
3539 B
3540 A
3541 A
3542 B
3543 A
3544 C
3545 C
3546 B
3547 A
3548 C
3549 B
3550 C
3551 B
3552 A
3553 B
3554 A
3555 C
3556 C
3557 B
3558 B
3559 A
3560 B
3561 A
3562 A
3563 C
3564 B
3565 C
3566 C
3567 D
3568 A
3569 C
3570 B
3571 B
3572 B
3573 B
3574 B
3575 B
3576 B
3577 A
3578 A
3579 A
3580 A
3581 B
3582 A
3583 B
3584 B
3585 B
3586 A
3587 A
3588 C
3589 A
3590 C
3591 A
3592 A
3593 A
3594 B
3595 A
3596 A
3597 A
3598 B
3599 B
3600 A
3601 B
3602 C
3603 B
3604 C
3605 B
3606 A
3607 C
3608 A
3609 C
3610 B
3611 A
3612 B
3613 B
3614 B
3615 C
3616 C
3617 B
3618 A
3619 B
3620 A
3621 A
3622 B
3623 B
3624 A
3625 A
3626 C
3627 C
3628 A
3629 C
3630 C
3631 B
3632 A
3633 A
3634 B
3635 C
3636 C
3637 A
3638 C
3639 A
3640 A
3641 A
3642 A
3643 B
3644 A
3645 A
3646 A
3647 A
3648 A
3649 A
3650 A
3651 A
3652 A
3653 A
3654 C
3655 A
3656 C
3657 A
3658 A
3659 C
3660 A
3661 B
3662 C
3663 B
3664 A
3665 A

3666 B
3667 B
3668 B
3669 A
3670 A
3671 A
3672 A
3673 B
3674 A
3675 B
3676 B
3677 A
3678 A
3679 C
3680 B
3681 A
3682 B
3683 A
3684 B
3685 C
3686 B
3687 B
3688 A
3689 B
3690 B
3691 B
3692 B
3693 B
3694 A
3695 A
3696 B
3697 B
3698 B
3699 C
3700 A
3701 A
3702 A
3703 A
3704 A
3705 B
3706 B
3707 B
3708 A
3709 C
3710 C
3711 A
3712 C
3713 B
3714 A
3715 A
3716 C
3717 B
3718 A
3719 A
3720 A
3721 C
3722 B
3723 C
3724 B
3725 C
3726 B
3727 B
3728 C
3729 B
3730 B
3731 C
3732 B
3733 B
3734 B
3735 B
3736 A
3737 A
3738 A
3739 A
3740 C
3741 B
3742 C
3743 B
3744 B
3745 A
3746 B
3747 B
3748 B
3749 B
3750 C
3751 B
3752 B
3753 C
3754 A
3755 B
3756 A
3757 C
3758 A
3759 C
3760 C
3761 B
3762 D
3763 A
3764 A
3765 A
3766 D
3767 B
3768 C
3769 B
3770 A
3771 B
3772 B
3773 A
3774 C
3775 A
3776 C
3777 A
3778 A
3779 B
3780 A
3781 A
3782 A
3783 B
3784 B
3785 B
3786 A
3787 C
3788 C
3789 B
3790 B
3791 C
3792 B
3793 C
3794 B
3795 A
3796 C
3797 A
3798 C
3799 C
3800 C
3801 C
3802 C
3803 A
3804 B
3805 C
3806 A
3807 B
3808 C
3809 B
3810 B

3811 B
3812 B
3813 A
3814 A
3815 A
3816 C
3817 B
3818 A
3819 A
3820 A
3821 A
3822 C
3823 A
3824 C
3825 C
3826 A
3827 B
3828 A
3829 B
3830 C
3831 C
3832 B
3833 C
3834 C
3835 B
3836 C
3837 A
3838 B
3839 B
3840 B
3841 C
3842 E
3843 B
3844 C
3845 A
3846 A
3847 C
3848 A
3849 B
3850 B
3851 B
3852 A
3853 C
3854 C
3855 C
3856 B
3857 B
3858 C
3859 A
3860 A
3861 B
3862 C
3863 A
3864 C
3865 B
3866 B
3867 A
3868 C
3869 B
3870 B
3871 C
3872 B
3873 A
3874 C
3875 A
3876 C
3877 C
3878 A
3879 B
3880 A
3881 B
3882 A
3883 C
3884 B
3885 A
3886 B
3887 A
3888 A
3889 C
3890 A
3891 C
3892 B
3893 B
3894 B
3895 B
3896 A
3897 C
3898 C
3899 C
3900 C
3901 C
3902 A
3903 B
3904 C
3905 A
3906 B
3907 A
3908 A
3909 B
3910 B
3911 A
3912 C
3913 A
3914 C
3915 C
3916 B
3917 B
3918 C
3919 C
3920 C
3921 A
3922 A
3923 B
3924 B
3925 A
3926 B
3927 B
3928 C
3929 C
3930 D
3931 B
3932 A
3933 B
3934 B
3935 B
3936 B
3937 B
3938 A
3939 C
3940 B
3941 C
3942 D
3943 C
3944 A
3945 B
3946 B
3947 A
3948 A
3949 B
3950 A
3951 C
3952 A
3953 A
3954 B
3955 B

3956 A	3965 C	3974 A	3983 B	3992 B
3957 A	3966 C	3975 A	3984 B	3993 A
3958 B	3967 A	3976 B	3985 B	3994 B
3959 A	3968 A	3977 B	3986 A	3995 B
3960 A	3969 B	3978 D	3987 C	3996 C
3961 B	3970 B	3979 A	3988 B	3997 B
3962 B	3971 B	3980 A	3989 A	3998 C
3963 A	3972 B	3981 C	3990 B	3999 B
3964 A	3973 C	3982 C	3991 C	4000 A

ABBREVIATIONS

A—amps

ac—alternating current

ACCA—Air Conditioning Contractors of America

AEV—automatic expansion valve

AGA—American Gas Association

API — American Petroleum Institute

ASHRAE—American Society of Heating, Refrigeration, and Air Conditioning Engineers

ASME—American Society of Mechanical Engineers

AWS—American Welding Society

AEV or AXV—automatic expansion valve

bip—black iron pipe

BOCA—Building Officials and Code Administration

Btu—British thermal units

Btuh—British thermal units per hour

cfm—cubic feet per minute

cps — cycles per second

CSIR—capacitor start induction run

CSR—capacitor start-run

db—dry bulb temperature

dc—direct current

DOE—(United States) Department of Energy

DPDT—double pole double throw

DPST—double pole single throw

EDR—equivalent direct radiation

FM—factory mutual

fpm—feet per minute

FS—federal specifications

ft or ' —foot or feet

ft^2—square feet

ft^3—cubic feet

ga—gauge

gpm—gallons per minute

HDA—horizontal double acting

Hg—mercury

HRT—horizontal return tube

HTHW—high temperature hot water

ID—inside diameter

IIAR—International Institute of Ammonia Refrigeration

in or " wc—inches water column

in or " —inch(es)

kcal—kilocalories

kV—kilovolts

kW—kilowatts

kWh—kilowatts per hour

lb — pound(s)

LP—liquid propane

mV—millivolts

NF—National fine threads

NFPA—National Fire Protection Association

OD—outside diameter

ppm—parts per million

PSC—permanent split capacitor

psi—pounds per square inch

psia—pounds per square inch, absolute

psig—pounds per square inch gauge

R—refrigerant

rh—relative humidity

rpm—revolutions per minute

SMACNA—Sheet Metal and Air Conditioning Contractors' National Association

SPDT—single pole double throw

SPST—single pole single throw

SSU—Saybolt Seconds Universal

TX or TXV—thermostatic expansion valve

UL—Underwriters Laboratories

V—volts

VSA—vertical single acting

wb—wet bulb temperature

°C—degrees Centigrade

°F—degrees Fahrenheit